Vehicle Stability

MECHANICAL ENGINEERING
A Series of Textbooks and Reference Books

Founding Editor

L. L. Faulkner

Columbus Division, Battelle Memorial Institute
and Department of Mechanical Engineering
The Ohio State University
Columbus, Ohio

1. *Spring Designer's Handbook*, Harold Carlson
2. *Computer-Aided Graphics and Design*, Daniel L. Ryan
3. *Lubrication Fundamentals*, J. George Wills
4. *Solar Engineering for Domestic Buildings*, William A. Himmelman
5. *Applied Engineering Mechanics: Statics and Dynamics*, G. Boothroyd and C. Poli
6. *Centrifugal Pump Clinic*, Igor J. Karassik
7. *Computer-Aided Kinetics for Machine Design*, Daniel L. Ryan
8. *Plastics Products Design Handbook, Part A: Materials and Components; Part B: Processes and Design for Processes*, edited by Edward Miller
9. *Turbomachinery: Basic Theory and Applications*, Earl Logan, Jr.
10. *Vibrations of Shells and Plates*, Werner Soedel
11. *Flat and Corrugated Diaphragm Design Handbook*, Mario Di Giovanni
12. *Practical Stress Analysis in Engineering Design*, Alexander Blake
13. *An Introduction to the Design and Behavior of Bolted Joints*, John H. Bickford
14. *Optimal Engineering Design: Principles and Applications*, James N. Siddall
15. *Spring Manufacturing Handbook*, Harold Carlson
16. *Industrial Noise Control: Fundamentals and Applications*, edited by Lewis H. Bell
17. *Gears and Their Vibration: A Basic Approach to Understanding Gear Noise*, J. Derek Smith
18. *Chains for Power Transmission and Material Handling: Design and Applications Handbook*, American Chain Association
19. *Corrosion and Corrosion Protection Handbook*, edited by Philip A. Schweitzer
20. *Gear Drive Systems: Design and Application*, Peter Lynwander
21. *Controlling In-Plant Airborne Contaminants: Systems Design and Calculations*, John D. Constance
22. *CAD/CAM Systems Planning and Implementation*, Charles S. Knox
23. *Probabilistic Engineering Design: Principles and Applications*, James N. Siddall
24. *Traction Drives: Selection and Application*, Frederick W. Heilich III and Eugene E. Shube
25. *Finite Element Methods: An Introduction*, Ronald L. Huston and Chris E. Passerello

26. *Mechanical Fastening of Plastics: An Engineering Handbook*, Brayton Lincoln, Kenneth J. Gomes, and James F. Braden
27. *Lubrication in Practice: Second Edition*, edited by W. S. Robertson
28. *Principles of Automated Drafting*, Daniel L. Ryan
29. *Practical Seal Design*, edited by Leonard J. Martini
30. *Engineering Documentation for CAD/CAM Applications*, Charles S. Knox
31. *Design Dimensioning with Computer Graphics Applications*, Jerome C. Lange
32. *Mechanism Analysis: Simplified Graphical and Analytical Techniques*, Lyndon O. Barton
33. *CAD/CAM Systems: Justification, Implementation, Productivity Measurement*, Edward J. Preston, George W. Crawford, and Mark E. Coticchia
34. *Steam Plant Calculations Manual*, V. Ganapathy
35. *Design Assurance for Engineers and Managers*, John A. Burgess
36. *Heat Transfer Fluids and Systems for Process and Energy Applications*, Jasbir Singh
37. *Potential Flows: Computer Graphic Solutions*, Robert H. Kirchhoff
38. *Computer-Aided Graphics and Design: Second Edition*, Daniel L. Ryan
39. *Electronically Controlled Proportional Valves: Selection and Application*, Michael J. Tonyan, edited by Tobi Goldoftas
40. *Pressure Gauge Handbook*, AMETEK, U.S. Gauge Division, edited by Philip W. Harland
41. *Fabric Filtration for Combustion Sources: Fundamentals and Basic Technology*, R. P. Donovan
42. *Design of Mechanical Joints*, Alexander Blake
43. *CAD/CAM Dictionary*, Edward J. Preston, George W. Crawford, and Mark E. Coticchia
44. *Machinery Adhesives for Locking, Retaining, and Sealing*, Girard S. Haviland
45. *Couplings and Joints: Design, Selection, and Application*, Jon R. Mancuso
46. *Shaft Alignment Handbook*, John Piotrowski
47. *BASIC Programs for Steam Plant Engineers: Boilers, Combustion, Fluid Flow, and Heat Transfer*, V. Ganapathy
48. *Solving Mechanical Design Problems with Computer Graphics*, Jerome C. Lange
49. *Plastics Gearing: Selection and Application*, Clifford E. Adams
50. *Clutches and Brakes: Design and Selection*, William C. Orthwein
51. *Transducers in Mechanical and Electronic Design*, Harry L. Trietley
52. *Metallurgical Applications of Shock-Wave and High-Strain-Rate Phenomena*, edited by Lawrence E. Murr, Karl P. Staudhammer, and Marc A. Meyers
53. *Magnesium Products Design*, Robert S. Busk
54. *How to Integrate CAD/CAM Systems: Management and Technology*, William D. Engelke
55. *Cam Design and Manufacture: Second Edition*; with cam design software for the IBM PC and compatibles, disk included, Preben W. Jensen
56. *Solid-State AC Motor Controls: Selection and Application*, Sylvester Campbell
57. *Fundamentals of Robotics*, David D. Ardayfio
58. *Belt Selection and Application for Engineers*, edited by Wallace D. Erickson
59. *Developing Three-Dimensional CAD Software with the IBM PC*, C. Stan Wei
60. *Organizing Data for CIM Applications*, Charles S. Knox, with contributions by Thomas C. Boos, Ross S. Culverhouse, and Paul F. Muchnicki

61. *Computer-Aided Simulation in Railway Dynamics*, by Rao V. Dukkipati and Joseph R. Amyot
62. *Fiber-Reinforced Composites: Materials, Manufacturing, and Design*, P. K. Mallick
63. *Photoelectric Sensors and Controls: Selection and Application*, Scott M. Juds
64. *Finite Element Analysis with Personal Computers*, Edward R. Champion, Jr., and J. Michael Ensminger
65. *Ultrasonics: Fundamentals, Technology, Applications: Second Edition, Revised and Expanded*, Dale Ensminger
66. *Applied Finite Element Modeling: Practical Problem Solving for Engineers*, Jeffrey M. Steele
67. *Measurement and Instrumentation in Engineering: Principles and Basic Laboratory Experiments*, Francis S. Tse and Ivan E. Morse
68. *Centrifugal Pump Clinic: Second Edition, Revised and Expanded*, Igor J. Karassik
69. *Practical Stress Analysis in Engineering Design: Second Edition, Revised and Expanded*, Alexander Blake
70. *An Introduction to the Design and Behavior of Bolted Joints: Second Edition, Revised and Expanded*, John H. Bickford
71. *High Vacuum Technology: A Practical Guide*, Marsbed H. Hablanian
72. *Pressure Sensors: Selection and Application*, Duane Tandeske
73. *Zinc Handbook: Properties, Processing, and Use in Design*, Frank Porter
74. *Thermal Fatigue of Metals*, Andrzej Weronski and Tadeusz Hejwowski
75. *Classical and Modern Mechanisms for Engineers and Inventors*, Preben W. Jensen
76. *Handbook of Electronic Package Design*, edited by Michael Pecht
77. *Shock-Wave and High-Strain-Rate Phenomena in Materials*, edited by Marc A. Meyers, Lawrence E. Murr, and Karl P. Staudhammer
78. *Industrial Refrigeration: Principles, Design and Applications*, P. C. Koelet
79. *Applied Combustion*, Eugene L. Keating
80. *Engine Oils and Automotive Lubrication*, edited by Wilfried J. Bartz
81. *Mechanism Analysis: Simplified and Graphical Techniques, Second Edition, Revised and Expanded*, Lyndon O. Barton
82. *Fundamental Fluid Mechanics for the Practicing Engineer*, James W. Murdock
83. *Fiber-Reinforced Composites: Materials, Manufacturing, and Design, Second Edition, Revised and Expanded*, P. K. Mallick
84. *Numerical Methods for Engineering Applications*, Edward R. Champion, Jr.
85. *Turbomachinery: Basic Theory and Applications, Second Edition, Revised and Expanded*, Earl Logan, Jr.
86. *Vibrations of Shells and Plates: Second Edition, Revised and Expanded*, Werner Soedel
87. *Steam Plant Calculations Manual: Second Edition, Revised and Expanded*, V. Ganapathy
88. *Industrial Noise Control: Fundamentals and Applications, Second Edition, Revised and Expanded*, Lewis H. Bell and Douglas H. Bell
89. *Finite Elements: Their Design and Performance*, Richard H. MacNeal
90. *Mechanical Properties of Polymers and Composites: Second Edition, Revised and Expanded*, Lawrence E. Nielsen and Robert F. Landel
91. *Mechanical Wear Prediction and Prevention*, Raymond G. Bayer

92. *Mechanical Power Transmission Components*, edited by David W. South and Jon R. Mancuso
93. *Handbook of Turbomachinery*, edited by Earl Logan, Jr.
94. *Engineering Documentation Control Practices and Procedures*, Ray E. Monahan
95. *Refractory Linings Thermomechanical Design and Applications*, Charles A. Schacht
96. *Geometric Dimensioning and Tolerancing: Applications and Techniques for Use in Design, Manufacturing, and Inspection*, James D. Meadows
97. *An Introduction to the Design and Behavior of Bolted Joints: Third Edition, Revised and Expanded*, John H. Bickford
98. *Shaft Alignment Handbook: Second Edition, Revised and Expanded*, John Piotrowski
99. *Computer-Aided Design of Polymer-Matrix Composite Structures*, edited by Suong Van Hoa
100. *Friction Science and Technology*, Peter J. Blau
101. *Introduction to Plastics and Composites: Mechanical Properties and Engineering Applications*, Edward Miller
102. *Practical Fracture Mechanics in Design*, Alexander Blake
103. *Pump Characteristics and Applications*, Michael W. Volk
104. *Optical Principles and Technology for Engineers*, James E. Stewart
105. *Optimizing the Shape of Mechanical Elements and Structures*, A. A. Seireg and Jorge Rodriguez
106. *Kinematics and Dynamics of Machinery*, Vladimír Stejskal and Michael Valášek
107. *Shaft Seals for Dynamic Applications*, Les Horve
108. *Reliability-Based Mechanical Design*, edited by Thomas A. Cruse
109. *Mechanical Fastening, Joining, and Assembly*, James A. Speck
110. *Turbomachinery Fluid Dynamics and Heat Transfer*, edited by Chunill Hah
111. *High-Vacuum Technology: A Practical Guide, Second Edition, Revised and Expanded*, Marsbed H. Hablanian
112. *Geometric Dimensioning and Tolerancing: Workbook and Answerbook*, James D. Meadows
113. *Handbook of Materials Selection for Engineering Applications*, edited by G. T. Murray
114. *Handbook of Thermoplastic Piping System Design*, Thomas Sixsmith and Reinhard Hanselka
115. *Practical Guide to Finite Elements: A Solid Mechanics Approach*, Steven M. Lepi
116. *Applied Computational Fluid Dynamics*, edited by Vijay K. Garg
117. *Fluid Sealing Technology*, Heinz K. Muller and Bernard S. Nau
118. *Friction and Lubrication in Mechanical Design*, A. A. Seireg
119. *Influence Functions and Matrices*, Yuri A. Melnikov
120. *Mechanical Analysis of Electronic Packaging Systems*, Stephen A. McKeown
121. *Couplings and Joints: Design, Selection, and Application, Second Edition, Revised and Expanded,* Jon R. Mancuso
122. *Thermodynamics: Processes and Applications*, Earl Logan, Jr.
123. *Gear Noise and Vibration*, J. Derek Smith
124. *Practical Fluid Mechanics for Engineering Applications,* John J. Bloomer
125. *Handbook of Hydraulic Fluid Technology*, edited by George E. Totten
126. *Heat Exchanger Design Handbook*, T. Kuppan

127. *Designing for Product Sound Quality*, Richard H. Lyon
128. *Probability Applications in Mechanical Design*, Franklin E. Fisher and Joy R. Fisher
129. *Nickel Alloys,* edited by Ulrich Heubner
130. *Rotating Machinery Vibration: Problem Analysis and Troubleshooting*, Maurice L. Adams, Jr.
131. *Formulas for Dynamic Analysis,* Ronald L. Huston and C. Q. Liu
132. *Handbook of Machinery Dynamics*, Lynn L. Faulkner and Earl Logan, Jr.
133. *Rapid Prototyping Technology: Selection and Application,* Kenneth G. Cooper
134. *Reciprocating Machinery Dynamics: Design and Analysis*, Abdulla S. Rangwala
135. *Maintenance Excellence: Optimizing Equipment Life-Cycle Decisions,* edited by John D. Campbell and Andrew K. S. Jardine
136. *Practical Guide to Industrial Boiler Systems*, Ralph L. Vandagriff
137. *Lubrication Fundamentals: Second Edition, Revised and Expanded*, D. M. Pirro and A. A. Wessol
138. *Mechanical Life Cycle Handbook: Good Environmental Design and Manufacturing,* edited by Mahendra S. Hundal
139. *Micromachining of Engineering Materials,* edited by Joseph McGeough
140. *Control Strategies for Dynamic Systems: Design and Implementation,* John H. Lumkes, Jr.
141. *Practical Guide to Pressure Vessel Manufacturing,* Sunil Pullarcot
142. *Nondestructive Evaluation: Theory, Techniques, and Applications,* edited by Peter J. Shull
143. *Diesel Engine Engineering: Thermodynamics, Dynamics, Design, and Control,* Andrei Makartchouk
144. *Handbook of Machine Tool Analysis,* Ioan D. Marinescu, Constantin Ispas, and Dan Boboc
145. *Implementing Concurrent Engineering in Small Companies,* Susan Carlson Skalak
146. *Practical Guide to the Packaging of Electronics: Thermal and Mechanical Design and Analysis*, Ali Jamnia
147. *Bearing Design in Machinery: Engineering Tribology and Lubrication,* Avraham Harnoy
148. *Mechanical Reliability Improvement: Probability and Statistics for Experimental Testing,* R. E. Little
149. *Industrial Boilers and Heat Recovery Steam Generators: Design, Applications, and Calculations*, V. Ganapathy
150. *The CAD Guidebook: A Basic Manual for Understanding and Improving Computer-Aided Design,* Stephen J. Schoonmaker
151. *Industrial Noise Control and Acoustics,* Randall F. Barron
152. *Mechanical Properties of Engineered Materials,* Wolé Soboyejo
153. *Reliability Verification, Testing, and Analysis in Engineering Design,* Gary S. Wasserman
154. *Fundamental Mechanics of Fluids: Third Edition*, I. G. Currie
155. *Intermediate Heat Transfer*, Kau-Fui Vincent Wong
156. *HVAC Water Chillers and Cooling Towers: Fundamentals, Application, and Operation*, Herbert W. Stanford III
157. *Gear Noise and Vibration: Second Edition, Revised and Expanded,* J. Derek Smith

158. *Handbook of Turbomachinery: Second Edition, Revised and Expanded,* edited by Earl Logan, Jr., and Ramendra Roy
159. *Piping and Pipeline Engineering: Design, Construction, Maintenance, Integrity, and Repair,* George A. Antaki
160. *Turbomachinery: Design and Theory*, Rama S. R. Gorla and Aijaz Ahmed Khan
161. *Target Costing: Market-Driven Product Design,* M. Bradford Clifton, Henry M. B. Bird, Robert E. Albano, and Wesley P. Townsend
162. *Fluidized Bed Combustion,* Simeon N. Oka
163. *Theory of Dimensioning: An Introduction to Parameterizing Geometric Models,* Vijay Srinivasan
164. *Handbook of Mechanical Alloy Design*, edited by George E. Totten, Lin Xie, and Kiyoshi Funatani
165. *Structural Analysis of Polymeric Composite Materials,* Mark E. Tuttle
166. *Modeling and Simulation for Material Selection and Mechanical Design,* edited by George E. Totten, Lin Xie, and Kiyoshi Funatani
167. *Handbook of Pneumatic Conveying Engineering,* David Mills, Mark G. Jones, and Vijay K. Agarwal
168. *Clutches and Brakes: Design and Selection, Second Edition*, William C. Orthwein
169. *Fundamentals of Fluid Film Lubrication: Second Edition,* Bernard J. Hamrock, Steven R. Schmid, and Bo O. Jacobson
170. *Handbook of Lead-Free Solder Technology for Microelectronic Assemblies*, edited by Karl J. Puttlitz and Kathleen A. Stalter
171. *Vehicle Stability*, Dean Karnopp

Additional Volumes in Preparation

Mechanical Wear Fundamentals and Testing: Second Edition, Revised and Expanded, Raymond G. Bayer

Engineering Design for Wear: Second Edition, Revised and Expanded, Raymond G. Bayer

Progressing Cavity Pumps, Downhole Pumps, and Mudmotors, Lev Nelik

Mechanical Engineering Software

Spring Design with an IBM PC, Al Dietrich

Mechanical Design Failure Analysis: With Failure Analysis System Software for the IBM PC, David G. Ullman

Vehicle Stability

Dean Karnopp
University of California, Davis
Davis, California, U.S.A.

CRC Press
Taylor & Francis Group
Boca Raton London New York

CRC Press is an imprint of the
Taylor & Francis Group, an **informa** business

Library of Congress Cataloging-in-Publication Data
A catalog record for this book is available from the Library of Congress.

ISBN: 0-8247-5711-4

This book is printed on acid-free paper.

Headquarters
Marcel Dekker 270 Madison Avenue, New York, NY 10016, U.S.A.
tel: 212-696-9000; fax: 212-685-4540

The publisher offers discounts on this book when ordered in bulk quantities. For more information, write to Special Sales/Professional Marketing at the headquarters address above.

Current printing (last digit):

10 9 8 7 6 5 4 3

PRINTED IN THE UNITED STATES OF AMERICA

Preface

This book is mainly the result of two activities that have given me a great deal of pleasure and satisfaction over a period of more than 30 years. The first activity was the initiation and teaching of a course with the same title as the book for seniors and first-year graduate students in mechanical and aeronautical engineering at the University of California, Davis. The second was a course, "Vehicle Dynamics and Active Control," given by me and my colleague Professor Donald Margolis numerous times in the United States and in several European countries. This short professional course was given under the auspices of the University Consortium for Continuing Education and was intended primarily for engineers in the automotive industry.

Although the short course contained much of the material in the academic course and in the present book, it was specialized in that it dealt with automotive topics. On the other hand, the short course was in one sense more general in that it dealt with several aspects of automotive vehicle dynamics besides stability. For example, where this book deals with steering response and the horizontal dynamics of automobiles, the short course also dealt with automotive suspension systems and vertical dynamics.

An aspect of the professional course that was of particular interest to working engineers was the discussion of active means of influencing the dynamics of automobiles. It is now fairly common to find active or semiactive suspension systems, active steering systems, and electronically controlled braking systems, as well as electronically controlled engines and transmissions. Chapter 11 discusses some of the active means used to stabilize vehicles such as cars and airplanes. The chapter focuses on model reference active steering control since this type of active vehicle control relates most directly to the topics discussed in prior chapters.

The unique feature of this book is its treatment of some of the stability aspects of a variety of vehicle types. This requires basic mathematical models describing the dynamic behavior of the various types of vehicles to be studied. Because of this, there was a temptation to call this book *Vehicle Dynamics* rather than *Vehicle Stability*. However, not all the dynamic effects in several types of vehicles can be described in a book of reasonable length. In the end it seemed better to restrict the discussion mainly to certain aspects of the dynamics of vehicles that are relevant for stability studies. Inevitably, some topics related to the handling properties and control features of vehicles are, out of necessity, discussed.

By emphasizing stability analyses using relatively simple linearized mathematical models, it is possible to use similar mathematical models for a variety of vehicle types. This allows the similarities and critical differences in the stability properties of different vehicle types to be easily appreciated. Although the emphasis is on linearized mathematical models, the nonlinear relations relating to force generation are discussed in several cases, so that it would be possible to extend the models beyond the range of small perturbation inherent in the analysis of stability. Through the use of computer simulation, one can discover the behavior of unstable vehicles when the perturbation variables grow to an extent that the linearized equations are no longer valid.

The book and the academic course on which it is based provide an opportunity for an engineer or engineering student to exercise a number of techniques from engineering analysis that are often learned separately and without benefit of interesting applications. In the course of studying the stability problems that arise in vehicles, the reader will encounter topics from kinematics, rigid body dynamics, system dynamics, automatic control, stability theory, and aerodynamics. Specialized topics dealing specifically with vehicle dynamics such as the force generation by pneumatic tires, railway wheels, and wings are also treated.

Since all the studies of vehicle stability begin with the formulation of mathematical models of the vehicles, there is a great deal of emphasis on the use of methods for formulating equations of motion. Chapter 2 deals with the description of rigid body motion. In most textbooks on dynamics, there is little emphasis on formulation of equations of motion in the forms one encounters in technical papers on vehicle dynamics. In particular, the use of a body-fixed coordinate system that often proves to be very useful in describing the dynamics of vehicles is usually not discussed. Chapter 2 describes the fundamental principles of mechanics for rigid bodies as well as the way these principles work out when a body-fixed coordinate system is used.

The more conventional derivations of the equations of motion using Newton's laws directly or Lagrange's equations with inertial coordinate systems are illustrated by a series of examples of increasing levels of complexity throughout the book. This gives the reader the opportunity to compare several ways to formulate equations of motion for a vehicle.

For those with some familiarity with bond graph methods for system dynamics, Chapter 2 contains a section that shows how rigid body dynamics using a body-fixed coordinate frame can be represented in graphical form using bond graphs. The Appendix gives complete bond graph representations for an automobile model used in Chapter 6 and for a simplified aircraft model used in Chapter 9. This opens up the possibility of using computer-automated equation formulation and simulation for nonlinear versions of the linearized models used for stability analysis.

The academic course has been very popular over the years for several reasons. First, many people are inherently interested in the dynamics of vehicles such as cars, bicycles, motorcycles, airplanes, and trains. Second, the idea that vehicles can exhibit unstable and dangerous behavior for no obvious reason is in itself fascinating. These instabilities are particularly obvious in racing situations or in speed record attempts, but in everyday life it is common to see trailers swaying back and forth or to see cars slewing around on icy roads and to wonder why this happens. Third, for those with some background in applied mathematics, it is always satisfying to see that simple mathematical models can often illuminate dynamic behavior that is otherwise baffling.

On the other hand, the book makes no real attempt to be a practical guide to the design or modification of vehicles. The reader will learn how an aircraft designer goes about designing a statically stable aircraft but will not find here a complete discussion of the practical knowledge needed to become an expert. The reader may gain an appreciation of the terms understeer, oversteer, and critical speeds, for example, but there are no rules of thumb for modifying an autocross racer. The bibliography does include books and papers that may prove helpful in building up a practical knowledge base relating to a particular vehicle stability problem.

For those interested in using the book as a text, it is highly recommended that experiments and demonstrations be used in parallel with the classroom lectures. The University of California, Davis, is fortunate to be located near the California Highway Patrol Academy, and their driving instructors have been generous in giving demonstrations of automobile dynamics on their high-speed track and their skid pads. The students are always impressed that the instructors use the same words to describe the handling of patrol cars that the engineers use, such as "understeer" and "oversteer," but without using the mathematical definitions that engineers prefer. Also, a trip to a local dirt track provides a demonstration of the racer's terms "tight" and "loose" as well as some spectacular demonstrations of unstable behavior.

Furthermore, the university has its own airport. This has permitted the students to experience aircraft dynamics personally as well as analytically, as in the discussions in Chapter 9. The relation between stability and control is much more obvious to a passenger in a small airplane than to one in a commercial aircraft. A good pilot can easily demonstrate several oscillatory modes of motion (except possibly for the low-frequency phugoid mode, these are all stable for production light airplanes) and, if the passengers are willing, can show some divergent instabilities for extreme attitudes.

In addition, it is possible to design laboratory experiments to illustrate many of the analyses in the book. As examples, we have designed a demonstration of trailer instability using a sanding belt and a stationary model trailer as well as a small trailer attached to a three-wheeled cycle. These models show that changes in center of mass location and moment of inertia do indeed influence stability, just as the theory in Chapter 5 predicts. Model gliders have been designed to illustrate static stability and instability, as discussed in Chapter 9. Even a rear-steering bicycle was fabricated to illustrate the control difficulties described in Chapter 7.

A number of exercises have been given at the back of the book that may be assigned if the book is used as a text. Some of these problems are included to help students appreciate assumptions behind derivations given in the book. A true understanding of analytical results often requires one to reproduce parts of the manipulations. Other problems extend the analyses of the corresponding chapter to new situations or relate topics in one chapter to other chapters. Still other problems are of a much more extensive nature and can form the basis of small projects. They are intended to illustrate how mathematical models of varying degrees of complexity can be used to suggest design rules for assuring stability of vehicles.

I hope that the readers of this book will be as fascinated with vehicle stability as I am and will be inspired to learn even more about the topic.

Dean Karnopp

Contents

Vehicle Stability

1
Elementary Vehicles

I. INTRODUCTION

Vehicles such as cars, trains, ships and airplanes are intended to move people and goods from place to place in an efficient and safe manner. This book deals with certain aspects of the motion of vehicles usually described using the terms "stability and control." Although most people have a good intuitive idea what these terms mean, this book will deal with mathematical models of vehicles and with more precise and technical meaning of the terms. As long as the mathematical models reasonably represent the real vehicles, the results of analysis of the models can yield insight into the actual problems that vehicles sometimes exhibit and, in the best case, can suggest ways to cure vehicle problems by modification of the physical aspects of the vehicle.

All vehicles represent interesting and often complex dynamic systems that require careful analysis and design to make sure that they behave properly. The stability aspect of vehicle motion has to do with assuring that the vehicle does not depart spontaneously from a desired path. It is possible for an automobile to start to spin out at high speed or a trailer to begin to oscillate back and forth in ever-wider swings seemingly without provocation. Most of this book deals with how the physical parameters of a vehicle influence its dynamic characteristics in general and its stability properties in particular.

The control aspect of vehicle motion has to do with the ability of a human operator or an automatic control system to guide the vehicle along a desired trajectory. In the case of a human operator, this means that the dynamic properties of such vehicles should be tailored to allow humans to control them with reasonable ease and precision. A car or an airplane that requires a great deal of attention to keep it from deviating from a desired path would probably not be considered satisfactory. Modern studies of the Wright brother's 1903 Flyer indicate that, while the brothers learned to control the airplane, it apparently was unstable enough that modern pilots are reluctant to fly an exact replica. The Wright brothers, of course, did not have the benefit of the understanding of aircraft stability that aeronautical engineers now have. Modern light planes can now be designed to be stable enough that flying them is not the daunting task that it was for the Wrights.

For vehicles using electronic control systems, the dynamic properties of the vehicle must be considered in the design of the controller to assure that the controlled vehicles are stable. Increasingly, human operators exercise supervisory control of vehicles with automatic control systems. In some cases, the control system may even stabilize an inherently unstable vehicle so that it is not difficult for the human operator to control its trajectory. This is the case for some modern fighter aircraft. In other cases, the active control system simply aids the human operator in controlling the vehicle. Aircraft autopilots, fly-by-wire systems, anti-lock braking systems and electronic stability enhancement systems for automobiles are all examples of systems that modify the stability properties of vehicles with active means to increase the ease with which they can be controlled. Active stability enhancement techniques will be discussed in Chapter 11 after the stability properties of several types of vehicles have been analyzed.

In some cases, vehicle motion is neither actively controlled by a human operator nor by an automatic control system but yet the vehicle may exhibit very undesirable dynamic behavior under certain conditions. A trailer being pulled by a car, for example, should follow the path of the car in a stable fashion. As we will see in Chapter 5, a trailer that is not properly designed or loaded may however exhibit growing oscillations at high speeds, which could lead to a serious accident. Trains that use tapered wheelsets are intended to self-center on the tracks without any active control, but above a critical speed, an increasing oscillatory motion called hunting can develop that may lead to derailment in extreme cases. For many vehicles, this type of unstable behavior may suddenly appear as a critical speed is exceeded. This type of unstable behavior is particularly insidious since the vehicle will appear to act in a perfectly normal manner until the first time the critical speed is exceeded at which time the unstable motion can have serious consequences.

This book will concentrate on the stability of the motion of vehicles, although some related aspects of vehicle dynamics will also be treated. These aspects of vehicle dynamics fall into the more general category of "vehicle handling," or "vehicle controllability" and are of particular importance under extreme conditions associated with emergency maneuvers.

Obviously vehicle stability is of interest to anyone involved in the design or use of vehicles, but the topic of stability is of a more general interest. Since the dynamic behavior of vehicles such as cars, trailers and airplanes is to some extent familiar to almost everybody, these systems can be used to introduce a number of concepts in system dynamics, stability and control. To many people, these concepts seem abstract and difficult to understand when presented as topics in applied mathematics without some familiar physical examples. Everyday experience with a variety of vehicles can provide examples of these otherwise abstract mathematical concepts.

Engineers involved in the design and construction of vehicles typically use mathematical models of vehicles in order to understand the fundamental dynamic problems of vehicles and to devise means for controlling vehicle motion. Unfortunately, it can be a formidable task to find accurate mathematical descriptions of the dynamics of a wide variety of vehicles. Not only do the descriptions involve nonlinear differential equations that seem to have little similarity from vehicle type to vehicle type, but, in particular, the characterization of the force producing elements can be quite disparate. One can easily imagine that rubber tires, steel wheels, boat hulls or airplane wings act in quite different ways to influence the motions of vehicles operating on land, on water or in the air.

On the other hand, all vehicles have some aspects in common. They are all usefully described for many purposes as essentially rigid bodies acted upon by forces that control their motion. Some of the forces are under control of a human operator, some may be under active control of an automatic control system but all are influenced by very nature of the force generating mechanisms inherent to the particular vehicle type. This means that not very many people claim to be experts in the dynamics of a large number of types of vehicles.

When describing the stability of vehicle motion, however, the treatment of the various types of force generating elements exhibits a great deal of similarity. In stability analysis, it is often sufficient to consider small deviations from a steady state of motion. The basic idea is that a stable vehicle will tend to return to the steady motion if it has been disturbed while an unstable vehicle will deviate further from the steady state after a disturbance. The mathematical description of the vehicle dynamics for stability analysis typically uses a linearized differential equation form based on the nonlinear differential equations that generally apply. The linearized equations show more similarity among vehicle types than the more accurate nonlinear equations. Thus a focus on stability allows one to appreciate that there are interesting similarities and differences among the dynamic properties of a variety of vehicle types without being confronted with the complexities of nonlinear differential equation models.

In what follows, stability analyses will be performed for extremely simplified vehicle models to illustrate the approach. Then gradually, more realistic vehicle models will be introduced and it will become clear that despite some analogous effects among vehicle types, ultimately, the differences among the force generating mechanisms for various vehicle types determine their behavior particularly under more extreme conditions than are considered in stability analyses.

In this introductory chapter, two examples will be analyzed that require essentially no discussion of force generating elements such as tires or wings, but that introduce the basic ideas of vehicle stability analysis. The first example is actually kinematic rather than dynamic in the usual sense, since Newton's laws are not needed. The second example is truly dynamic but kinematic constraints take the place of force generating elements. Succeeding Chapters 2, 3 and 4 provide a basis for the more complete dynamic analyses to follow in Chapters 5–11.

II. TAPERED WHEELSET ON RAILS

Although the ancient invention of the wheel was a great step forward for the transportation of heavy loads, when soft or rough ground was to be covered, wheels still required significant thrust to move under load. The idea of a railway was to provide a hard, smooth surface on which the wheels could roll and thus to reduce the effort required to move a load.

The first rail vehicles were used to transport ore out of mines. Cylindrical wheels with flanges on the inside to keep the wheels from rolling off the rails were used. To the casual observer, these early wheelsets resemble closely those used today. In fact, modern wheelsets differ in an important way from those early versions.

Originally it was found that the flanges on the wheels were often rubbing in contact with the sides of the rails. This not only increased the resistance of the wheels to rolling but, more importantly, also caused the flanges to wear quite rapidly. Eventually it was discovered that if slightly tapered rather than cylindrical wheels were used, the wheel sets would automatically tend to self-center and the flanges would hardly wear at all since they rarely touched the sides of the rails. Since this time almost all rail cars use tapered wheels.

The actual analysis of the stability of high-speed trains is quite complicated since multiple wheels, trucks on which the wheels are mounted and the car bodies are involved. Also, somewhat surprisingly, steel wheels rolling on steel rails at high speeds do not simply roll as rigid bodies without slipping as one might imagine. There are lateral and longitudinal "creeping" motions between the wheels and the rails that make the stability of the entire vehicle dependent upon the speed. Above a so-called critical speed a railcar will begin to "hunt" back and forth between the rails. At high enough speeds, the flanges will contact the rails and in the extreme case, the wheels may derail. The dynamic analysis of rail vehicles is discussed in more detail in Chapter 10.

The following analysis will be as simple as possible. Only a single wheelset is involved and the analysis is not even dynamic but rather kinematic. The two wheels will be assumed to roll without slip, which is a reasonable assumption at low speeds. The point of the exercise is to show that at very low speeds when the accelerations and hence the forces are very low, a wheelset with tapered wheels will tend to steer itself automatically towards the center of the rails. If the taper were to be in the opposite sense, the wheelset would be unstable and tent to veer toward one side of the rails or the other until the flanges would contact the rails. This example will serve to show how a small perturbation from a steady motion leads to linearized equations of motion that can be analyzed for their stability properties.

Figure 1.1 shows the wheelset in two conditions. At the left in Fig. 1.1, the wheelset is in a centered position rolling at a constant forward speed on straight rails. The wheelset consists of two wheels rigidly attached to a common axle so both wheel rotate about their common axle at the same rate. The wheels are assumed to have point contact with the two straight lines that represent the rails. In the centered position, the radius of each wheel at the contact point on the rails is the same, r_0.

The second part of the figure to the right shows the wheelset in a slightly perturbed position. In this part of the figure, the wheelset has moved off its centered position and its axle has assumed a slight angle with respect to a line perpendicular to the rails. The forward speed is assumed to remain constant.

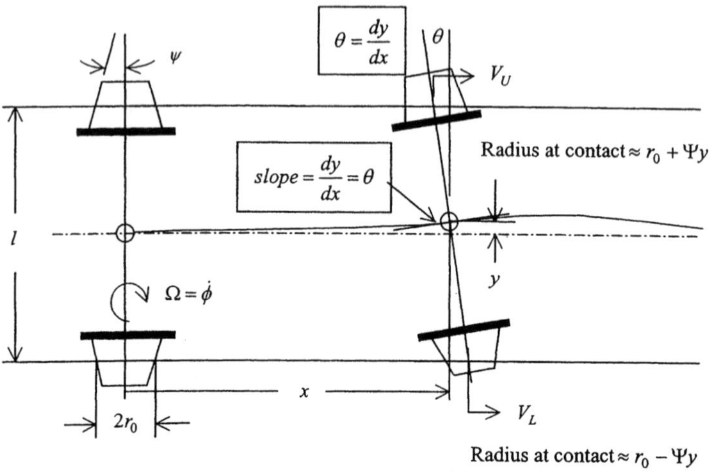

FIG. 1.1 A tapered wheelset.

In all of the examples to follow, we will distinguish between *parameters*, which are constants in the analysis, and *variables*, which change over time and are used to describe the motion of the system. In this case, the parameters are l, the separation of the rails, Ψ, the taper angle, r_0, the rolling radius when the wheelset is centered and Ω, the angular velocity of the wheelset about its axle.

The main variables for this problem are y, the deviation of the center of the wheelset from the center of the rails and θ, the angle between the wheelset axis and the perpendicular to the rails. Other variables that may be of interest include x, the distance the wheelset has moved down the track and ϕ, the angle around the axle that the wheelset has turned when it has rolled a distance x. Still other variables such as the velocities of the upper and lower wheels in the sketch of Fig. 1.1, V_U and V_L, will be eliminated in the final equations describing the system.

Generally for a vehicle stability analysis, a possible steady motion will be defined and then small deviations from this steady motion will be assumed. This steady motion will often be called in this book the *basic motion*. The essence of a stability analysis is to determine whether the deviations from the steady basic motion will tend to increase or decrease in time. If the deviations increase in time, the system is called unstable. If the deviations tend to decrease and the vehicle system returns to the basic motion, the system is called stable. These concepts will be made more precise later.

In this case the basic motion occurs when the wheelset is perfectly centered with $\theta \cong dy/dx$, and y both vanishing. Since Ω is assumed to be constant, $\dot{x} = r_0\Omega = V_U = V_L$ also are constant for the basic motion.

The deviated motion is called the *perturbed motion* and is characterized by the assumption that the deviation variables are small in some sense. In the present case, it is assumed that the variable y is small with respect to l and that the angle θ is also small. (Remember that angles are dimensionless and small in this case means small with respect to 2π radians. We will encounter angles that must be considered small frequently in the succeeding chapters. Small angle approximations for trigonometric functions will then be used.)

For an unstable system, the small, perturbed variables such as y and θ will grow spontaneously in time. For a stable motion, the perturbed variables decrease in time and the motion approaches the basic motion. In this example, y and θ will tend to return toward zero if the wheelset is stable.

For the perturbed motion, with y and θ small but not zero, the upper and lower wheel velocities are slightly different because the rolling radii of the two wheels are no longer the same. The taper angle is assumed to be small so the change in radius is approximately Ψy. See Fig. 1.1. The rotational rate Ω of the wheelset is assumed to be constant even when the wheelset is not perfectly centered. The forward speeds of the upper and lower wheels are then

$$V_U = (r_0 + \Psi y)\Omega \tag{1.1}$$

$$V_L = (r_0 - \Psi y)\Omega \tag{1.2}$$

Note that when $y = 0$, the two wheel speeds are equal and have the value $r_0\Omega$ which is the forward speed of the wheelset.

In fact, the forward velocity of the center point of the wheelset, which was also $r_0\Omega$ for the basic motion, does not change appreciably for the perturbed motion. Simple kinematic considerations show that the speed of the center point of the wheelset is essentially the average of the motion of the two ends of the axle.

$$\dot{x} = (V_U + V_L)/2 = r_0\Omega \tag{1.3}$$

Now one can compute the rate of change of the angle θ, which has to do with the difference between the velocities of the two wheels. This is again a matter of simple kinematics.

$$\dot{\theta} = (V_L - V_U)/l = -2\Psi y\Omega/l \tag{1.4}$$

The rate of change of y can also be found using the idea that $\theta \cong dy/dx$ and Eq. (1.3),

$$\frac{dy}{dt} \equiv \dot{y} \cong \frac{dy}{dx}\frac{dx}{dt} = \frac{dy}{dx}\dot{x} \cong \theta(r_0\Omega) \tag{1.5}$$

By differentiating Eq. (1.5) with respect to time and substituting $\dot{\theta}$ from Eq. (1.4), a single second order equation for $y(t)$ results.

$$\ddot{y} + \left(\frac{r_0\Omega^2 2\Psi}{l}\right)y = 0 \tag{1.6}$$

Alternatively, Eqs. (1.4 and 1.5), can also be expressed in state space form, Ogata, 1970.

$$\begin{bmatrix} \dot{\theta} \\ \dot{y} \end{bmatrix} = \begin{bmatrix} 0 & -2\Psi\Omega/l \\ r_0\Omega & 0 \end{bmatrix} \begin{bmatrix} \theta \\ y \end{bmatrix} \tag{1.7}$$

In either form, Eq. (1.6) or Eq. (1.7), these linear differential equations are easily solved. The solution and stability analysis of linear equations will be discussed in some detail in Chapter 3.

In this section, we will simply note the analogy of Eq. (1.6) to the familiar equation of motion for a mass-spring vibratory system,

$$\ddot{x} + (k/m)x = 0 \tag{1.8}$$

in which x is the position of the mass, k is a spring constant and m is the mass.

Common experience with a mass attached to a normal spring shows that after a disturbance, the spring pulls the mass back towards the equilibrium position, $x = 0$. Thereafter the mass oscillates back and forth and, if there is any friction at all, the mass will eventually come closer and closer to the equilibrium position. This is almost obviously a stable situation.

The general solution of Eq. (1.8), for the position $x(t)$, assuming both m and k are positive parameters, is $x = A\sin(\omega_n t + const.)$ and $\omega_n = \sqrt{k/m}$ is the undamped natural frequency. Note that in Eq. (1.8) no term representing friction is present so the oscillation persists.

This analogy between Eq. (1.6) and Eq. (1.8) makes clear that the tapered wheelset (with the taper sense positive as shown in the Fig. 1.1) does tend to steer itself toward the center when disturbed and will follow a sinusoidal path down the track after being disturbed. Note that the solution of Eq. (1.8) assumed is correct if $(k/m)$is positive. The analogy holds because the corresponding term $(r_0\Omega^2 2\Psi/l)$ in Eq. (1.6) is also positive for the situation shown in Fig. 1.1.

It is easy and instructive to build a demonstration device that will show the self-centering properties of a wheelset using, for example, wooden dowels for the rails and paper cups taped together to make the wheelset with tapered wheels.

If the taper angle, Ψ were to be negative, the wheelset equation would be analogous to a mass-spring system but with a spring with a negative spring constant rather than a positive one. A spring with a negative spring constant would create a force tending to push the mass away from the equilibrium position $x = 0$ rather than tending to pull it back as a normal spring does.

Such a system would not return to an equilibrium position after a disturbance but would rather accelerate ever farther away from equilibrium. Thus one can see that a wheelset with a negative taper angle will tend to run off center until stopped by one of the flanges. In this case, one would surely conclude that the motion of a wheelset with a negative taper angle is unstable. (The demonstration device with the paper cups taped together such that the taper is the opposite of the taper shown in Fig. 1.1 will demonstrate this clearly.)

On the other hand, the wheelset with a positive taper angle while it is not unstable also is not strictly stable in this analysis since it continues to oscillate sinusoidally as it progresses down the rails. Just as a mass-spring oscillator with no damping would oscillate forever, the wheelset after an initial disturbance would not return to the basic motion but rather would continue to wander back and forth at a constant amplitude as it traveled along.

In the case of the wheelset with positive taper angle, the natural frequency is found by analogy to the mass spring system to be

$$\omega_n = \Omega\sqrt{(2r_0\Psi/l)} \tag{1.9}$$

The period of the oscillation is

$$T = \frac{2\pi}{\omega_n} = \frac{2\pi}{\Omega}\sqrt{\frac{l}{2r_0\Psi}} \tag{1.10}$$

and the wavelength of the oscillating motion is just the distance traveled in the time of one oscillation,

$$\lambda = \dot{x}T = r_0\Omega T = 2\pi r_0\sqrt{\frac{l}{2r_0\Psi}} \tag{1.11}$$

It is interesting to note that the wavelength is independent of Ω and hence the speed at which the wheelset is rolling along the track.

For cylindrical wheels with $\Psi = 0$, there would be neither a self-centering effect nor a tendency to steer away from the center and $\ddot{y}$ would equal zero. In this case, the wheelset would roll in a straight line until one of the flanges encountered the side of one of the rails if the initial value of θ were not precisely zero.

This example assumed rolling without any relative motion between the wheel and the rail at the contact point, which turns out to be unrealistic at high speed. What this simple example has shown is merely that tapered wheels tend to steer themselves toward the center of the tracks, but this mathematical model does not reveal whether the resulting oscillation would actually die down or build up in time. The tiniest change in the assumption of rolling without slip changes the purely sinusoidal motion to one that decays or grows slightly in amplitude.

This leaves open the possibility that the tapered wheelset, even with the positive taper angle, might actually be either truly stable or actually unstable. A more complicated model of the wheel-rail interaction is required to answer this question and it will be presented in Chapter 10.

Although this first example was analyzed in the time domain, Eq. (1.11) suggests that time is not inherently a part of this problem. The sinusoidal motion is the same in space regardless of the speed of the wheelset so it might be more logical the problem as a function of space, x, rather than time, t. The next introductory example actually uses a dynamic model based on rigid body dynamics that is inherently time-based. In this sense it is more typical of the vehicle models to follow.

III. THE DYNAMICS OF A SHOPPING CART

A large part of the rest of this book will be devoted to ground vehicles using pneumatic tires. It is not easy to start immediately to analyze such vehicles without beginning with a fairly long discussion of the means of characterizing the generation of lateral forces by tires. This is the subject of Chapter 4. On the other hand, it is possible to introduce some of the basic ideas of vehicle stability if a ground vehicle can be idealized in such a way the tire characteristics do not have to be described in any detail. In the introductory case to be discussed below, an idealized shopping cart, we will replace actual tire characteristics with simple kinematic constraints. This idealization actually works quite well for the hard rubber tires typically used on shopping carts up to the point at which the wheels actually are forced to slide sideways. This type of idealization allows one to focus on the dynamic model of the vehicle itself without worrying about details of the tire force generating mechanism.

Most courses in dynamics mainly treat inertial coordinate systems when applying the laws of mechanics to rigid bodies. As will be demonstrated in Chapter 5, this approach is logical when studying the stability of simple trailers but there is another way to write equations for freely moving vehicles such as automobiles and airplanes that turns out to be simpler and is commonly used in vehicle dynamics studies. The description of the vehicle motion involves the use of a body-fixed coordinate system, i.e., a coordinate system attached to the body and moving with it. This type of description is not commonly discussed in typical mechanics textbooks so it will be presented in a general framework in Chapter 2. In the present example, the two ways of writing dynamic equations will be presented in a simplified form.

Finally, the shopping cart example introduces the analysis of stability for dynamic vehicle models in the simplest possible way. The final dynamic equation of motion is only of first order so the eigenvalue problem that is at the heart of stability analysis is almost trivial. The general concepts and theory behind stability analysis for linearized systems is discussed in some detail in Chapter 3.

A supermarket shopping cart has casters in the front that are supposed to swivel so that the front can be pushed easily in any direction and a back axle with fixed wheels which are supposed to roll easily and to resist sideways motion. Anyone who has actually used a shopping cart realizes that real carts often deviate significantly from these ideals. The casters at the front often don't pivot well so the cart is hard to turn and one is often forced to skid the back wheels to make a sharp turn. Furthermore, the wheels often don't roll easily so quite a push is required to keep the cart moving. On the other hand, a mathematically ideal cart serves as a good introduction to the type of analysis used to study ground vehicles in general.

After analyzing the stability properties of the ideal cart, one can experiment with a non-ideal real cart and find that the basic conclusions do hold true in the main even when the idealizations are not strictly true. (This might be better done in a parking lot than in the aisles of a supermarket.) This example provides a prevue of the much more complex analysis of the lateral stability of automobiles in Chapter 6.

The interaction of the tires of the shopping cart with the ground will be highly simplified. It will be assumed that the front wheels generate no side force at all because of the casters. The front of the cart can be moved sideways with no side force required because the casters are supposed to swivel without friction and the wheels are assumed to roll also without friction.

At the rear, the wheels are assumed to roll straight ahead and to allow no sideways motion at all. Obviously, if the rear wheels are pushed sideways hard enough they will slide sideways, but if the side forces are small enough, the wheels roll essentially in the direction that they are pointed. It will also be assumed at the rear that there is negligible resistance to rolling.

A. Inertial Coordinate System

The first analysis will consider motion of the cart to be described in an x-y inertial reference frame. Figure 1.2 shows a view of the shopping cart seen from above. The coordinates x and y locate the center of mass of the cart with respect to the ground and ϕ is the angle of the cart centerline with respect to a line on the ground parallel to the x-axis. The x-y axes are

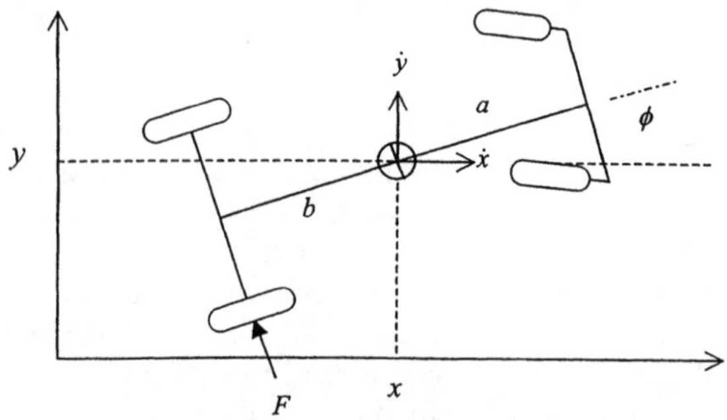

FIG. 1.2 Shopping cart described in an inertial coordinate frame.

supposed to be neither accelerating nor rotating and thus they are an inertial frame in which Newton's laws are easily written.

In Fig. 1.2, the parameters of the cart are a and b, the distance from the center of mass to the front and rear axles respectively, m, the mass and I_z, the moment of inertia about the mass center and with respect to the vertical axis. The basic motion will be considered to be along the x-axis so the variables needed to describe the motion are primarily $y(t)$ and $\phi(t)$. (In principle, $x(t)$ is also needed in general, but for the basic and the perturbed motion, it is assumed that $\dot{x} \cong U$, a constant, so $x(t) = Ut +$ constant. Thus the speed U will play the role of a parameter in the analysis.) In the basic motion, $\dot{y}$, ϕ and $\dot{\phi}$ all vanish and the cart proceeds along the x-direction at constant speed U.

For the perturbed motion, $\dot{y}$ is assumed to be small compared with U, and ϕ and $\dot{\phi}$ are also small. (To be precise, $b\dot{\phi}$ is small compared to U.) This means that the small angle approximations $\cos\phi \cong 1$ and $\sin\phi \cong \phi$ can be used when considering the perturbed motion. The side force at the rear is F and, since the rear wheels are assumed to roll without friction, there is no force in the direction the rear wheels are rolling. At the front axle there is no force at all in the x-y plane because of the casters. There are, of course, vertical forces at both axles that are necessary to support the weight of the cart, but they play no role in the lateral dynamics of the cart.

Because the cart is described in an inertial coordinate system and is executing plane motion, three equations of motion are easily written. They are force equals mass times acceleration in two directions and moment equals rate of change of angular momentum around the z-axis. Using the fact that the angle ϕ is assumed to be very small for the perturbed motion, the equations of motion are

$$m\ddot{x} = -F\sin\phi \cong -F\phi \cong 0 \tag{1.12}$$

$$m\ddot{y} = F\cos\phi \cong F \tag{1.13}$$

$$I_z\ddot{\phi} = -Fb \tag{1.14}$$

In Eq. (1.12), the acceleration in the x-direction is seen to be not exactly zero. However, we consider that the forward velocity is initially large and does not change much as long as the angle ϕ remains small. From Eqs. (1.13 and 1.14) we see that the acceleration in the y-direction and the angular acceleration are large compared to the x-direction acceleration when ϕ is small.

Considering the velocity of the center of the rear axle one can derive the condition of zero sideways velocity for the rear wheels. The basic kinematic velocity relation for two points on the same rigid body (see, for example, Crandall et al., 1968) is

$$\vec{v}_B = \vec{v}_A + \vec{\omega} \times \vec{r}_{AB} \tag{1.15}$$

where $\vec{\omega}$ is the angular velocity of the body and $\vec{r}_{AB}$ is the vector distance between A and B.

In the case at hand, let $\vec{v}_A$ represent the velocity of the mass center and let $\vec{v}_B$ represent the velocity of the center of the rear axle. Then $\vec{\omega}$ is a vector perpendicular to x and y in the z-direction. The magnitude of $\vec{\omega}$ is $\dot{\phi}$. The magnitude of $\vec{r}_{AB}$ is just the distance, b, between the center of gravity and the center of the rear axle.

Figure 1.3 shows the velocity components involved when Eq. (1.15) is evaluated. The component $b\dot{\phi}$ represents the term $\vec{\omega} \times \vec{r}_{AB}$ in Eq. (1.15).

From Fig. 1.3 one can see that the side velocity (with respect to the center line of the cart) at the center of the rear axle is

$$\dot{y}\cos\phi - b\dot{\phi} - U\sin\phi \tag{1.16}$$

Using the small angle approximations, the final relation needed is simply a statement that this sideways velocity of the center of the rear axle should vanish.

$$\dot{y} - b\dot{\phi} - U\phi = 0 \tag{1.17}$$

(It is true but perhaps not completely obvious at first that if the two rear wheels cannot move sideways but can only roll in the direction that they are pointed instantaneously, then any point on the rear axle also cannot have any sideways velocity. This means that the kinematic constraint of Eq. (1.17) derived for the center of the axle properly constrains the variables such that the two wheels also have no side velocity.)

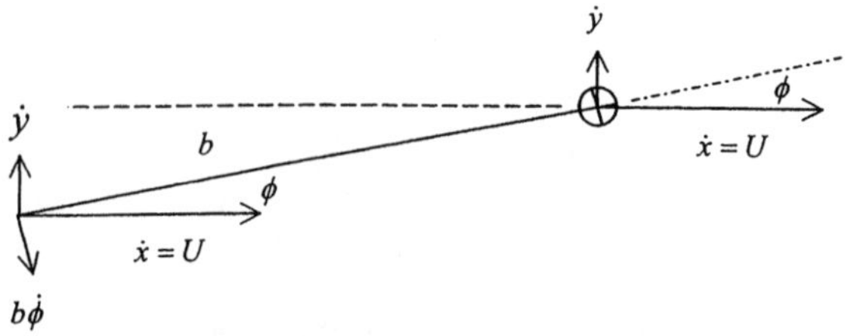

FIG. 1.3 Velocity components at the center of the rear axle.

Now combining Eqs. (1.13 and 1.14), one can eliminate F yielding a single dynamic equation. Then, after differentiating Eq. (1.17) with respect to time, $\ddot{y}$ can be eliminated from this dynamic equation to yield a single equation for ϕ. The result is

$$(I_z + mb^2)\ddot{\phi} + mbU\dot{\phi} = 0 \tag{1.18}$$

Equation (1.18) may appear to be of second order since it involves a term containing $\ddot{\phi}$, but because ϕ itself is missing, a first order version of Eq. (1.18) can be studied instead.

The angle ϕ actually has no particular significance. It is just the angle the cart makes with the x-axis, which itself is a line on the ground in an arbitrary direction. Since the angle ϕ itself does not appear in the Eq. (1.18), it is logical to consider the angular rate $\dot{\phi}$ as the basic variable rather than ϕ. The angle ϕ is called the *yaw* angle and it is common to call the angular rate $\dot{\phi}$ the *yaw rate* and to give it the symbol r in vehicle dynamics. Standard symbols usually used in vehicle dynamics will be presented later in Chapter 2.

In terms of yaw rate, the Eq. (1.18) can be rewritten in first order form. Noting that $\dot{\phi} = r$, $\ddot{\phi} = \dot{r}$, the result is

$$\dot{r} + \{mbU/(I_z + mb^2)\}r = 0 \tag{1.19}$$

This equation is a linear ordinary differential equation with constant coefficients and is of the general form

$$\dot{r} = A\mathrm{r} \tag{1.20}$$

with $A = -mbU/(I_z + mb^2)$.

The assumption that it is possible for stability analysis to consider small perturbations from a basic motion generally leads to linear differential equations. In the present case, the nonlinear equations, Eqs.(1.12, 1.13, and 1.16), became linear because of a small perturbation assumption.

The analysis of the stability of the first order equation, Eq. (1.20), is elementary. The solution to this linear equation can be assumed to have an exponential form such as $r = \mathrm{R}e^{st}$. Then $\dot{r} = s\mathrm{R}e^{st}$, which when substituted into Eq. (1.20), yields $s\mathrm{R}e^{st} = A\mathrm{R}e^{st}$ or

$$(s - A)\mathrm{R}e^{st} = 0 \tag{1.21}$$

This result is the basis of an *eigenvalue* analysis. The general theory of eigenvalues and their use in stability analysis will be presented in Chapter 3. For now it is enough to note that when three factors must multiply to zero, as in Eq. (1.21), the equation will be satisfied if any one of the three factors vanishes.

For example, if R were to be zero, the product in Eq. (1.21) would certainly be zero. This represents the so-called *trivial solution*. The assumed solution would then simply imply $r(t) = 0$, which should have been an obvious possibility from the beginning. Another possible factor to vanish in Eq. (1.21) is e^{st}. Not only is the vanishing of this factor not really possible, but also even if it were, the result would again be the trivial solution. The only important condition is the vanishing of the final factor in Eq. (1.21), which happens when

$$s = A \tag{1.22}$$

Thus we have determined that the *eigenvalue* s takes on the value A and the only nontrivial solution has the form

$$r = \mathrm{R}e^{At} \tag{1.23}$$

in which R is an arbitrary constant determined by the initial value of r at time $t = 0$. Eigenvalues will be discussed at length in Chapter 3.

Figure 1.4 shows the nature of the solutions for the cases when A is greater than and less than zero. When A is negative, the system is stable and the yaw rate will return to zero after a disturbance. If A is positive, the yaw rate increases in time, the cart begins to spin faster and faster and the system is unstable.

It the case of the shopping cart with

$$A = -mbU/(I_z + mb^2) \tag{1.24}$$

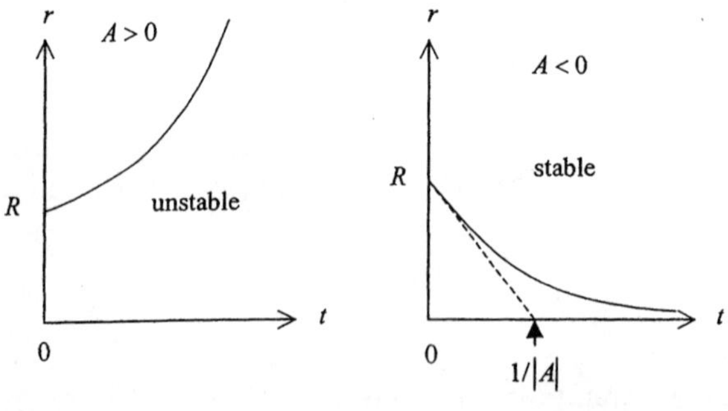

FIG. 1.4 Responses for first order unstable and stable systems.

it is clear that A will normally be negative since all the parameters are positive and there is a negative sign in front of the expression. The motion will then be stable as long as U is positive as shown in Fig. 1.3 since all other parameters are inherently positive.

This means that if a shopping cart is given a push in the normal direction but with an initial yaw rate, the yaw rate will decline towards zero exponentially as time goes on. Ultimately, the cart will roll in essentially a straight line.

On the other hand, Eq. (1.17) remains valid whether U is considered to be positive in the sense shown in Fig. 1.3 or negative and the same is true of Eq. (1.24). If the cart is pushed backwards, one can simply consider U to be negative. Then the combined parameter A in Eq. (1.24) will be positive and if the yaw rate is given any initial value however small, the yaw rate will increase exponentially in time. In fact, the cart will eventually turn around and travel in the forward direction even though the linearized dynamic equation we have been using cannot predict this since the small angle approximations no longer are valid after the cart has turned through a large angle.

Another interesting aspect of first order equation response is the speed with which the stable version returns to zero. If the response is written in the form

$$r = Re^{-t/\tau} \tag{1.25}$$

where τ is defined to be the *time constant*,

$$\tau = 1/|A| \tag{1.26}$$

it is clear that the time constant is just the time at which the initial response decays to $1/e$ times the initial value.

For the shopping cart,

$$\tau = (I_z + mb^2)/mbU \tag{1.27}$$

Figure 1.4 shows that the time constant is readily shown on a plot of the response in the stable case by extending the slope of the response from an initial point (assuming $t = 0$ at the initial point) to the zero line. The time constant is the time when the extended slope line intersects the zero line.

From Eq.(1.27) it can be seen that the time constant depends on the physical parameters of the cart and the speed. Increasing the speed decreases the time constant and increases the quickness with which the raw rate declines toward zero for the stable case. Pushing the cart faster in the backwards direction also quickens the unstable growth of the yaw rate.

B. Body-Fixed Coordinate System

The use of an inertial coordinate system to describe the dynamics of a vehicle may seem reasonable at first, but in fact many analyses of vehicles use a moving coordinate system attached to the vehicle itself. The vehicle motion is described by considering linear and angular velocities rather than positions. (In the previous analysis of the shopping cart, the angular position of the cart turned out to be less important than the angular rate.)

To introduce this idea we will now repeat the previous stability analysis using a coordinate system attached to the center of mass of the cart and rotating with it. This will require a restatement of the laws of dynamics taking into consideration the non-inertial moving coordinate system. The general case of rigid body motion described in a coordinate system attached to the body itself and executing three-dimensional motion will be presented in Chapter 2. Here we will present the simpler case of plane motion appropriate for the shopping cart.

Figure 1.5 shows the moving coordinate system and the velocity components associated with it. In this description of the motion, the basic motion consists of a constant forward motion with velocity, U which again will function as a parameter. The variables for the cart are now the side or lateral velocity V and the yaw rate r. This notation is in conformity with the general notation to be introduced in Chapter 2. For the basic motion, the side velocity and angular velocity are zero, $V = r = 0$.

The perturbed motion has small lateral velocity and yaw rate, $V \ll U$, $r \ll U/b$. Again the parameters are a, b, m, I_z, U.

The x-y coordinate system moves with the cart and the side force Y at the rear axle always points exactly in the y-direction.

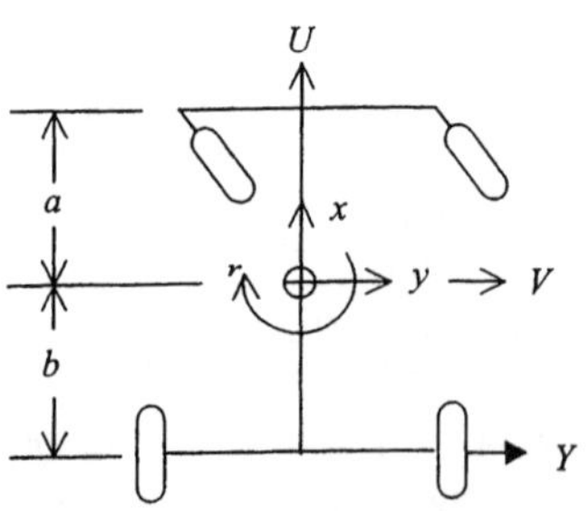

$x-y-z$ coordinates are attached at the center of mass and move with the cart

U,V are velocity components with respect to x,y

r is the yaw rate (angular velocity around the z-axis)

Y is the force on the rear axle

FIG. 1.5 Coordinate system attached to the shopping cart.

We are now in a position to write equations relating force to the mass times the acceleration of the center of mass and the moment to the change of angular momentum. Because the x-y-z is rotating with the body, we must properly find the absolute acceleration in terms of components in a rotating frame.

The fundamental way to find the absolute rate of change of any vector $\vec{v}$, measured in a frame rotating with angular velocity $\vec{\omega}$ is given by the well-known formula

$$\frac{d\vec{v}}{dt} = \left.\frac{d\vec{v}}{dt}\right|_{\text{rel}} + \vec{\omega} \times \vec{v} \tag{1.28}$$

Again see Crandall et al., 1968, for example.

The first term on the right of Eq. (1.28) represents the rate of change of the vector as seen in the coordinate system itself and the second term corrects for the effects of the rotation of the coordinate system. In the case at hand, we are interested in the acceleration so $\vec{v}$ represents the velocity with components U and V and the only component of $\vec{\omega}$ is the z-component, r. In Chapter 2, general formulas for accelerations in vehicle centered coordinate systems will be derived but for now the relevant expressions resulting from the use of the general formula will simply be stated. The acceleration components in the x- and y-directions upon application of Eq. (1.28) are

$$a_x = \dot{U} - rV \tag{1.29}$$

$$a_y = \dot{V} + rU \tag{1.30}$$

The equations of motion then are

$$m(\dot{U} - rV) = X = 0 \tag{1.31}$$

$$m(\dot{V} + rU) = Y \tag{1.32}$$

$$I_z\dot{r} = -bY \tag{1.33}$$

where the moment about the center of mass in the positive z-direction is due only to the force on the back axle and is negative when Y is positive as shown in the sketch. These equations have the same meaning as Eqs.(1.12–1.14) but they are expressed in the body-fixed coordinate system.

From Eq. (1.31), one can see that even when the x-force is zero, the forward velocity U is not really constant. In fact, $\dot{U} = rV$. For the basic motion both r and V are zero and U is exactly constant. For the perturbed motion both V and r are assumed to be small, $\dot{U} \cong 0$ and $\dot{U}$ is nearly but not exactly constant.

To complete the analysis, we still have to put in the fact that the side velocity at the rear axle is zero. Using the same kinematic law used previously, Eq. (1.15), the side velocity at the center point of the rear axle is the side velocity of the center of mass plus the side velocity due to rotation. Considering the Fig. 1.5, the side velocity is $V-br$ and the constraint that this velocity is assumed to vanish is

$$V - br = 0 \tag{1.34}$$

This relation corresponds to Eq. (1.17) in the previous formulation using the inertial coordinate system. (Once again it can be verified that if the center of the axle is constrained to have no side velocity, then there is no side velocity at either wheel attached to the axle.)

Using the body-fixed coordinate system there is no need to assume that the yaw angle is small as was required to derive Eq. (1.17). In fact the yaw angle will not enter the analysis at all with the body-fixed coordinate system.

Now, considering U to be essentially constant, the Eqs. (1.32 and 1.33) and the zero side velocity relation Eq. (1.34) form a set of three equations for the three variables V, r and Y. It is left as an exercise to show that if V and Y are eliminated from these three equations the result is the same equation of motion as before, Eq. (1.19). Thus the use of the coordinate system attached to the cart produces exactly the same equation of motion as was produced using the inertial coordinate system. (It would be a great problem for Newtonian mechanics if two alternative ways to set up an equation of motion for a system did not produce an equivalent if not identical equation.) The stability analysis is then identical for vehicle motion whether an inertial reference frame or a moving frame is used.

It turns out that the use of a coordinate system moving with the vehicle is in many cases the most convenient way to describe the dynamics of vehicles. Freely moving vehicles such as cars and airplanes are often described with the help of body fixed coordinate system such as the one just used for the shopping cart. Unfortunately this type of coordinate system is rarely discussed in basic dynamics texts so it is worthwhile to devote some time in Chapter 2 to deriving general equations of motion for a rigid body using this type of coordinate system.

2
Rigid Body Motion

I. INTRODUCTION

The first step in the analysis of vehicle stability is the construction of a suitable mathematical model of the vehicle. This always involves the classical kinematic and dynamic laws of mechanics as well as basic assumptions about the type of model and the characterization of the force generating elements. Although one could argue that an accurate mathematical model of a vehicle should include nonlinear and distributed parameter effects, an overly complicated model may well be as useless as an overly simplified one if the point of the model is to understand the stability properties of a certain type of vehicle.

If the model is so complex that only a large computer analysis is possible, then it may be difficult to discern any patterns in the parameter values that lead either to stability or instability. In this book, we will attempt to gain insight into how and why the dynamics of vehicles can lead to unstable behavior. Thus the concentration here will be on the simplest possible models that can contribute to understanding of the phenomenon. More accurate and therefore more complex models are certainly appropriate for the detailed design of specific vehicles.

Many vehicles are usefully described as a single rigid body moving in space for the purposes of analyzing the stability of motion. Of course, airplane wings do flap and convertible car cowls do shake, but in many cases the gross motion of these vehicles is not much affected by these vibratory motions so the rigid body model is a useful simplification. It is a somewhat surprising fact of mechanics that even the description of the motion of a single rigid body is not necessarily a simple matter.

Other vehicles are composed of several connected essentially rigid bodies. Tractor-trailer trucks and railroad trains are vehicle systems of this sort. Such composite vehicles bring their own complications to the mathematical modeling process, but they represent a straightforward extension of single rigid body models. In this chapter, we will concentrate on the description of the motion of a single rigid body moving in three-dimensional space or in a two-dimensional plane.

II. INERTIAL FRAME DESCRIPTION

Most elementary expositions of the principles of dynamics concentrate on writing equations in an inertial coordinate frame (Crandall et al., 1968). In this case, the motion is described with respect to a coordinate frame in which Newton's laws are supposed to be valid. It turns out to be surprisingly difficult to describe precisely what an inertial frame should be, but when considering vehicles moving near earth, a frame attached to the surface of the earth is often assumed to be an inertial frame. As the earth is rotating this is clearly only an approximation. Most vehicles exhibit much greater accelerations than those associated with the movements of the surface of the earth so the errors connected with this assumption are generally negligible. Roughly speaking, other inertial coordinate frames should neither rotate nor accelerate with respect to a basic inertial frame such as one attached to the surface of the earth.

Rigid bodies moving in an inertial frame are said to have six degrees-of-freedom, meaning that six quantities are required to specify the position of the body in the three-dimensional space. For example, three position variables could be defined to locate the center of mass, and three angles, such as Euler angles, could be defined to locate the body in an angular sense. Even for a single body, it is not a simple job to define the position variables and to write the corresponding equations of motion.

Often, one can study restricted aspects of motion in which the motion is restricted to a plane and only one or two degrees-of-freedom are required. The shopping cart of Chapter 1 was assumed to execute only plane motion. In such cases, the motion description and the writing of the equations of motion in an inertial frame are accomplished with relatively little difficulty. The only disadvantage to this approach was that the equation of motion appeared to be of second order when it proved to be actually of first order as far as a stability analysis was concerned. The problem of describing three-dimensional motion in an inertial coordinate system is considerably more complex.

The study of trailer dynamics in Chapter 5 uses inertial frames for an entirely different reason. In this case, the towing vehicle is assumed to be moving in a straight line in inertial space and it is the motion of the trailer with respect to this inertial space that is important. In such constrained motion cases it is natural to use an inertial coordinate frame to describe the vehicle motion.

However, in the field of vehicle dynamics, an alternative description of motion is commonly used to describe the motion of vehicles free to move arbitrarily in space. Examples of such vehicles are automobiles and airplanes. It is often the case that forces on the vehicle actually depend neither upon where the vehicle finds itself in inertial space nor on the vehicle's angular orientation with respect to inertial space. If this is the case, instead of using an inertial coordinate frame to describe the vehicle motion, the motion is more conveniently described in a coordinate frame emanating from the center of mass of the vehicle and moving with it. An example of the use of this type of description has already been given in Chapter 1 for the shopping cart.

The moving body-fixed coordinate frame is rarely discussed in standard expositions of dynamics. For this reason, a brief outline of the dynamic equations for a rigid body in a body-fixed frame is given below. Although it will be seen that the most general form of the dynamic equations in a body-centered coordinate frame can be quite complex, in many cases including those presented in subsequent chapters useful simplifications based on symmetries and restrictions on the motions to be studied can be made. When these simplifications are made, the equations of motion become simple and straightforward in the body-fixed coordinate system and the equations are often easier to understand than the equivalent equations formulated using an inertial coordinate frames.

It is not absolutely necessary to study the most general equations presented in this chapter in great detail at this point as only these simplified versions of the equations will be used in subsequent chapters. It is important, however, to understand how the basic principles of dynamics are represented in this specific moving reference frame because the expressions for accelerations assume somewhat nonintuitive forms in the body-fixed coordinate frame. The results presented in this chapter will be referred to later when analyzing specific vehicles.

As an illustration of the usefulness of the three-dimensional equations of motion, the spin stabilization of satellites will be briefly discussed in this chapter.

III. BODY-FIXED COORDINATE FRAME DESCRIPTION

In many cases the description of vehicle dynamics is best done in a frame moving with the vehicle itself. This is a rotating and thus noninertial frame so the equations of motion take on different forms from those usually presented in dynamics textbooks. In their most general form, the equations even for a single rigid body are quite complex as will be seen below, but often one does not need all possible degrees-of-freedom to study a particular phenomenon and vehicles typically have symmetries which reduce the complexities of the equations considerably.

Fig. 2.1 shows a coordinate frame attached at the vehicle center of mass and moving with the vehicle. A standard notation is used for the velocity of the center of mass and angular velocity components as indicated in Fig. 2.1. When appropriate, this notation will be used in subsequent chapters. The x axis points in the forward direction and the y axis point to the right from the point of view of the pilot or driver in Fig. 2.1. For a right-handed system the z axis then points down. Some authors of vehicle dynamics papers prefer the z axis to point up rather than down. In this case the y axis points to the left.

It is important to use a right-handed coordinate system because the dynamic equations involve cross products and, if a left-handed coordinate system is inadvertently used, a number of terms will have incorrect signs.

The velocity of the center of mass, $\vec{v}$, and the angular velocity of the body, $\vec{\omega}$, can be expressed as follows:

$$\vec{v} = U\vec{i} + V\vec{j} + W\vec{k} \tag{2.1}$$

$$\vec{\omega} = p\vec{i} + q\vec{j} + r\vec{k} \tag{2.2}$$

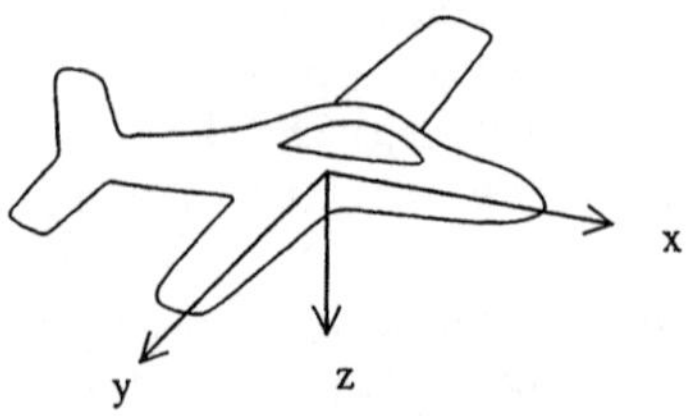

Coordinates	x y z
Unit vectors	$\vec{i}$ $\vec{j}$ $\vec{k}$
Velocity components	U V W
Angular velocity Components	p q r

FIG. 2.1 Standard coordinate system for vehicle dynamic quantities.

Alternatively, these quantities can be expressed as column vectors.

$$[v] = \begin{bmatrix} U \\ V \\ W \end{bmatrix} \tag{2.3}$$

$$[\omega] = \begin{bmatrix} p \\ q \\ r \end{bmatrix} \tag{2.4}$$

It may be worthwhile to note here that the notation for angular velocity components is not as easy to remember as one might at first think. Referring to Fig. 2.1, it is clear that p, q, and r are angular rates around the x, y, and z axes, respectively. But the names given to these quantities do not correspond very well with their symbols: p is called the *r*oll rate, q is called the *p*itch rate, and r is called the yaw rate. It might be nicer if the first letters of the names "pitch" and "roll" corresponded to p and r, but this is unfortunately not the case.

Another notable feature of the body-fixed coordinate system is that while velocities and accelerations can be expressed in the coordinate system, the position of the body is not represented at all. In contrast, when inertial coordinates are used, it is common to first describe the position of a body and then proceed to find expressions for velocities and accelerations. This was done for the shopping cart in Chapter 1 in the section in which inertial coordinates were used. As this example showed, there are cases in which the position or motion path is not important for a stability analysis although positions and the motion path may be of interest for other reasons.

In case it is of interest to discover the position of the body, by computer simulation, for example, it is common either to use an inertial coordinate system from the beginning or to use a body-centered frame and to introduce another inertial frame in which the position of the body can be found by means of integration of velocity components. An illustration of this technique is found in Chapter 7. The extra equations that need to be integrated to find the path taken by the tilting vehicle being analyzed are found in Eq. (7.6).

The net force, $\vec{F}$, on the body and the net moment or torque on the body, $\vec{\tau}$, are defined as vectors using the symbols X, Y, Z and L, M, N for components as indicated in Eqs. (2.5) and (2.6).

$$\vec{F} \equiv X\vec{i} + Y\vec{j} + Z\vec{k} \tag{2.5}$$

$$\vec{\tau} \equiv L\vec{i} + M\vec{j} + N\vec{k} \tag{2.6}$$

These general force and moment components can always be defined, of course, no matter how the forces and moments are determined, but in certain cases, these definitions are particularly useful. This is the case when the force components remain in the x–y–z directions as the vehicle moves. For example, it may be that the thrust force from an airplane propeller can be assumed always to act in the x direction or that the lateral force on a rear tire of a car always acts in the y direction. On the other hand, forces or moments that depend on the linear or rotational position of the body are harder to represent in the forms of Eqs. (2.5 and 2.6) because the body-centered coordinate system itself is incapable of representing the position of the body.

The linear momentum, $\vec{P}$, is a vector. It is proportional to the velocity of the center of mass with the scalar mass, m, being the proportionality constant.

$$\vec{P} = m\vec{v} \tag{2.7}$$

The angular momentum is also a vector, but it is best represented as a column vector related to the angular velocity.

$$[H_c] = [I_c][\omega] \tag{2.8}$$

The relationship between the angular momentum and the angular velocity in Eq. (2.8) involves the centroidal inertia matrix, which has nine components.

$$[I_c] = \begin{bmatrix} I_{xx} & I_{xy} & I_{xz} \\ I_{yx} & I_{yy} & I_{yz} \\ I_{zx} & I_{zy} & I_{zz} \end{bmatrix} \tag{2.9}$$

The components are called moments of inertia (along the diagonal of the matrix) and products of inertia (for the off-diagonal terms) and are defined by the well-known relationships

$$I_{xx} = \int (y^2 + z^2)dm, \quad I_{xy} = -\int (xy)dm, \text{ etc.} \tag{2.10}$$

Refer to any standard dynamics text such as Crandall et al. (1968).

The defining relationships in Eq. (2.10) imply that the matrix is symmetrical so that only six of the nine components are independent. Because most vehicles are nearly symmetrical about certain planes, at least some of the products of inertia are commonly assumed to be zero. If it happens that the x–y–z axes are the so-called principal axes for the body, then all the off-diagonal products of inertia in Eq. (2.9) vanish.

All rigid bodies have three orthogonal directions in the body that are the principal directions. If the x–y–z axes are aligned with the principal directions, then the axes become principal axes. When it is reasonable to assume that the x–y–z axes are principal axes, the equations of motion take on a highly simplified form.

A. Basic Dynamic Principles

The basic dynamic principles for a rigid body can be stated in a form that is valid for any coordinate frame.

$$\vec{F} = d\vec{P}/dt = m\vec{a} \tag{2.11}$$

$$\vec{\tau} = d\vec{H}_c/dt, \ [H_c] = [I_c][\omega] \tag{2.12}$$

Note that because Eq. (2.11) involves the scalar mass, it can be written as a simple vector equation. On the other hand, Eq. (2.12) is complicated by the tensor or matrix nature of $[I_c]$ in Eq. (2.9). As will be seen when Eqs. (2.11) and (2.12) are written out in component form, the dynamic equations for rotation in three dimensions are considerably more complicated than those for linear motion.

B. General Kinematic Considerations

When the velocity components are defined with respect to a coordinate frame that is itself rotating, a standard technique must be used to find the absolute acceleration in terms of the acceleration with respect to the frame plus the effect of the frame rotation. This relation is typically presented in basic dynamics texts, but rarely applied to generate equations of motion in body-centered frames as will be done here.

Let an arbitrary vector $\vec{A}$ be expressed in a coordinate frame rotating with respect to an inertial frame, $\vec{A} = A_x\vec{i} + A_y\vec{j} + A_z\vec{k}$, and let $\vec{\omega}$ be the angular velocity of the rotating frame, $0xyz$. Then the derivative of the vector as it would be seen in an inertial frame can be related to the derivative in the rotating frame by the relation

$$d\vec{A}/dt = (\partial\vec{A}/dt)_{rel} + \vec{\omega} \times \vec{A} \tag{2.13}$$

The first term on the right of Eq. (2.13) is what an observer in the moving frame would think the change in the vector $\vec{A}$ would be. Expressed in terms of column vectors, the relative change in the vector $\vec{A}$ as it would appear in the rotating coordinate frame is simply

$$\left[\frac{\partial \vec{A}}{\partial t_{rel}}\right] = \begin{bmatrix} \dot{A}_x \\ \dot{A}_y \\ \dot{A}_z \end{bmatrix} \tag{2.14}$$

The second term on the right of Eq. (2.13) can be evaluated using the standard determinate form for a cross product and the definition of $\vec{\omega}$ from Eq. (2.4).

$$\left[\vec{\omega} \times \vec{A}\right] = \det \begin{vmatrix} \vec{i} & \vec{j} & \vec{k} \\ p & q & r \\ A_x & A_y & A_z \end{vmatrix} = \begin{matrix} \vec{i}(qA_z - rA_y) \\ -\vec{j}(pA_z - rA_x) \\ +\vec{k}(pA_y - qA_x) \end{matrix} \tag{2.15}$$

The final result is best expressed in a column vector form.

$$\left[\frac{d\vec{A}}{dt}\right] = \begin{bmatrix} \dot{A}_x + qA_z - rA_y \\ \dot{A}_y + rA_x - pA_z \\ \dot{A}_z + pA_y - qA_x \end{bmatrix} \tag{2.16}$$

Now this result will be used in the dynamic equations. First, Eq. (2.16) will be applied to the velocity of the center of mass to find the acceleration of the center of mass. Equation (2.11) becomes the following in the body-fixed system:

$$[F] = \begin{bmatrix} X \\ Y \\ Z \end{bmatrix} = m \begin{bmatrix} \dot{U} + qW - rV \\ \dot{V} + rU - pW \\ \dot{W} + pV - qU \end{bmatrix} \tag{2.17}$$

When Eq. (2.16) is applied to the angular momentum law, Eq. (2.12), the result is quite complex in general. First, the angular momentum is expressed in column vector form using Eqs. (2.8) and (2.9).

$$[H_c] = \begin{bmatrix} H_x \\ H_y \\ H_z \end{bmatrix} = [I_c][\omega] = \begin{bmatrix} I_{xx}p + I_{xy}q + I_{xz}r \\ I_{yx}p + I_{yy}q + I_{yz}r \\ I_{zx}p + I_{zy}q + I_{zz}r \end{bmatrix} \tag{2.18}$$

Then Eq. (2.16) can be applied to the angular momentum column vector. The result is

$$[\tau] = \begin{bmatrix} L \\ M \\ N \end{bmatrix} = \begin{bmatrix} I_{xx}\dot{p} + I_{xy}\dot{q} + I_{xz}\dot{r} \\ I_{yx}\dot{p} + I_{yy}\dot{q} + I_{yz}\dot{r} \\ I_{zx}\dot{p} + I_{yz}\dot{q} + I_{zz}\dot{r} \end{bmatrix}$$

$$+ \begin{bmatrix} q(I_{zx}p + I_{zy}q + I_{zz}r) - r(I_{yx}p + I_{yy}q + I_{yz}r) \\ r(I_{xx}p + I_{xy}q + I_{xz}r) - p(I_{zx}p + I_{zy}q + I_{zz}r) \\ p(I_{yx}p + I_{yy}q + I_{yz}r) - q(I_{xx}p + I_{xy}q + I_{xz}r) \end{bmatrix} \tag{2.19}$$

This is the general dynamic equation relating the moments about the x-, y-, and z axes to the rates of change of the angular velocity components. The contrast between Eqs. (2.17) and (2.19) is striking and is due to the differences between the otherwise analogous Eqs. (2.11) and (2.12).

The situation is often not quite as complicated as these general equations might suggest. For many vehicles, the x–y–z axes are lined up at least partially with the principal inertial axes of the vehicle. Recall that this means that, if this is the case, some (or perhaps even all) of the products of inertia vanish, which simplifies Eqs. (2.18) and (2.19).

For example if the x–z plane is a plane of symmetry, then $I_{xy} = I_{yx} = I_{yz} = I_{zy} = 0$. Then Eq. (2.18) becomes

$$\begin{bmatrix} H_x \\ H_y \\ H_z \end{bmatrix} = \begin{bmatrix} I_{xx}p + I_{xz}r \\ I_{yy}q \\ I_{zz}r + I_{zx}p \end{bmatrix} \tag{2.20}$$

and Eq. (2.19) simplifies to

$$\begin{bmatrix} L \\ M \\ N \end{bmatrix} = \begin{bmatrix} I_{xx}\dot{p} + I_{xz}\dot{r} \\ I_{yy}\dot{q} \\ I_{zx}\dot{p} + I_{zz}\dot{r} \end{bmatrix}$$

$$+ \begin{bmatrix} q(I_{xz}p + I_{zz}r) - (I_{yy}q) \\ r(I_{xx}p + I_{zx}r) - p(I_{zx}p + I_{zz}r) \\ p(I_{yy}q) - q(I_{xx}p + I_{xz}r) \end{bmatrix} \tag{2.21}$$

It is also common that one does not always need to study complete three-dimensional motion. For land vehicles, for example, the heave velocity W and the pitch angular velocity q are sometimes ignored. Assuming the symmetry discussed above leading to Eqs. (2.20) and (2.21) and further assuming W and q are zero, a smaller number of fairly simple dynamic equations result when compared with the general case.

$$X = m(\dot{U} - rV) \tag{2.22}$$

$$Y = m(\dot{V} + rU) \tag{2.23}$$

$$L = I_{xx}\dot{p} + I_{xz}\dot{r} \tag{2.24}$$

$$N = I_{zz}\dot{r} + I_{zx}\dot{p} \tag{2.25}$$

Equations about as complex as Eqs. (2.22)–(2.25) will be used to study the stability of automobiles (by neglecting roll and heave motions) and the longitudinal stability of airplanes. Both cases involve plane motion rather than general three-dimensional motion.

It is also worth noting that most vehicles have obvious planes of symmetry and are usually loaded nearly symmetrically, so it is rare that the most general forms of the equations of motion, Eqs. (2.17) and (2.19), are used in analytical stability studies.

IV. SPIN STABILIZATION OF SATELLITES

Although most of the use of body-fixed coordinates to describe vehicle dynamics in succeeding chapters will deal with plane motion, the general three-dimensional motion equations have a number of interesting applications. Although it may be stretching the point to consider a satellite in space a vehicle, a satellite does transport instruments and communication gear. The satellite itself represents a nearly rigid body moving in three dimensions and it has some interesting stability problems. Furthermore, the general equations, Eqs. (2.17) and (2.19), are well suited to describing satellite motion.

In outer space, the forces and torques exerted on compact bodies are extremely small and for many purposes can be neglected. (There is, of course, a large gravity force that provides the acceleration associated with the orbit of the satellite, but we are assuming the inertial coordinate system is moving with the satellite and in this system the satellite appears to be "weightless.")

If the forces X, Y, and Z are assumed to vanish, and if the angular velocity components p, q, and r are also assumed to be zero, then Eq. (2.17) predicts that the linear velocity components U, V, and W will be constant. This means that a satellite traveling near a space ship in the same basic orbit would have a constant velocity with respect to the space ship. This result dealing with linear velocities contains no surprises, but a study of rotation is more interesting and complex than one might expect.

If one assumes that the moments L, M, and N are zero and that the angular rates p, q, and r are initially zero, then Eq. (2.19) is satisfied when $\dot{p}$, $\dot{q}$, and $\dot{r}$ are zero. That is, a nonrotating satellite remains nonrotating in space.

On the other hand, if even a small moment or angular impulse from a particle hitting the satellite should occur, the satellite will acquire an angular velocity and it will slowly change attitude. This may disturb the orientation of sensitive antennas. This change in orientation can be corrected if the satellite has thrusters, but the less the thrusters need to be used the better because the amount of rocket fuel on board is strictly limited.

One idea to reduce the influence of torque disturbances is spin stabilization. The idea is to spin the satellite around a principal axis. The satellite then acts somewhat like a gyroscope and is more resistant to changes in the direction of the spin axis because of the small disturbance torques than it would be if it were not spinning.

To analyze spin stabilization, let us first assume that the x–y–z axis system is a principal axis system. Every rigid body has three principal axis directions oriented somewhere in the body and they are always at right angles to each other (Crandall et al., 1968). With this assumption, Eq. (2.19) simplifies considerably because

$$I_{xy} = I_{yx} = I_{xz} = I_{zx} = I_{yz} = I_{zy} = 0 \tag{2.26}$$

It is easy to see from Eq. (2.19) that the satellite can spin around any one of the three principal axes if the moments are zero and if the angular velocities about the other two axes also vanish. For example, suppose the satellite is spinning around the x axis with

$$q = r = o \tag{2.27}$$

Equation (2.19) then yields

$$\begin{bmatrix} I_{xx}\dot{p} \\ I_{yy}\dot{q} \\ I_{zz}\dot{r} \end{bmatrix} = \begin{bmatrix} 0 \\ 0 \\ 0 \end{bmatrix} \tag{2.28}$$

This result shows that if Eq. (2.27) is true at some initial time then

$$p = \Omega = \text{constant} \tag{2.29}$$

and Eq. (2.27) will remain true as time goes on. It is easy to show that similar results hold for spin about any one of the principal axes.

We now consider spin about a principal axis as a basic motion and consider a perturbed motion in which angular velocities about the remaining two axes have small but nonzero values. This will lead to an analysis of the stability of spin stabilization.

The assumption will be that the spin is about the x axis,

$$p = \Omega, \; q = \Delta q, \; r = \Delta r \tag{2.30}$$

where Δq and Δr are supposed to be small with respect to the spin speed Ω.

The first equation in Eq. (2.19) under the assumptions of Eq. (2.30) becomes

$$I_{xx}\dot{p} + \Delta q \Delta r (I_{zz} - I_{yy}) = 0 \tag{2.31}$$

Under the assumption that the product of two small quantities $\Delta q \Delta r$ is negligible, we see that p is nearly zero and that Eq. (2.29) is approximately true for the perturbed as well as for the basic motion because $\dot{p}$ in Eq. (2.31) nearly vanishes.

The last two equations in Eq. (2.19) yield the following two equations for Δq and Δr after some minor manipulation:

$$\Delta\dot{q} + \frac{(I_{xx} - I_{zz})\Omega}{I_{yy}} \Delta r = 0 \tag{2.32}$$

$$\Delta\dot{r} + \frac{(I_{yy} - I_{xx})\Omega}{I_{zz}} \Delta q = 0 \tag{2.33}$$

An interesting pattern arises when these two equations are put in a vector–matrix form.

$$\begin{bmatrix} \Delta\dot{q} \\ \Delta\dot{r} \end{bmatrix} + \begin{bmatrix} 0 & \frac{(I_{xx}-I_{zz})\Omega}{I_{yy}} \\ \frac{(I_{yy}-I_{xx})\Omega}{I_{zz}} & 0 \end{bmatrix} \begin{bmatrix} \Delta q \\ \Delta r \end{bmatrix} = \begin{bmatrix} 0 \\ 0 \end{bmatrix} \tag{2.32a, 2.33a}$$

These two equations are in a form that will be discussed in some detail in Chapter 3 but similar equations have already been seen for the tapered wheelset in Chapter 1 [Eqs. (1.6) and (1.7)].

These two first-order equations, Eqs. (2.32) and (2.33), can be converted into a single second-order form similar to Eq. (1.6) quite easily.

If Eq. (2.32) is differentiated once with respect to time, then $\Delta\dot{r}$ from Eq. (2.33) can be substituted to find a single equation for Δq.

$$\Delta\ddot{q} + \frac{(I_{xx} - I_{zz})(I_{xx} - I_{yy})\Omega^2}{I_{yy}I_{zz}}\Delta q = 0 \tag{2.34}$$

The same equation arises if Eqs. (2.32) and (2.33) are manipulated to eliminate Δq in favor of Δr.

This form of the system equation resembles both the equation for the tapered wheelset, Eq. (1.6), and the equation for a mass spring system, Eq. (1.8). In Chapter 1 we argued that a mass spring system with positive spring constant and mass parameters is not unstable and has an oscillatory response. Without going into eigenvalue analysis (the topic of the next chapter), we can again argue that if the term

$$\frac{(I_{xx} - I_{zz})(I_{xx} - I_{yy})\Omega^2}{I_{yy}I_{zz}} \tag{2.35}$$

is positive, then the behavior for Δq must be analogous to the behavior of a mass spring system with a positive spring constant, i.e., Δq will have a sinusoidal response with a radian frequency equal to the square root of the term in Eq. (2.35). (In fact, Δr will also have a sinusoidal response.)

There are two situations for which the term in Eq. (2.35) will be positive: The first is if I_{xx} is the largest of the three principal moments of inertia, and the second is if I_{xx} is the smallest. In the first case, the two terms in parentheses are both positive. In the second case, the terms are both negative but their product is positive. In the case when I_{xx} is intermediate in magnitude, then the term in Eqs. (2.34) and (2.35) will be negative.

Using the analogy between Eq. (2.34) and the equation for a mass spring system, Eq. (1.8), one can say the satellite spin stabilized around a principal axis with either the largest or the smallest moment of inertia will not be unstable. The perturbation variables Δq and Δr may oscillate sinusoidally in time, but they will not grow ever larger. (This was the case for the tapered wheelset with the taper in the positive sense.)

On the other hand, if the satellite is spin stabilized around the intermediate axis, the term in Eq. (2.35) will be negative and the perturbation variables will act like the variables of a mass spring system with a negative spring constant. A mass-spring system having a spring with a negative spring constant will deviate more and more from an equilibrium position if it is even slightly disturbed. From this analogy, one can see that if x is an intermediate principal axis, Δq and Δr will grow in time and the spin stabilization system is thus unstable. In Chapter 3, we will see that the growth is initially exponential in nature.

In this analysis we have linearized nonlinear equations for the purpose of studying the stability of a basic motion. For an unstable system, the perturbation variables grow to a size such that the linearized equation is no longer an accurate representation of the system. Practically, this means that while the perturbation spins do grow for a time, they really do not grow infinitely large in time. The satellite has only a finite amount of energy so what ultimately happens is that the initial spin energy associated with the spin about the x axis is distributed into spin energy associated with angular velocities about the y- and z axes. If one attempts to spin stabilize a satellite about an intermediate principal axis, the result is that the satellite will soon have a complicated and unsatisfactory tumbling motion.

This result can even be demonstrated on earth. A compact object (such as a blackboard eraser) thrown into the air behaves for a while much like a torque-free rigid body. The only moments about the center of mass arise from the object's interaction with the air and for a while their effect is small. If the object has obvious principal axis directions and if it is clear which moment of inertia is the largest and which is the smallest, then it is easy to throw the object up with a spin primarily around the largest or the smallest principal axes. An attempt to spin the object about the intermediate principal axis will always result in a tumbling motion.

There are rumors that an early satellite was planned to be spin stabilized about the intermediate axis as the engineers knew that any rigid body can spin steadily about any principal axis. What they did not realize was the spinning about the intermediate axis is unstable (although applied mechanicians knew this long before satellites existed). The result was that the satellite tumbled uncontrollably and spin stabilization would have been better called spin unstabilization.

V. BOND GRAPHS FOR RIGID BODY DYNAMICS

The equations developed above can be represented in terms of bond graphs (Karnopp, 1976; Karnopp, et al., 2000), and these have been used for more complex models of vehicle dynamic systems (Pacejka, 1986; Margolis and Ascari, 1989). The equations appear in a slightly different form from Eqs. (2.17) and (2.19) because bond graphs fundamentally use linear and angular momentum variables rather than linear and angular accelerations. In many cases, it is useful to have a graphical representation of a system model to study how system components interact. Bond graphs are particularly useful when actuators are combined with passive components to control the dynamics response of vehicles. Chapter 11 will provide an introduction to active control of vehicle stability.

The angular momentum has been defined in Eq. (2.8) and written out in component form in Eq. (2.18). The linear momentum has been defined in Eq. (2.7) as a vector, but we now define the momentum components in a column vector.

$$\begin{bmatrix} P_x \\ P_y \\ P_z \end{bmatrix} = m \begin{bmatrix} U \\ V \\ W \end{bmatrix} \tag{2.36}$$

The basic dynamic principles, Eqs. (2.11) and (2.12), can then be expressed in terms of momentum variables using the kinematic rule of Eq. (2.13) for a moving coordinate system

$$\vec{F} = d\vec{P}/dt = (\partial\vec{P}/\partial t)_{rel} + \vec{\omega} \times \vec{P} \tag{2.37}$$

$$\vec{\tau} = d\vec{H}_c/dt = (\partial\vec{H}_c/\partial t)_{rel} + \vec{\omega} \times \vec{H}_c \tag{2.38}$$

When these equations are written out in components, the results are equivalent to Eqs. (2.17) and (2.19).

$$\begin{bmatrix} X \\ Y \\ Z \end{bmatrix} = \begin{bmatrix} \dot{P}_x + (mq)W - (mr)V \\ \dot{P}_y + (mr)U - (mp)W \\ \dot{P}_z + (mp)V - (mq)U \end{bmatrix} \tag{2.39}$$

$$\begin{bmatrix} L \\ M \\ N \end{bmatrix} = \begin{bmatrix} \dot{H}_x + (H_z)q - (H_y)r \\ \dot{H}_y + (H_x)r - (H_z)p \\ \dot{H}_z + (H_y)p - (H_x)q \end{bmatrix} \tag{2.40}$$

Although the linear momentum equation, Eq. (2.39), may seem more difficult to understand than the equivalent linear acceleration equation, Eq. (2.17), the angular momentum equation, at least, is considerably simpler than the equivalent angular acceleration equation, Eq. (2.19).

When the momentum components are used as state variables, the velocity components are found from the momentum components by inverting Eqs. (2.36) and (2.18).

In bond graph terms, elements that relate velocities (or flows in bond graph terms) with momenta are called inertia or I-elements. Fig. 2.2 shows how Eqs. (2.36) and (2.18) are represented as bond graph inertia elements.

Note that on the left side of Fig. 2.2, the linear momentum relation is represented by three separate inertia elements each with the scalar mass as a parameter in accordance with Eq. (2.36). The bond graph variables appended near the bonds are forces and velocities (efforts and flows), with the convention that the forces are represented as the derivatives of momenta (Karnopp et al., 2000). These momentum derivatives are components of $(\partial\vec{P}/\partial t)_{rel}$ in Eqs. (2.37) and (2.39).

On the right side of Fig. 2.2, the angular momentum relationship has to be represented by a multiport inertial element because of the nature of the tensor or matrix involved in (Eqs. (2.8), (2.9), and (2.18). If x, y, and z happened to be principal axes, the inertia matrix would be diagonal and the multiport representation could be reduced to three separate inertia elements with the three principal moments of inertia as parameters. The effort and flow variables are moments and angular velocities with the moments denoted as the derivatives of angular momenta. The derivatives of the angular momenta are components of $(\partial\vec{H}_c/\partial t)_{rel}$ in Eqs. (2.38) and (2.40).

Although the inertia elements shown in Fig. 2.2 are strictly linear in Newtonian mechanics, the equations of motion for the rigid body, (Eqs. (2.39) and (2.40) or (2.17) and (2.19), are nonlinear due to the correction terms that arise because of the use of a body-fixed coordinate system. When these correction terms are incorporated in a bond graph, a complete representation of the motion of a rigid body in body-fixed coordinates is given as shown in Fig. 2.3.

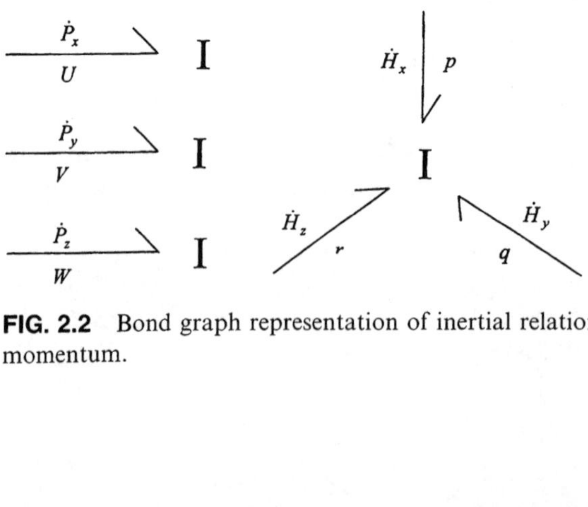

FIG. 2.2 Bond graph representation of inertial relations for linear and angular momentum.

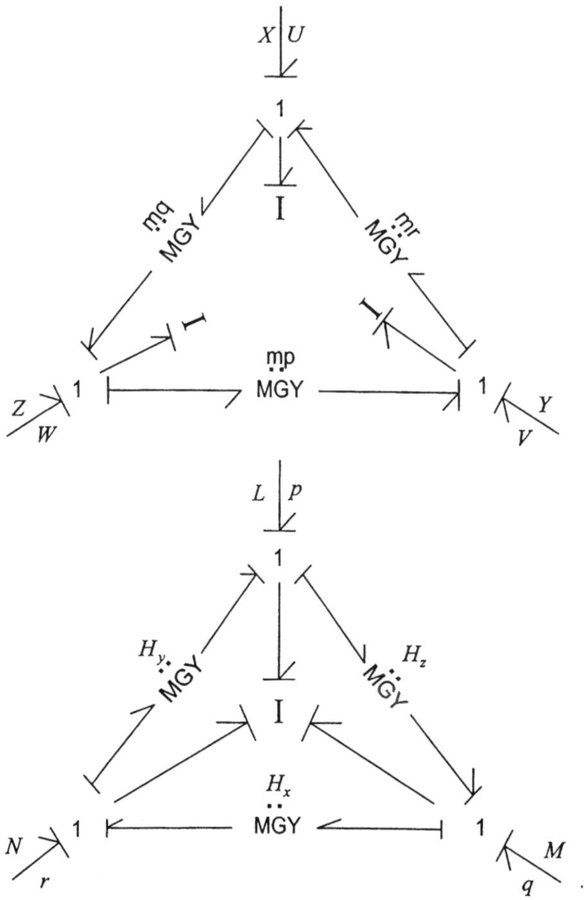

FIG. 2.3 General bond graph for a rigid body with body-fixed coordinate frame variables.

The two bond graphs represent the translational and rotational dynamics of a rigid body moving in three-dimensional motion. The causal marks that were absent in Fig. 2.2 have been added in Fig. 3. The state variables are the three linear momentum and the three angular momentum variables. The input forcing quantities are the three forces, X, Y, and Z, and the three moments, L, M, and N. The output variables and the linear and angular velocities.

It may appear that the translational and rotational bond graphs are decoupled. This is only partly true. Because of the choice of the center of mass as the point used for the acceleration law, Eq. (2.11), and the angular momentum law, Eq. (2.12), the basic dynamic equations are uncoupled. However, the use of a rotating coordinate frame requires that the components of the angular velocity vector must be used to compute the changes in the linear and angular momenta as shown in Eqs. (2.37)–(2.40). The gyrator rings in both parts of Fig. 2.3 are modulated by the angular velocity components so they are indeed coupled.

The gyrator rings shown in Fig. 2.3 are sometimes called Euler junction structures because they reflect the manipulations necessary to find absolute accelerations or changes in momenta when using a rotating coordinate system. Euler's equations for the angular velocities of a rigid body use the fundamental kinematic law, Eq. (2.3), and the result is just Eq. (2.19), although this result is typically expressed for the special case of principal axes.

The bond graphs shown in Fig. 2.3 can be simplified considerably if the x–y–z axes are principal or if not all of the six linear and angular velocities need to be considered. Many of the vehicle examples to be discussed in later chapters consider only plane motion and in this case a number of the bonds in Fig. 2.3 disappear (Karnopp, 1976). In other cases, vehicle studies have used the complete bond graph for general three-dimensional motion (Karnopp et al., 2000).

As an example of the simplification that occurs when some motions are assumed to vanish, consider the case in which both q and W are zero. This assumption led to Eqs. (2.22)–(2.25) as simplified versions of Eqs. (2.17) and (2.19). In this case Eqs. (2.39) and (2.40) become

$$\begin{aligned} X &= \dot{P}_x - (mr)V \\ Y &= \dot{P}_y + (mr)U \end{aligned} \tag{2.41}$$

$$\begin{aligned} L &= \dot{H}_x - (H_y)r \\ N &= \dot{H}_z + (H_y)p \end{aligned} \tag{2.42}$$

The linear and angular velocities are found from simplified versions of Eqs. (2.36) and (2.18).

$$m\begin{bmatrix} U \\ V \end{bmatrix} = \begin{bmatrix} P_x \\ P_y \end{bmatrix} \tag{2.43}$$

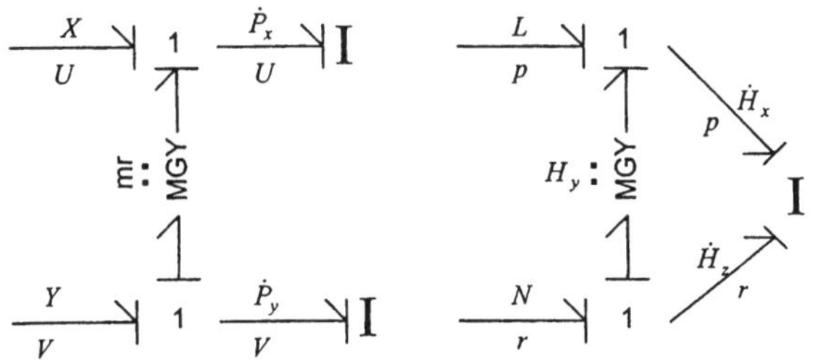

FIG. 2.4 Four-port representation of rigid body dynamics when heave (W) and pitch (q), are zero.

$$\begin{bmatrix} I_{xx} & I_{xz} \\ I_{zx} & I_{zz} \end{bmatrix} \begin{bmatrix} p \\ r \end{bmatrix} = \begin{bmatrix} H_x \\ H_z \end{bmatrix} \tag{2.44}$$

The quantity H_y that appears in Eq. (2.42) can also be found in Eq. (2.18).

$$H_y = I_{yx}p + I_{yz}r \tag{2.45}$$

Fig. 2.4 shows the simplified bond graph that results when q and W are assumed to vanish.

An even simpler bond graph applies if the roll velocity, p, is also assumed to be zero. Then the body is in plane motion and the bond graph at the right of Fig. 2.4 degenerates into a single one-port inertia element relating the yaw rate r to the angular momentum about the z axis. Only one of the original six modulated gyrators in Fig. 2.3 would then remain. A number of models in Chapters 5, 6, and 9 will be concerned with the simple case of plane motion.

The Appendix contains examples of bond graph representations of models developed for vehicles in Chapters 6 and 9. As these are plane motion models, the rigid body parts of the bond graphs are somewhat simplified versions (Fig. 2.4) and the rest of the bond graph represents the means in which the total forces and moments are generated by the vehicles.

3
Stability of Motion—Concepts and Analysis

I. INTRODUCTION

Before dealing with the specific topic of vehicle stability, it is useful first to discuss some general concepts of stability. Mathematicians sometimes joke that there is no stability to the definition of stability. This is because, for nonlinear differential equations, numbers of definitions of stability and instability have been developed. Each definition has a precise use and each differs from the others enough that one cannot say that they are all essentially equivalent. However, for the purposes of this book, the situation is not as hopeless as it might at first seem.

The concern here is mainly the stability of motion of vehicles traveling in a straight line or in a steady turn. The idea will be to try to predict whether a small perturbation from a basic motion will tend to die out or will tend to grow. The analysis will be based on linearized differential equations and the analysis of the stability properties of linear differential equations is fairly straightforward. The appropriate analytical techniques are discussed in this chapter.

In addition, engineers realize that no mathematical model is ever exactly representative of a real device. Therefore they do not expect that the results of analysis or even of simulation of a mathematical model will exactly predict how a real device will behave. An engineer's use of analytical and computational techniques is an attempt to reduce the necessity of experimentation and to foresee trends rather than to obtain precise results in the physical world. A good engineer is never completely surprised to find that some new phenomenon arises when a prototype is tested that was not predicted by analytical or even complicated computational results.

The use of mainly linear techniques to study vehicle stability has two main benefits. First, the analysis of linearized systems is often justified by experimental results as far as the question of the growth or decay of small perturbations. What this type of analysis cannot do is to predict the behavior of unstable vehicles after the perturbations have grown large. These questions can be addressed by experiment or simulation using nonlinear component characteristics.

Secondly, the study of fairly low-order linearized models often allows on to see trends toward stability or instability in terms of physical parameters that are hard to see in experimental or simulation results. This allows one to develop some intuition about stability that is very helpful in preliminary design of vehicles. Simple mathematical models that are amenable to analysis also aid in the search for solutions to stability problems found experimentally.

Almost everyone has at least a vague idea what the terms stability and instability mean, but these ideas are often not precise enough for use in studying vehicle stability. Even engineers dealing with vehicle dynamics often use the terms in relatively imprecise ways that are nevertheless often useful. For example, in vehicle stability discussions, one often encounters the concepts of static and dynamic stability. The first section of this chapter will consider these concepts in a general context. Static stability considerations arise logically in the studies of aircraft in Chapter 9, while dynamic stability studies typically must be used when studying ground vehicles. Chapter 9 will provide an argument about why the use of the special case of static stability is often sufficient for aircraft while the more general concept of dynamic stability is necessary for most other vehicles.

II. STATIC AND DYNAMIC STABILITY

The basic ideas of stability and instability are often introduced by considering simple physical systems that exhibit stability or instability in an intuitive way when displaced from an equilibrium position. For example, a marble lying on a surface and acted upon by gravity can illustrate the concept of *static stability*. Three cases are illustrated in Fig. 3.1, distinguished by the surface the marble is resting on that can be flat, curved downward, or curved upward.

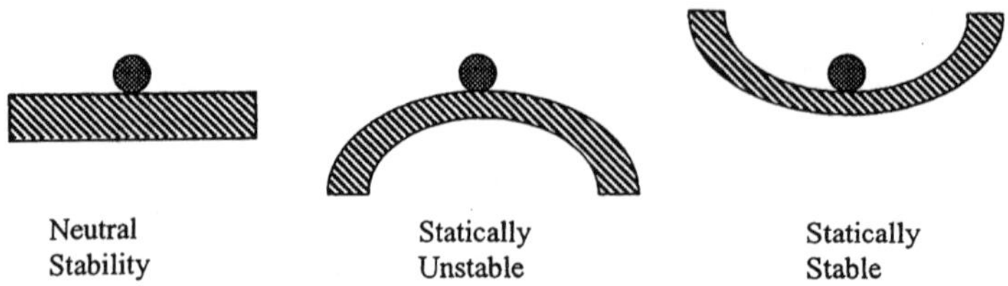

FIG. 3.1 Static stability of a marble resting on a surface.

The idea of the first case in Fig. 3.1 is that a marble resting on a flat, horizontal surface could stand still at any point along the surface. In this case, every point on the surface is an equilibrium point. The marble would neither tend to return to any point if moved a distance away nor would it tend to move further from the original point. Thus any point on the surface is called an equilibrium point of *neutral static stability*.

In the second case in Fig. 3.1, the marble rests at the top of a hill. The marble could remain there if it were exactly at the top of the hill, but if it were to be moved even a small distance toward either side, it would tend to roll further away from the equilibrium point down the hill. The point at the very top of the hill is called a *statically unstable* equilibrium point.

Finally, if the marble finds itself at the bottom point of a depression as shown in the last case in Fig. 3.1, it is clear that the marble will tend to roll back toward the bottom point if it is displaced away from the bottom point. This is then called a *statically stable* equilibrium point.

Note that the idea of static stability only has to do with the *tendency* of the marble to move after it has been displaced from equilibrium, not with the question of whether or not the marble would settle down back to the equilibrium point after some time.

In terms of a simple differential equation such as that for a mass–spring system, as shown in Fig. 3.2,

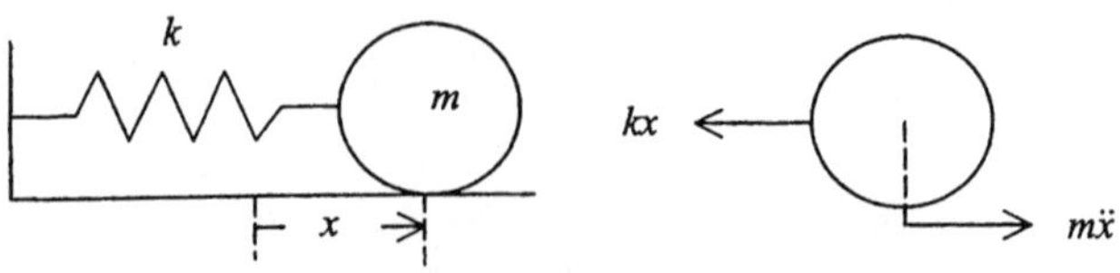

FIG. 3.2 A simple mass–spring system with the free body diagram.

$$m\ddot{x} + kx = 0 \tag{3.1}$$

where m is a mass (an inherently positive parameter), k is a spring constant, and x is a position variable, static stability has only to do with the sign of k.

For the case $k = 0$, the system is neutrally statically stable. The spring exerts no force at all on the mass no matter what the position happens to be.

If $k > 0$, the system is statically stable because the spring force is in a direction tending to pull the mass back toward the equilibrium point at $x = 0$ for any positive or negative initial value of x.

If $k < 0$, the system is statically unstable because the spring force, kx, which is zero only when $x = 0$, is in a direction that tends to push the mass farther from zero for any initial position. (The force points in the opposite direction to that shown in Fig. 3.2.)

It is easy to imagine that a statically unstable system is certainly unstable in the wider dynamic sense. In the case of the mass acted upon only by a spring with a negative spring constant, there is simply no reason for the mass to return to the equilibrium point. (Springs with negative spring constants are quite easy to construct. Many toggle or snap-through devices have such springs. In such devices, the central equilibrium position is unstable by design and the device is intended to run to a plus or minus limit stop.)

However, a statically stable system is not necessarily *dynamically stable* because the statically stable system has only been shown to have a force tending to bring the system back toward equilibrium. It is conceivable that upon returning to the equilibrium point, the system could keep moving past the equilibrium point and not reverse its motion until it was farther away on the other side than it was initially. This type of overshooting behavior could continue, meaning that the statically stable system would never settle down to the equilibrium point and thus would not deserve to be called dynamically stable.

Consider an extended version of the mass–spring equation [Eq. (3.1)] but now including a damping term, $b\dot{x}$.

$$m\ddot{x} + b\dot{x} + kx = 0 \tag{3.2}$$

Static stability considerations still apply to k (the spring constant), but now the sign of the damping constant, b, is also important.

If $b > 0$, the mass is acted upon by a linear damping mechanism that removes energy from the mass. This is fundamentally a stabilizing effect that means that for a positive spring constant, the mass will either oscillate around the equilibrium point with decreasing amplitude of the cycles or, if the damping is large enough, the mass will approach the equilibrium without any oscillations. In this case, the system is statically stable and is also dynamically stable.

On the other hand, even if b is positive, a negative value for the spring constant k will mean that the system is both statically and dynamically unstable. The damper cannot stabilize a statically unstable system. The positive damping effect will slow down the deviation from equilibrium but the spring will still push the mass away. In Fig. 3.1, the statically unstable case remains unstable even when the marble moves in a viscous fluid.

If $b < 0$, the "negative damping" effect actually adds energy to the system and the system will be dynamically unstable regardless of whether it is statically stable or unstable. In the statically stable but dynamically unstable case, the system will *tend* to return to its equilibrium point after being disturbed, but it will overshoot and begin an ever-increasing oscillation around the equilibrium point. Fig. 3.3 shows the qualitative difference between static instability (a divergence) and dynamic instability (a diverging oscillation). For the mass–spring–damper system, the divergence shown to the left in Fig. 3.3 corresponds to the negative spring constant case while the divergent oscillation shown to the right corresponds to the positive spring constant but negative damping constant case.

As another physical example, consider a pendulum. When it is hanging straight down, it is clearly in equilibrium. If it is moved away from the straight down position and released, it will tend to return to equilibrium so, in this case, it is statically stable. Furthermore, if the

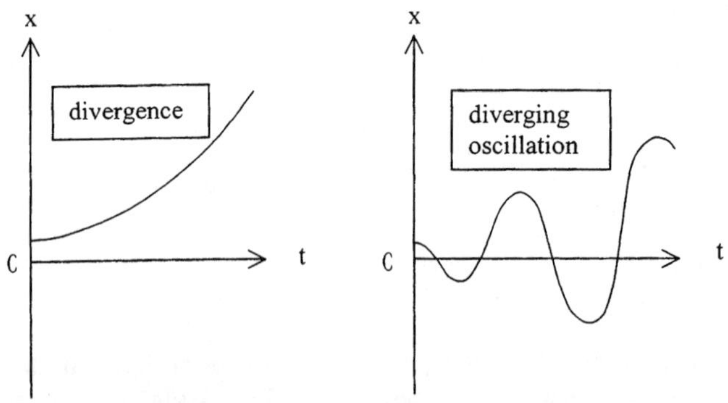

FIG. 3.3 Two types of unstable behavior.

pendulum had some viscous friction at the pivot point, the oscillations would reduce in amplitude in time, indicating that the pendulum would be also dynamically stable.

On the other hand, if some device put a moment on the pendulum around the pivot similar to the moment provided by a viscous damper but with opposite sign, instead of reducing the amplitude of the pendulum swing in time, the device would increase the swing amplitude. The pendulum would then be statically stable but dynamically unstable.

It may seem that the idea of a "negative damping coefficient" device is completely artificial, and, of course, it is. But the wobble of a motorcycle front fork, the flutter of an airplane wing, or the galloping of a transmission line in a strong wind are all examples of dynamically unstable systems that behave as if the damping in the system were negative. In all these cases, one can identify a source of energy that can feed into the oscillation and increase its amplitude in time.

Finally, an upside-down pendulum is statically unstable and will clearly fall over if displaced from the straight upright position. In this case, the pendulum is dynamically unstable too, even when there is a viscous damping moment at the pivot. The viscous damping moment slows down the fall of the upside down pendulum but cannot bring the pendulum back to the upright position.

What this means is that the often-used concept of static stability is not always sufficient for the determining vehicle stability. Some stability problems can be avoided by eliminating static instabilities. For example, the wheels of cars, bicycles, and motorcycles are designed to have a caster effect, which means that they have a self-centering action and that they are statically stable. On the other hand, even with a positive caster effect, the wheels can exhibit shimmy or wobble, which is a dynamic instability phenomenon. In many cases, vehicle dynamic instabilities that arise in statically stable systems are difficult to understand. The causes of these divergent oscillations may be numerous and the remedies are often far from obvious. Some of the problems associated with wheels designed to behave like casters will be discussed in Chapter 8.

III. EIGENVALUE CALCULATIONS AND THE ROUTH CRITERION

Although it is useful to introduce concepts of stability using examples that can be described with first- or second-order differential equations, most systems require higher-order equations to accurately describe their dynamics. In some cases, judging the stability of vehicles requires mathematical models of very high order and there is no alternative to numerical computer studies. In any case, the elementary ideas of static and dynamic stability need to be generalized when higher-order mathematical models of vehicles are to be analyzed.

In the this section, we will present means for studying the stability of linearized mathematical models of vehicles that are of low-enough order such that analytical methods are practical. When the number of parameters is not too high, analytical methods can give insights into the basic causes of unstable behavior, which are often hard to find in purely numerical studies. Numerical studies sometimes degenerate into a frustrating trial-and-error procedure of varying parameter values to try to find a set that results in a stable vehicle. Experimental studies of stability can be even more frustrating in the absence of a guiding theory because of the time and expense involved in changing physical aspects of a vehicle in an attempt to solve a stability problem.

Eigenvalue analysis is the primary tool used in discussing the stability of linearized systems of equations. The term "eigenvalue" is a German-English hybrid word that is in common use. A completely English version of the word is "characteristic value," which is probably used less often just because it is longer. Practically every area of engineering makes use of eigenvalue analysis but it is of central importance for the study of vehicle stability.

No mater what form the equations of a linear system model may take, the eigenvalues are essentially identical as long as no error has been made in the equation formulation. (As will be shown for the case of automobiles in Chapter 6, some equation formulations may produce extra zero-valued eigenvalues but the finite-valued eigenvalues are identical for all correct equation sets.) The idea of eigenvalues is that they are related to the dynamic characterization for the system itself and are independent of the equations used to describe the system. The eigenvalues are independent of any external inputs that may be forcing the linear system.

In this section, we first discuss the typical forms mathematical models of vehicle dynamics may take. Then, the definition of eigenvalues and the means for determining them will be presented. Next, the criterion for system stability given the eigenvalues will be discussed. Finally, Routh's method for predicting stability or instability without actually finding the eigenvalues is presented.

The importance of Routh's method lies in the ability to find algebraic conditions for stability in terms of literal coefficients and thus in terms of the physical parameters of vehicles. Naturally, very complex, high-order models have so many parameters that it is hopeless to use analytical methods to find general patterns in the parameter space that will guarantee stability; however, for fairly low-order models, Routh's method is very effective. Often, insight into the causes of instability can be obtained through the use of low-order vehicle models and the application of Routh's criterion.

The stability analysis of linear system is based on equations with no external forcing terms. That is, the stability of a linear system is a property of the so-called free response and has nothing to do with the response to forcing functions that may be imposed on the system. The basic stability of an automobile, for example, is studied for no steering input although a skilled driver may be able to stabilize the car by manipulating the steering wheel. This shows that there is a distinction between the stability of a car itself and the stability of the car and driver combination.

Nonlinear systems may have stability problems resulting from the way they are forced and, because all vehicles are in principle nonlinear, we must keep in mind that the stability analysis based on linearized models may not reveal all the dynamic problems a real vehicle may have. On the other hand, if the linear analysis of a model of a vehicle reveals that a vehicle will exhibit unstable behavior, it is highly likely that the real vehicle would also exhibit unstable dynamic behavior.

A. Mathematical Forms for Vehicle Dynamic Equations

There are several common ways in which linearized mathematical models of vehicles are presented. Usually, the forms that the equations take result from the formulation technique used. In principle, all equation forms can be converted to other equivalent forms, but the manipulations required are sometimes quite complex, and, because the eigenvalues will be identical for all correct equation sets, it is useful to discuss a number of typical forms related to the formulation techniques in common use.

The first form to be considered is a single nth-order differential equation.

$$a_0 \frac{d^n x}{dt^n} + a_1 \frac{d^{n-1} x}{dt^{n-1}} + \ldots + a_{n-1} \frac{dx}{dt} + a_n x = 0 \tag{3.3}$$

in which it will be assumed that $a_n \neq 0$ so that the system is really of order n. In this formulation, the variable x can be any variable in the vehicle model. The right-hand side is zero because for eigenvalue analysis, no forcing terms are assumed. [Different choices for the variable x would give different right-hand sides for Eq. (3.3) if there were forcing functions involved, but because there are no forcing functions considered for eigenvalue analysis, Eq. (3.3) basically applies for all variable choices.]

A second common form involves n equations of first-order form. Some formulation techniques directly produce this type of formulation. This is the case, for example, for bond graph methods (Karnopp et al., 2000). Examples of this form have already been encountered in examples treated in Chapters 1 and 2.

$$\begin{aligned} \dot{x}_1 &= a_{11}x_1 + a_{12}x_2 + \ldots + a_{1n}x_n \\ \dot{x}_2 &= a_{21}x_1 + a_{22}x_2 + \ldots + a_{2n}x_n \\ &\cdot \\ &\cdot \\ &\cdot \\ \dot{x}_n &= a_{n1}x_n + a_{n2}x_2 + \ldots + a_{nn}x_n \end{aligned} \tag{3.4}$$

In vector–matrix form, Eqs. (3.4) can be represented in a compact way.

$$[\dot{x}] = [A][x] \tag{3.5}$$

with

$$[\dot{x}] = \begin{bmatrix} \dot{x}_1 \\ \dot{x}_2 \\ \cdot \\ \cdot \\ \cdot \\ \dot{x}_n \end{bmatrix}, \quad [x] = \begin{bmatrix} x_1 \\ x_2 \\ \cdot \\ \cdot \\ \cdot \\ x_n \end{bmatrix}, \quad [A] = \begin{bmatrix} a_{11} & a_{12} & \ldots & a_{1n} \\ a_{21} & a_{22} & \ldots & a_{2n} \\ \cdot & & & \\ \cdot & & & \\ \cdot & & & \\ a_{n1} & a_{n2} & \ldots & a_{nn} \end{bmatrix} \tag{3.6}$$

In Eq. (3.6), x_1, x_2, $\ldots x_n$ are called state variables for the system. Once again, there are a number of choices that can be made for the state variables and the matrix $[A]$ will be different for each choice. In fact, the matrix $[A]$ even assumes different forms for different ordering of the state variables in the column vector $[x]$. However, if no error has been made, all the $[A]$ matrices for a particular system model will yield exactly the same eigenvalues.

A third form of dynamic equation that will be encountered in Chapter 5, for example, involves sets of second-order equations. Such equations arise naturally when Lagrange's equations are used to describe vehicle motion. Also, when using inertial coordinate systems as discussed in Chapter 2, the use of Newton's laws often produce sets of second-order equations.

The second-order equation sets resemble those used in vibration studies but, while vibratory systems are generally not unstable, the vehicle dynamic equations will often have the possibility of being unstable for certain conditions among the physical parameters.

In vector–matrix form, the equations can be written

$$[M][\ddot{x}] + [B][\dot{x}] + [K][x] = [0] \tag{3.7}$$

Here the vector $[x]$ contains n generalized coordinates and the vehicle model is said to have n degrees of freedom. The $[M]$, $[B]$, and $[K]$ matrices would be called mass, damping, and spring constant matrices for a vibration problem but, for vehicle dynamic models, this nomenclature for the $[B]$ and $[K]$ matrices is not entirely appropriate because some of the terms in these matrices do not come from damping or spring elements. In Chapter 5, it will be clear that the characterization of the force-generating properties of pneumatic tires can result in terms in the $[B]$ and $[K]$ matrices.

For unforced linear systems, represented in any of the forms of Eqs. (3.3)–(3.7), the general solution can always be assumed to be a sum of constants times an exponential in time, e^{st}. This form of solution will only work for special values of the generally complex number s. The values of s that allow the assumed solutions to fit the differential equations are called the *eigenvalues* of the equation sets.

The eigenvalues are found by using the fact that all system variables can be assumed to have a form such as $x(t) = Xe^{st}$. The derivatives then have the form $\dot{x}(t) = sXe^{st}$, $\ddot{x}(t) = s^2 Xe^{st}$, and so on. If this assumption is used with the nth-order equation form [Eq. (3.3)], the result of substituting $x(t)$ and all derivatives into the dynamic equation and collecting terms is

$$(a_0 s^n + a_1 s^{n-1} + \ldots + a_{n-1} s + a_n)Xe^{st} = 0 \tag{3.8}$$

In Eq. (3.8), there are three factors that must multiply to zero to satisfy the equation. Thus there appear to be three distinct ways to satisfy Eq. (3.8). But if either X or e^{st} were to vanish, then $x(t)$ as well as all its derivatives would also vanish. This solution, which is not very useful, is called the *trivial solution*, and should have been an obvious solution from the beginning. [Eq. (3.3) is trivially satisfied if x and all of its derivatives are zero.]

Only if the third factor in Eq. (3.8) vanishes do we obtain a useful result.

$$a_0 s^n + a_1 s^{n-1} + \ldots + a_{n-1} s + a_n = 0 \tag{3.9}$$

This is called the *characteristic equation* and values of s that satisfy the equation are the system eigenvalues (or characteristic values in pure English). The theory of linear algebra gives two useful facts about solutions to the characteristic equation [Eq. (3.9)]. First, an nth-order algebraic equation such as Eq. (3.9) will have exactly n solutions. Second, if the coefficients of the equation are real, the solutions for the eigenvalues will either be real or will come in complex conjugate pairs. In the case of physical systems such as vehicles, the coefficients will indeed be real-valued parameters, so the n eigenvalues will be either real or complex conjugate numbers.

For the vector-matrix form of the equations [Eq. (3.5)], the argument for determining the eigenvalues yields equivalent results but is slightly different. The assumption is first made that

$$[x] = [X]e^{st} \tag{3.10}$$

where $[X]$ is a vector of constants that at this point is not determined. After substitution of Eq. (3.10) into Eq. (3.4) or Eq. (3.5), this leads to the relation

$$s[X]e^{st} = [A][X]e^{st} \tag{3.11}$$

or, after combining terms to the equation

$$[sI - A][X]e^{st} = [0] \tag{3.12}$$

In this equation,

$$[I] = \begin{bmatrix} 1 & 0 & & 0 \\ 0 & 1 & & 0 \\ \cdot & & & \\ \cdot & & & \\ \cdot & & & \\ 0 & 0 & & 1 \end{bmatrix} \tag{3.13}$$

is the unit matrix that has 1's along the main diagonal and 0's elsewhere.

Note that if all the components of $[X]$ were zero (or if e^{st} actually could be zero), then Eq. (3.12) would be satisfied but one would have the trivial solution $[x(t)] = [0]$ in which all state variables and their first derivatives are zero. If the trivial solution is to be rejected, and after factoring out the (nonzero) e^{st} in Eq. (3.12), the result is an algebraic problem in which there are n linear algebraic equations for the n components of $[X]$. The assumption of Eq. (3.10) has changed a problem in differential equations to a problem in algebraic equations.

If the determinant of the equation coefficients in Eq. (3.12) is not zero,

$$\det[sI - A] \neq 0 \tag{3.14}$$

then linear algebraic theory states that the only possible solution for $[X]$ is unique. But that would mean that the trivial solution (which always exists) is the only solution. To allow solutions for $[X]$ other than [0], we require the opposite of Eq. (3.14),

$$\det[sI - A] = 0 \tag{3.15}$$

Equation (3.15) is actually the characteristic equation for determining the system eigenvalues for equations in the form of Eq. (3.5). When the determinant is written out, Eq. (3.15) takes on the same form as Eq. (3.9).

As an example, consider a second-order set of equations in the form of Eq. (3.5) with

$$[A] = \begin{bmatrix} a_{11} & a_{12} \\ a_{21} & a_{22} \end{bmatrix} \tag{3.16}$$

Then, $[sI - A]$ becomes

$$[sI - A] = \begin{bmatrix} s - a_{11} & -a_{12} \\ -a_{21} & s{-}_{22} \end{bmatrix} \tag{3.17}$$

and Eq. (3.15) can be evaluated.

$$\det[sI - A] = s^2 + (-a_{11} - a_{22})s + (a_{11}a_{22} - a_{12}a_{21}) = 0 \tag{3.18}$$

Because the characteristic equation must be the same no matter in which form the system equations are written, we can see an equivalence between the coefficients of the characteristic equation as written for a single second-order differential equation [Eq. (3.3)] and for this set of two first-order equations [Eq. (3.5)] assuming that both equations refer to the same system model.

$$a_0 s^2 + a_1 s + a_2 = 0 \tag{3.19}$$

$$a_0 = 1, \; a_1 = -a_{11} - a_{22}, \quad a_2 = a_{11}a_{22} - a_{12}a_{21} \tag{3.20}$$

Finally, if the system equations are in the form of Eq. (3.7), the assumed solution of Eq. (3.10) results in the equation

$$\{[M]s^2 + [B]s + [K]\}[X]e^{st} = [0] \tag{3.21}$$

The same arguments that lead to the characteristic equation in the form of Eq. (3.15) can now be applied to Eq. (3.21), namely that is if the trivial solution is not to be the only solution, then the determinant of the coefficient matrix in Eq. (3.21) must vanish. The result is that for system equations in the form of Eq. (3.7), the characteristic equation for determining the eigenvalues can be expressed as

$$\det\{[M]s^2 + [B]s + [K]\} = 0 \tag{3.22}$$

For an n degree of freedom system, Eq. (3.22) will be a $2n$th-order polynomial in the variable s. As an example, a 2 degree of freedom version of Eq. (3.22) would have the following general form:

$$\det\left\{\begin{matrix} m_{11}s^2 + b_{11}s + k_{11} & m_{12}s^2 + b_{12}s + k_{12} \\ m_{12}s^2 + b_{12}s + k_{12} & m_{22}s^2 + b_{22}s + k_{22} \end{matrix}\right\} = 0 \tag{3.23}$$

If the determinant in Eq. (3.23) is written out, a fourth-order polynomial in s results. A number of examples of the use of Eq. (3.22) will be found for vehicle models in succeeding chapters.

B. Computing Eigenvalues

Assuming that one could solve the characteristic equation, the result for an nth-order system would be n eigenvalues, $s = s_1, s_2, s_3, \ldots, s_n$. Each eigenvalue will contribute a part of the free response of the system with a time response of the form of a constant times the function e^{st}, where s is the eigenvalue in question. (The actual values of the constants could be determined by the system initial conditions, but for stability analysis, it is the nature of the eigenvalues that is important not the value of the constants for specific initial conditions.)

If the eigenvalue is real and if it is negative, the contribution to the free response by the eigenvalue will be a decreasing exponential function of time. This means that the contribution to all system variables from this eigenvalue will decrease in time and all variables will tend to return to their zero equilibrium values. If it is real and positive, the contribution will be an exponentially increasing function of time and the system variables will deviate more and more from their zero equilibrium values. Thus the system will be unstable if *any* one or more of the eigenvalues is real and positive.

When two eigenvalues come in a complex conjugate pair, it can be shown that they together contribute components to the free response, which are sinusoidal but multiplied by an exponential that is e^{rt}, where r is the real part of either of the two eigenvalues. (Complex conjugates have identical real parts, of course.) Thus if the real part is negative, the contribution is an oscillating function of time with decreasing amplitude. If the real part is positive, the amplitude of the oscillatory component increases in time. If a system has complex conjugate eigenvalues with positive real parts, then the system is unstable.

Roughly speaking, the static instability discussed in the first part of this chapter is correlated with a positive real eigenvalue and the dynamic instability is related to a complex conjugate pair of eigenvalues with positive real parts. Fig. 3.3 shows these two types of unstable behavior.

It is important to note that for system stability, *all* eigenvalues must either be real and negative or have negative real parts if they are complex conjugates. Unstable systems will have one or more eigenvalues with positive real parts. If one considers the system eigenvalues in a complex s plane, it is clear that stable systems have *all* their n eigenvalues in the left half plane (where the real parts are negative). A system is unstable if even a single eigenvalue exists in the right half of the s plane.

Neutral stability is associated with eigenvalues that, if real, are zero or, if complex conjugate pairs, have zero real parts. In the first case, a zero eigenvalue contributes a constant to the system free response. In the second case, the pair contributes a sinusoidal component of the response with constant amplitude. The wheelset in Chapter 1 was actually neutrally stable for positive wheel taper and unstable for the taper of the opposite sense. A spin-stabilized satellite discussed in Chapter 2 is neutrally stable if it is spinning around a principle axis with the largest or the smallest moment of inertia but is unstable if it is spinning around the intermediate principal axis.

For any system whose parameters have been given numerical values, it is possible to compute the numerical values of the eigenvalues and thus to determine whether the system is stable or not. However, the connection between changes in physical parameters and changes in the stability properties of a system can be very complicated. This often leads to a tedious trial-and-error process in trying to improve the dynamic behavior of a complex system such as a vehicle.

For models of modest order, it is possible to develop an analytical criterion for stability without actually finding the system eigenvalues. Essentially, this means that it is possible to determine whether any eigenvalues of a system have positive real parts by a fairly simple procedure. This procedure, called Routh's method, is presented next.

C. Routh's Stability Criterion

The straightforward way to decide on the stability of a system is to linearize the dynamic system equations, derive the characteristic equation, and then solve for the eigenvalues. Then, the system is stable if all the eigenvalues have negative real parts. If the coefficients in the characteristic equation [Eq. (3.9)], $a_0, a_1, \ldots, a_{n-1}, a_n$, are given numerical values, then it is easy to use a computer routine to find the roots of the equation, i.e., the eigenvalues. However, if the characteristic equation coefficients are given in terms of literal vehicle parameters such as masses, dimensions, moments of inertia, and the like, it is not possible to find general expressions for the eigenvalues for systems higher than second order.

The Routh criterion provides a technique for deciding on the stability of the system without actually computing the eigenvalues. It turns out to be possible to decide whether all of the eigenvalues have negative real parts without being able to actually find expressions for the eigenvalues. This allows the determination of sets of relations among the vehicle parameters that will assure that the vehicle is stable. Later we will see that using the Routh criterion it is possible, for example, to show that some vehicles will become unstable when traveling at a greater speed than a critical speed. With numerical studies, such a determination can only be made based on repetitive calculations.

The Routh criterion for system stability starts out by requiring that all the coefficients in Eq. (3.9), $a_0, a_1, \ldots, a_{n-1}, a_n$, be positive. [Of course, if all the coefficients in Eq. (3.9) were negative, multiplying through by -1 would make them all positive.] If any coefficient is negative, at least one eigenvalue has a positive real part so the system is unstable. If one coefficient is zero, the system has eigenvalues with zero real parts. This means that the system response contains either a constant response component or undamped oscillatory components and could be considered neutrally stable in the sense that if disturbed, the system will neither settle down to zero nor exhibit exponentially increasing responses.

For systems of order 1 or 2, all that is necessary for stability is that the coefficients in Eq. (3.9) all be positive. On the other hand, even if all the coefficients are positive, the system could still be unstable if the order of the system is greater than 2. The Routh procedure derives a criterion for assuring stability for arbitrarily high-order characteristic equations.

Assuming that all coefficients are positive, the procedure continues by giving rules for the construction of a table. The table is shown below as a pattern of rows and columns. The first two rows contain the coefficients from the characteristic equation [Eq. (3.9)], and the remaining rows are filled in by simple calculations that have some similarity to the calculations for finding the determinant of a 2 × 2 matrix.

$$\begin{array}{lllllll}
s^n & a_0 & a_2 & a_4 & a_6 & \cdot & \cdot & \cdot \\
s^{n-1} & a_1 & a_3 & a_5 & a_7 & \cdot & \cdot & \cdot \\
s^{n-2} & b_1 & b_2 & b_3 & b_4 & \cdot & \cdot & \cdot \\
s^{n-3} & c_1 & c_2 & c_3 & c_4 & \cdot & \cdot & \cdot \\
s^{n-4} & d_1 & d_2 & d_3 & d_4 & \cdot & \cdot & \cdot \\
\cdot & \cdot & \cdot \\
\cdot & \cdot & \cdot \\
\cdot & \cdot & \cdot \\
s^2 & p_1 & p_2 \\
s^1 & q_1 \\
s^0 & r_1
\end{array}$$

The quantities in the b-row are evaluated by a cross-multiplying scheme as shown below:

$$b_1 = (a_1 a_2 - a_0 a_3)/a_1$$
$$b_2 = (a_1 a_4 - a_0 a_5)/a_1$$
$$b_3 = (a_1 a_6 - a_0 a_7)/a_1$$
$$\cdot \quad \cdot \quad \cdot$$

Missing a-coefficients are counted as zero, so the calculation of the b's continues until the b's become zero. The c-row is calculated in a similar fashion from the two rows above.

$$c_1 = (b_1 a_3 - a_1 b_2)/b_1$$
$$c_2 = (b_1 a_5 - a_1 b_3)/b_1$$
$$c_3 = (b_1 a_7 - a_1 b_4)/b_1$$
$$\cdot \quad \cdot \quad \cdot$$

The d-row continues the pattern.

$$d_1 = (c_1 b_2 - b_1 c_2)/c_1$$
$$d_2 = (c_1 b_3 - b_1 c_3)/c_1$$
$$d_3 = (c_1 b_4 - b_1 c_4)/c_1$$

. . .

The process continues until the last row is completed. The rows become progressively shorter, forming a triangular pattern. The complete Routh criterion is then simply stated: The necessary and sufficient condition for a system to be stable is that all the coefficients in the characteristic equation be positive and that all the quantities in the first column of the Routh array $(a_0, a_1, b_1, c_1, d_1, \ldots)$ be positive.

Occasionally, it may be of interest to know that the number of eigenvalues with positive real parts is equal to the number of sign changes in the first column. This is not a particularly important information if one is interested only in whether the system is stable or not because the system will be unstable if even a single eigenvalue has a positive real part.

Another point of interest is that any entire row in the array can be multiplied or divided by any positive constant without changing the final result. This fact sometimes allows the calculations to be simplified as the array is being filled out.

As an example of the use if the procedure for constructing the Routh array and an application of the stability criterion, we consider a general third-order system. The stability of first- and second-order systems is already assured if the coefficients in the characteristic equation are positive but this is not the case for the third-order system. The final result is simple and easy to remember and will prove to be useful for any third-order system. The Routh array is shown below:

$$\begin{array}{lcc} s^3 & a_0 & a_2 \\ s^2 & a_1 & a_3 \\ s^1 & (a_1 a_2 - a_0 a_3)/a_1 & \\ s^0 & a_3 & \end{array}$$

The Routh stability criterion now states that the third-order system will be stable if the coefficients a_0, a_1, a_2, a_3 are all positive and, in addition, if

$$a_1 a_2 > a_0 a_3 \tag{3.24}$$

so that the third quantity in the first row is positive. The criterion of Eq. (3.24) will be used for finding stability criteria in terms of physical parameters for a number of simple vehicle models in the succeeding chapters.

The calculations required for filling out the Routh array increase in complexity quite rapidly as the order of the system increases. For numerical coefficients, it is possible to fill out the array fairly easily even for relatively high-order systems and indeed this was the main use of the criteria in the past. However, in the present, the alternative of using a computer program to find the eigenvalues for a system with numerical coefficients directly by solving the characteristic equation numerically is probably a more attractive alternative.

If the coefficients are in terms of literal system parameters, the Routh criterion can yield conditions for stability for systems of moderate order without resorting to numerical calculations that often do not yield much insight into the solution of an instability problem. The conditions stemming from Routh's criterion may increase understanding of the causes of instability and produce useful rules of thumb for design. Several examples of this will be seen in succeeding chapters.

It should be pointed out that this discussion of stability criteria is limited in several ways. First of all, it is true that the use of the Routh criterion in literal coefficients for high-order models of vehicles is impractical. Even if a computer program for algebraic manipulation were to be used to fill out the Routh array in terms of literal coefficients, the final results would be so complex that little useful information could be derived from them. High-order systems have so many parameter and the combinations that would be found in the Routh array would be so complex that one would probably have to resort to numerical calculation to decide if the first column contained only positive quantities. This would negate the idea of finding stability criteria in terms of the physical parameters rather than in terms of their numerical values.

Secondly, one must always keep in mind that it is not possible to learn everything about a nonlinear system from a linearized version of the system. Some discontinuous component characteristics cannot be linearized successfully at all. A good example of a phenomenon difficult to linearize is stick–slip friction. We have noted that an upside-down pendulum is statically unstable and cannot be stabilized by a linear viscous friction term in the equation of motion. But if the pendulum has a pivot point with the so-called dry friction, then it will not fall over for a certain range of initial positions. This is an example where linearization does not yield a complete description of the stability of the system.

Also, some unstable behavior for nonlinear systems only shows up for large perturbations but does not appear for the very small perturbations considered in the linearization process. Automobiles that seem to be stable under normal driving conditions can exhibit unstable behavior under extreme conditions due in part to the nonlinear characteristics to the tires.

For these and many other reasons, analytical studies always need to be supplemented with accurate computer simulations and physical experiments. The analytical results often give insight into the nature of unstable vehicle behavior and direction to the search for a cure to instability but there is ultimately no substitute for physical experimentation.

4
Pneumatic Tire Force Generation

I. INTRODUCTION

In Chapters 5, 6, and 8, a number of interesting stability problems associated with vehicles using pneumatic tires will be analyzed. Since the force-generating capability of pneumatic tires is a somewhat specialized topic, it will be discussed briefly in this chapter. As is the case in many specialized fields, there are certain traditional descriptive terms and concepts that are not always easy for a newcomer to the field to understand. It is the purpose of this chapter to provide background information about the properties of pneumatic tires for use in the construction of the vehicle mathematical models to be developed in later chapters.

II. TIRE–ROAD INTERACTION

One of the less well-known aspects of pneumatic tires is how they interact with road surfaces to generate the forces necessary to guide, propel, and brake road vehicles. Pneumatic tires are actually complex composite structures, and the frictional interaction between the rubber and the road surface is also far from simple. For these reasons, it is hard to predict the force-generating capabilities of tires theoretically with any accuracy. As a result, most descriptions of tire for generation are heavily based upon experiment and it might be fair to describe even the most sophisticated multidimensional computer models as exercises in curve fitting.

There are a number of reference books and technical papers that explain the force generating mechanism in some detail (Ellis, 1969, 1988; Gillespie, 1992; Kortuem and Lugner, 1994; Loeb et al., 1990; Milliken and Milliken, 1995; Nordeen, 1968; Pacejka and Sharp, 1991; Pacejka, 1989; Wong, 1978). A certain amount of actual data can be found in these references, although tire manufacturers are often reluctant to publish the results of their experimental investigations.

Accurate nonlinear mathematical models of tire–road interaction, verified by experiment, have been developed for use in computer simulation of vehicle dynamics (for example, Bakker et al., 1987, 1989). These models are useful for simulating complex vehicle models that are used to predict the response of proposed vehicle designs. They are, however, not very useful for the analytical treatment of the stability properties of low-order vehicle models.

Since the interest in this book is primarily in vehicle stability, simple linearized models of tire force generation will generally be used to analyze vehicle dynamics under the influence of small perturbations away from a basic state of motion. The main purpose of this chapter is to introduce the somewhat nonintuitive concepts of slip angle and longitudinal slip that are used in the description of the tire–road interaction. A primary aim is to provide some background to justify the "cornering coefficient" models of tire force generation that will be used in later chapters.

The word "slip" will be used in the term "slip angle" when discussing lateral or sideways tire forces and in the term "longitudinal slip" when longitudinal braking and traction forces are discussed. It is unfortunate that the word "slip" is naturally confused with the words "slide" or "skid." In extreme cases, tires do indeed exhibit gross sliding or skidding in which all parts of the tire contact patch have a relative motion with respect to the road surface. But when one speaks of a slip angle or of longitudinal slip, it can be the case that in the contact patch where the tire is in contact with the road surface, there are areas of the tire that are essentially locked to the road surface by frictional forces. In such a case, one should not think of the tire as slipping, sliding, or skidding.

The apparent "slips" are really due to the fact that pneumatic tires are deformable bodies. This deformation of the tire causes the wheel on which it is mounted neither to roll exactly in the direction it is pointed when it is subjected to a lateral force nor to rotate about its axle exactly as might be expected when subjected to braking or driving moments. The surprise is that this it true even when there are points of the tire in contact with the roadway surface that have no relative motion with respect to the road surface.

III. LATERAL FORCES

In the study of the stability of pneumatic tired vehicles, it is of primary importance to describe the generation of lateral forces by the tires. Obviously, these are the forces that allow a car to be steered or allow any wheeled vehicle to negotiate a turn. It is also obvious that no lateral force is necessary for a vehicle traveling on a straight and level road when it is not under the influence of any external loads. On the other hand, if one considers a small perturbation from a straight-line path, it is the lateral forces from the tires that will determine whether the vehicle will return to the original state of straight-line motion or will deviate ever more from it, i.e., whether the vehicle is stable or unstable.

A tire mounted on a wheel subjected to no lateral force at all will generally roll along a surface in the direction its wheel is pointed unless the tire is significantly unsymmetrical. Assuming that the wheel supports a load, there is a normal force between the road surface and the contact patch of the tire. If now a small lateral force is applied to the wheel, a corresponding lateral force will arise in the contact patch between the tire and the road surface due to friction. Although the lateral force may be too small to cause the tire to slide sideways over the surface, it will no longer be the case that the wheel continues to move in the direction it is pointed. Rather, it will acquire a sideways or lateral velocity in addition to its forward velocity. Thus the total velocity vector of the wheel will point in a different direction than the direction of the center plane of the wheel. This angle between the wheel and the direction the wheel moves when a lateral force is applied is called the slip angle, α. The slip angle is shown in Fig. 4.1.

For small lateral forces, the slip angle is small and it is due to the distortion of the tire. The most obvious aspect of this distortion is the

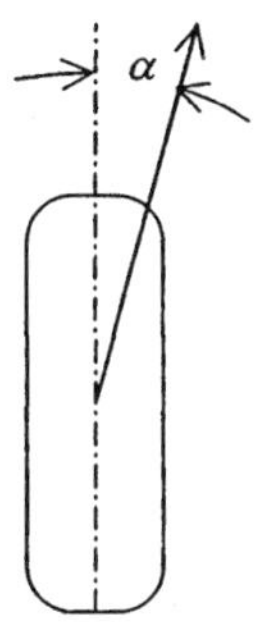

FIG. 4.1 The slip angle between a wheel centerline and the direction of travel.

deflection of the tire sidewall. The wheel appears to drift sideways as parts of the tire surface entering the contact patch are pushed to the side before they finally become locked to the road surface by the friction. (Parts of the tire exiting the rear of the contact move back toward their undistorted position as they lose the influence of the lateral friction force.) If the lateral force is increased sufficiently, there will come a point at which the entire contact patch begins to slide sideways. At this point, the slip angle can approach 90°.

In the analysis of the stability of vehicles using pneumatic tires, it is common to use a linearized model of the relation between the lateral force generated by the tire, F_y, and the slip angle, α. Over a certain range, the lateral force and the slip angle are found to be nearly proportional to each other. The linear tire force model will apply approximately for small values of the lateral force and for small values of the slip angle. As long as the lateral force remains small enough that the tire does not begin to skid, the slip angle remains quite small. For automobile tires, the slip angle typically remains less than about 10° even for quite vigorous cornering.

For small slip angles, a proportionality constant, the cornering coefficient, C_α, can be defined as the slope of the line relating lateral force to slip angle. At large slip angles, the tire begins to skid and the force vs. slip angle relation is significantly nonlinear. At a certain slip angle, the lateral force reaches a maximum, and for even larger values, the force tends to decline somewhat.

Although the slip angle is a main determinant of the side force, it is certainly not the only one. For example, if the wheel midplane is not at right angles to the ground plane but is tilted over at a so-called *camber angle*, this angle does influence side force generation. The effect is sometimes called *camber thrust*, but 1° of camber angle produces much less lateral force than 1° of slip angle. Thus changes in camber angle can often be neglected in a first analysis of vehicle stability. More importantly, if the wheel has a braking or driving torque, the lateral force generation can be significantly affected as will be demonstrated below.

A. Effect of Normal Force

Perhaps the most obvious influences on the side force generation for tires are the nature of the surface on which the tire rolls and the normal force, F_z, supported by the tire. It is common to assume that the side force at a given slip angle is roughly proportional to the normal force. This is consistent with the Coulomb friction model assumption that when the tire skids, the skidding force is approximately equal to a coefficient of friction times the normal force. The coefficient of friction depends on the road surface as well as on the tire material. Unfortunately, friction is a complicated phenomenon and cannot always be represented accurately using the coefficient of friction as a simple constant parameter as often is assumed in elementary treatments of friction.

For pneumatic tires, the coefficient of friction idea is often generalized to be the ratio of the lateral force to the normal force. This ratio then becomes a function of the slip angle. As the tire begins to skid, this ratio assumes a maximum value, but for very large slip angles, the ratio actually begins to drop off. Fig. 4.2 shows how a typical passenger tire generates a side force as a function of slip angle. Note that the condition of the roadway surface can have a major effect on the possible magnitude of the lateral force as anyone who has driven on an icy road can testify.

The generalized lateral coefficient of friction, μ_y, plotted in the graph is simply defined as the ratio of the side force to the normal force, F_y/F_z. Thus the lateral force can be written as

$$F_y = \mu_y F_z \tag{4.1}$$

The graph implies that the side force is strictly proportional to the normal force at a given slip angle, but this is only approximately true for a range of normal forces. When the normal force becomes very large, the ratio of side force to normal force at a given slip angle actually decreases. This

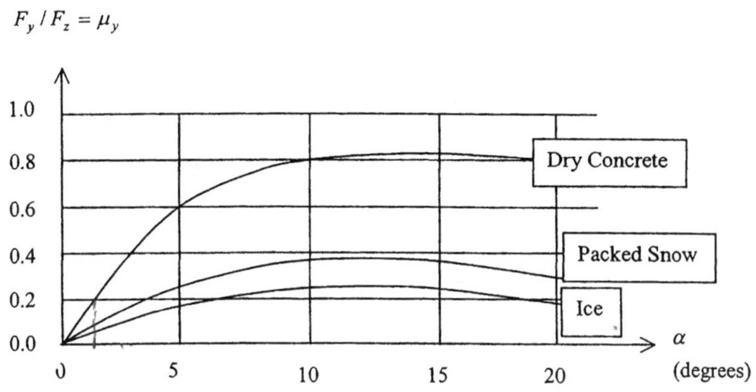

FIG. 4.2 The ratio of side force to normal force for a tire on several surfaces as a function of slip angle.

phenomenon is not represented in graphs such as the ones shown in Fig. 4.2.

The graphs in Fig. 4.2 give a good idea of the basic form of the relation between side force and slip angle. As the graphs show, there is no strictly linear portion of the curves, but up to about 5°, a linear approximation is reasonable. For many stability analyses, the tire is assumed to be operating in the vicinity of zero slip angle so a linear approximation to the relationship between the lateral force and the slip angle can be used. The *cornering coefficient*, C_α, is defined as the coefficient in the linearized version of the law relating the lateral force to the slip angle

$$F_y = C_\alpha \alpha \tag{4.2}$$

The cornering coefficient C_α is the slope of a graph of F_y as a function of α at the origin (with α in radians), assuming a constant normal force.

Using the assumption implied by Fig. 4.2 that the lateral force is proportional to the normal force, Eq. (4.1), one can find the cornering coefficient from the slope of the graphs in Fig. 4.1 near the origin, $\mathrm{d}\mu_y/\mathrm{d}\alpha$. In the linear region, the friction coefficient is approximately $(\mathrm{d}\mu_y/\mathrm{d}\alpha)\alpha$, and thus Eq. (4.1) becomes

$$F_y = (\mathrm{d}\mu_y/\mathrm{d}\alpha)\alpha F_z = \{(\mathrm{d}\mu_y/\mathrm{d}\alpha)F_z\}\alpha \tag{4.3}$$

A comparison of Eqs. (4.2) and (4.3) shows that

$$C_\alpha = \{(\mathrm{d}\mu_y/\mathrm{d}\alpha)F_z\} \tag{4.4}$$

Obviously, C_α depends on the road surface as can be seen in Fig. 4.2, but as mentioned above, the dependence on the normal force implied by the graphs in Fig. 4.2 and given in Eqs. (4.1), (4.3), and (4.4) is not strictly correct. As a result, the cornering coefficient is not strictly proportional to the normal force. As the normal force increases, the lateral force at a given slip angle begins to increase less than strictly proportionally at high normal force levels. Thus Fig. 4.2 and the relation, Eq. (4.4), derived from it should be regarded only as useful approximations for a restricted range of normal forces. See Fig. 4.3 for a plot of tire data showing the effect of changes in normal force.

Fig. 4.3 shows an important effect associated with changes in normal force. For a constant slip angle, there is a range of normal forces for which the lateral force is essentially proportional to the normal force as is assumed in Eqs. (4.1)–(4.4) and in Fig. 4.2. However, as the normal force is increased to ever-higher levels, the lateral force for a constant slip angle

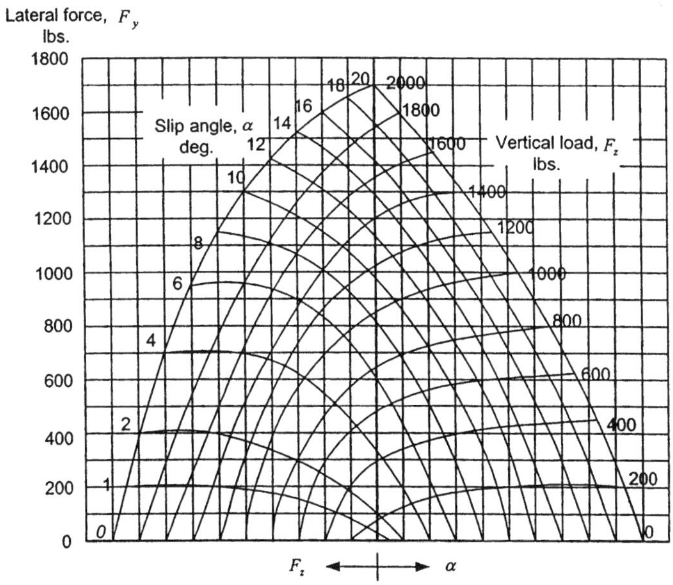

FIG. 4.3 Composite plot of lateral force, F_y, vs. slip angle, α, and normal force, F_z. (From Nordeen, 1968.)

does not continue to increase at the same rate. In fact, at some point, the lateral force ceases to increase at all with increasing normal force and even begins to decrease somewhat in some cases. This effect is particularly noticeable in Fig. 4.3 for slip angles from about 4° to 10°.

The curvature of the relation between lateral and normal force for constant slip angle will have implications for the vehicle models to be studied in Chapters 5 and 6. In many cases, it is convenient to consider a single equivalent wheel to represent the lateral forces generated by two wheels on a single axle. If the slip angles for both wheels are the same and if they both carry the same load, then it is easy to see that the total lateral force is just twice the lateral force for a single wheel. Furthermore, the axle cornering coefficient is twice the cornering coefficient for a single wheel as calculated, for example, by Eq. (4.4).

When a vehicle is negotiating a corner, there will be a transfer of load with the outside wheel taking more of the load and the inside wheel taking less. (Just how much load transfer takes place depends on a number of factors including the height of the vehicle center of mass, the distance between the wheels on the axle, and the details of the suspension system.) If the cornering coefficient was truly proportional to the normal force, the cornering coefficient for the axle could still be calculated as twice the coefficient for one wheel. Assuming that the total load supported by the axle is constant whether the vehicle is cornering or not, the increase in load on the outside wheel would be matched by the decrease in load carried by the inside wheel. The change in the cornering coefficient on one wheel would compensate the change in cornering coefficient on the other.

As can be seen in Fig. 4.3, the curvature of the lateral vs. normal force relation at a given slip angle means that the sum of the lateral forces for two wheels with equal normal forces is more than the sum of the lateral forces when one wheel has its normal force incremented by a certain amount and the other has its normal force decremented by the same amount. In simple terms, when load transfer occurs at an axle, the lateral force and the cornering coefficient at a given slip angle are less than they would be in the absence of load transfer. This effect is often used by suspension designers to adjust the handling properties of cars by varying the amount of load transfer taken by the front and rear axles through the use of antisway bars, the stiffness of the suspension springs, etc. What they are doing is influencing the effective cornering coefficients at the front and rear axles. How the cornering coefficients influence the stability of the car will be discussed in Chapter 6.

For the tire law sketched in Fig. 4.2, the maximum lateral force on dry concrete is a little more that 0.8 times the normal force. If the normal force were due solely to the weight supported by the tire, this would imply that the maximum lateral acceleration would be 0.8 times the acceleration of gravity.

Most passenger cars can barely achieve this level of lateral acceleration because of a number of effects including the fact that not all tires achieve their maximum levels of side force simultaneously in a typical case. Race cars may achieve much larger levels of lateral acceleration if they use tires using rubber compounds with higher maximum friction coefficients. Very high levels of lateral acceleration are possible particularly if cars have bodies that generate aerodynamic down-force so that the normal force on the tires is larger than the weight of the car. Maximum tire friction coefficients greater than unity are possible (usually at the expense of decreased tire life) and the aerodynamic down-force allows the tires to generate larger side forces without any increase in weight or mass, and thus lateral accelerations greater than one "g" are common for certain types of race cars.

IV. LONGITUDINAL FORCES

A free-rolling tire, i.e., a tire mounted on a wheel with perfect bearings so that no torque is applied to the wheel, establishes a relationship between the forward velocity, U, and the angular velocity, ω, in the form

$$U = r\omega \tag{4.5}$$

The parameter r is called the *dynamic rolling radius*. This radius corresponds roughly but not precisely to what one might measure as the tire radius. For a tire under load, there is no single radius. Under load, a pneumatic tire deflects so the distance from the wheel center to points on the tire circumference is not constant around the tire. Because of this, there must be some longitudinal slip between points on the tire circumference and the road surface as points on the tire move through the contact patch. Fig. 4.4 shows a wheel with a moment, τ, in the driving direction applied to the wheel. A braking moment would be in the opposite direction from the direction shown.

When a braking or driving torque is applied to the wheel, a longitudinal braking or traction force, F_x, is generated between the tire and the road surface. At the same time, the angular velocity and the forward velocity are no longer related exactly as they were for the free-rolling case. This is conventionally described by defining a nondimensional *longitudinal slip* that determines the longitudinal force in much the same way as the (nondimensional) slip angle determines the lateral force.

This longitudinal slip does not necessarily involve sliding or skidding of the entire contact patch. The longitudinal slip, much as the slip angle, is fundamentally due to the deformation of the tire. Complete sliding of the contact patch only occurs when a braked wheel stops rotating or locks and is forced to slide along the road or when a driven wheel begins to spin rapidly due to a large applied drive moment. Under less extreme con-

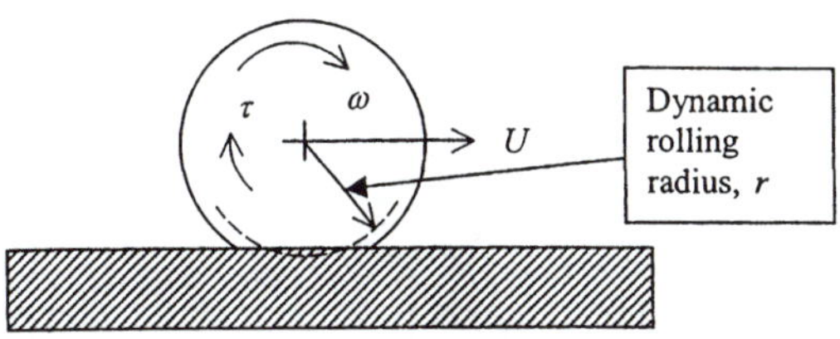

FIG. 4.4 Rolling wheel with drive moment.

ditions, the wheel velocity and rotation rate simply deviate somewhat from the relation Eq. (4.5).

During braking, the angular velocity is less than the free-rolling angular velocity and can even become zero if the wheel locks. The longitudinal slip during braking, s_B, is defined as

$$s_B = (U - r\omega)/U \tag{4.6}$$

Note that s_B is zero when there is no braking torque (the free-rolling case) because Eq. (4.5) holds and s_B becomes unity when the wheel locks ($\omega = 0$).

If a driving torque is applied to a wheel carrying a pneumatic tire, the angular velocity of the wheel is larger than it is for the free-rolling case. The wheel can even spin at standstill when the forward velocity is zero if a large-enough driving moment is applied. The definition of longitudinal slip as given in Eq. (4.6) is inconvenient for the driving moment case because it cannot be applied when $U = 0$. Thus a slightly different definition is used for longitudinal slip, s_D, when a driving moment is applied.

$$s_D = (r\omega - U)/r\omega \tag{4.7}$$

Again, the driving slip is zero in the free-rolling case because of Eq. (4.5) but becomes unity if the wheel spins at zero velocity ($U = 0$). The two different separate definitions of longitudinal slip, Eqs. (4.6) and (4.7), are necessary to handle the two situations in which either U or $r\omega$ could be zero without a division by zero.

The longitudinal traction force on the tire is defined to be in the positive direction in the driving torque case and, during braking, the force

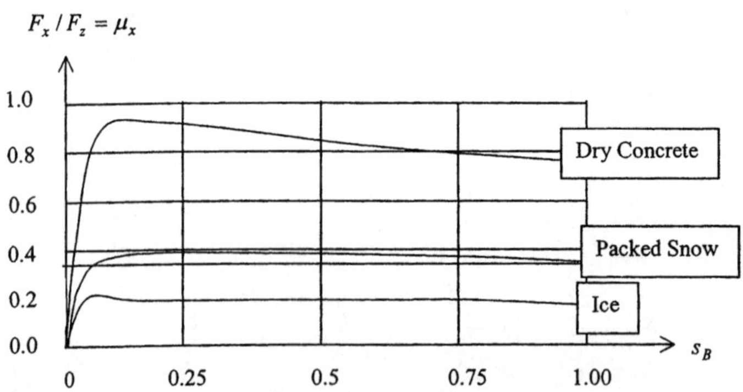

FIG. 4.5 Ratio of breaking force to normal force as a function of breaking slip.

is considered to be acting in the negative direction. In each case, however, the force depends on the appropriate version of longitudinal slip [Eqs. (4.6) and (4.7)]. Fig. 4.5 shows generally how the braking force depends on the braking slip for several surfaces. A similar plot can be made for the driving force as a function of driving longitudinal slip.

Again, the assumption, which is not strictly true, is that the longitudinal force is proportional to the normal force so that a longitudinal coefficient of friction can be defined, $\omega_x = F_x/F_z$. The similarity between the curves relating longitudinal force to longitudinal slip in Fig. 4.5 and the curves in Fig. 4.2 relating the lateral force to slip angle should be apparent.

V. COMBINED LATERAL AND LONGITUDINAL FORCES

Tires are often called upon to generate lateral and longitudinal forces simultaneously as when a car is turning and braking or accelerating at the same time. The interaction between lateral and longitudinal force generation is not favorable. The maximum lateral force is achieved when there is no longitudinal force and, conversely, the maximum longitudinal force is achieved when the lateral force is zero.

This can be appreciated by thinking of the maximum force that could be generated by a rubber tire while sliding in various directions. If the tire acted as predicted by the simple Coulomb friction model, the tire would generate the same magnitude friction force no matter what direction it was forced to slide. At a very large slip angle, the tire would generate a lateral force equal to the coefficient of friction times the normal force. A locked wheel or a wheel forced to spin rapidly would generate longitudinal forces of the same magnitude. A locked wheel forced to slide at a slip angle of 45° would produce a friction force of the same magnitude but in the negative direction of the sliding velocity. If this force was resolved into equal lateral and longitudinal components, each would have a magnitude of $1/\sqrt{2}$ times the friction coefficient times the normal force.

Although this coulomb friction model does not describe exactly how tires generate combined lateral and longitudinal forces, it leads to the useful approximate concept of the *friction circle* (Milliken and Milliken, 1995). The idea is that the maximum tire forces are essentially limited to a circle in the F_x–F_y plane assuming a constant normal force.

This means that the maximum lateral force that a tire can generate is reduced as it is required to simultaneously generate braking or traction forces, or, said in another way, the maximum braking force that a tire can generate is reduced in a cornering situation in which a lateral force must be produced simultaneously.

A number of ways have been developed to represent the behavior of tires subjected to a combination of slip angles and longitudinal slip in a mathematical form. All are quite complex and will not be discussed in detail here. The friction circle only describes the limit behavior and not how the tire behaves for slip angles and longitudinal slip values that yield forces within the circle.

Fig. 4.6 shows a plot of experimental data for a particular tire on a particular surface. The friction circle idea is clearly evident since the envelope of all the curves is approximately a circle. The behavior inside the circle, however, requires some effort to understand and is quite difficult to model in any simple way, although several quite general mathematical models are available in the literature and produce results similar to those shown in Fig. 4.6.

The sketches in Fig. 4.7 may help to understand the curves of constant slip angle but with varying longitudinal slip that appear in Fig. 4.6. The three cases show the tire operating at the fairly large constant slip angle of 20° but with a braking longitudinal slip of 100% and 0% and a driving longitudinal slip of 100%. They thus correspond to three points on the 20° curve in Fig. 4.6.

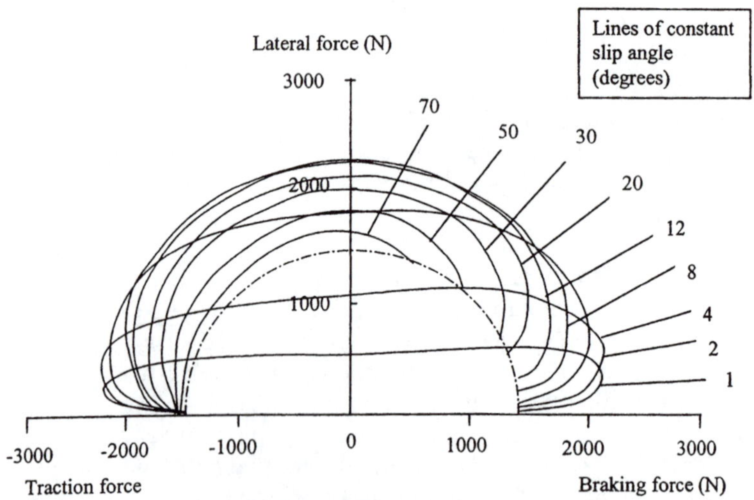

FIG. 4.6 Combined lateral and longitudinal forces.

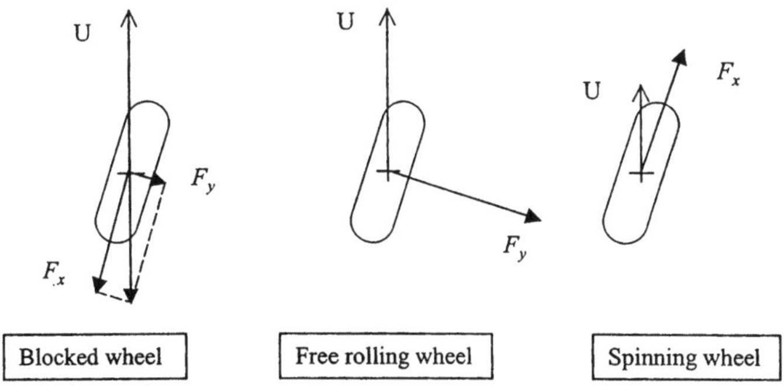

FIG. 4.7 Lateral and longitudinal tire forces for a slip angle of 20°: locked brakes; free rolling wheel; wheel spinning because of a high driving moment.

The first case on the left of Fig. 4.7 when the wheel is blocked corresponds to the end point on the right side of the 20° curve in Fig. 4.6. The middle sketch shows a free-rolling wheel with no longitudinal force and corresponds to a point at the top of the 20° curve in Fig. 4.6. Finally, the sketch at the right side of Fig. 4.7 when the driving longitudinal slip is practically 100% corresponds to the left-hand end point of the 20° curve in Fig. 4.6.

Note that the lateral force and the longitudinal braking or traction forces are defined with respect to the wheel. That is, a lateral component of force is defined to be perpendicular to the center plane of the wheel not perpendicular to the direction the wheel happens to be moving. In the three sketches, the wheel velocity is called U and the wheel is turned 20° from the direction of the velocity so that the slip angle is 20° in all three cases.

The locked wheel is simply sliding and the force is almost exactly in the opposite direction to the sliding velocity and roughly equal to a coefficient of sliding friction times the normal force. With respect to the wheel, there is a lateral force, F_y, and a braking force, F_x, and both, of course, are smaller in magnitude than the coefficient of friction times the normal force.

An important point is that the amplitude and direction of the total force on a blocked wheel have virtually nothing to do with the slip angle. The inner circle of the graph in Fig. 4.6 indicates this. At different slip angles, the lateral and longitudinal force components change with respect to the wheel, but the magnitude of the total force and its direction with respect to the velocity of the wheel remain constant. Thus there are two distinct reasons why antilock brakes are useful.

The first reason is that once a wheel stops turning or locks, a change in the steering angle does not change the direction of the force with respect to the direction the wheel is moving. This implies that steering control over the vehicle is lost. The direction of the force on a locked wheel has only to do with the direction that it is skidding and not the direction the wheel is pointing. Thus in this situation, steering the wheel does not change the magnitude or direction of the tire force.

The second reason is that at large slip angles or large longitudinal slips, the magnitude of the total force generated typically drops below its maximum value. This is clearly the case for the tire of Fig. 4.6. For the tire and the road surface of Fig. 4.6, a car with antilock brakes has the possibility of stopping in a shorter distance than if the brakes simply locked the wheels. Antilock braking systems have algorithms intended to modulate brake pressure in such a way as to converge on the value of longitudinal slip that will generate the maximum possible braking force.

The second sketch in Fig. 4.7 shows a free-rolling wheel with neither braking nor driving slip. In this case, only a lateral force is generated. In the case shown in Fig. 4.6, the magnitude of the lateral force is larger than the magnitude of the locked wheel force, but it is not quite on the outer friction circle. This means that 20° is a slip angle at which the lateral force with no longitudinal slip has already declined from the maximum value. A close inspection of Fig. 4.6 reveals that the slip angle that yields the maximum lateral force lies somewhere between 8° and 12°.

The third sketch in Fig. 4.7 represents the left end point of the 20° slip angle curve in Fig. 4.6 where the driving slip approaches 100%. The wheel is spinning so fast that the relative velocity between the wheel and the road surface is much larger than the wheel velocity U. In this case, the slip of the contact patch slip is nearly in the direction the wheel is pointing so the force is also nearly in the direction the wheel is pointed. The magnitude of this driving force is about the same as the magnitude of the force of a locked wheel. The force is also smaller than it would be at less than 100% driving slip.

Thus maximum acceleration would not occur for a car using this tire if the drive wheels spin. On the other hand, for a front-wheel drive car, even if the front wheel spins under power, some steering control will be present because the longitudinal force points nearly in the direction the wheel is pointed.

Although in the succeeding chapters that deal with ground vehicles the emphasis will be on the cornering coefficient expression, Eq. (4.2), that strictly applies only for small slip angles and small lateral forces, in Chapter 6, some effects of the nonlinear relation between lateral force and slip angle are discussed. Consideration of this nonlinear behavior is necessary when discussing stability in steady turns or ultimate cornering ability.

A consideration of Figs. 4.2 and 4.6 will allow at least a qualitative understanding of how braking or accelerating can affect lateral stability. For example, a drive or braking moment at a wheel essentially reduces the cornering coefficient for that wheel. This means that the distribution of braking forces can have an important effect on automobile stability. Furthermore, power effects on stability are different depending upon whether the car has front-wheel drive or rear-wheel drive.

The connection between longitudinal acceleration and lateral stability is quite a complex matter because not only do brake and drive moments affect individual tires as shown in Fig. 4.6, but also the longitudinal acceleration affects the normal forces at the front and rear axles. During acceleration, the normal forces at the rear tend to increase and the normal forces at the front tend to decrease. The shift in normal forces is easy to understand if the acceleration or deceleration is relatively steady, but for sudden changes in longitudinal acceleration, the vertical dynamics complicates the picture.

The suspension system stiffness and damping characteristics affect the normal forces at the wheels in every sort of transient situation. Whether it is a sudden change in longitudinal acceleration due to application of power or braking or a sudden steering input causing a change in lateral acceleration, the compliance and damping properties of the vehicle suspension determine in detail how the normal forces at the wheels change. The distribution of normal forces among the individual wheels changes the effective cornering coefficients and hence the stability properties of the entire automobile.

In Chapters 5 and 6, the vehicle models will not contain the suspension effects mentioned above in any detail. In some cases, one can imagine the qualitative effects of the vehicle suspension on stability during some maneuvers by considering the effect on axle cornering coefficients, i.e., the cornering coefficient considering the sum of the lateral forces from the two wheels on an axle. If both wheels at an axle have increased normal forces, the cornering coefficient for the axle increases. If the total load on the axle is constant but one wheel takes more of the load and the other takes less, the axle cornering coefficient generally decreases.

It is always possible to construct more complex vehicle models than the ones used in this book to determine how the vertical dynamics might affect the lateral dynamics that is the main interest here. The problem is that the more complex the model is, the more difficult it is to see general features of vehicle stability. There are numbers of technical papers that treat the dynamics and stability of particular vehicles and use detailed tire models that are capable of coupling the vertical dynamics with the lateral dynamics of a vehicle through the normal force effect. The models are, however, too complex and too particular to be included here.

5
Stability of Trailers

I. INTRODUCTION

In this chapter, several mathematical models of trailers will be developed and analyzed for their stability properties. All of the models are quite simple and easy to understand so they provide an opportunity to introduce equation formulation techniques such as the use of Lagrange's equations as an alternative to Newton's laws. In addition, these trailer models exhibit many of the concepts related to vehicle stability that will be encountered in more complex situations in succeeding chapters.

Most people have probably followed a car or truck towing a trailer that tended to oscillate back and forth after being hit by a gust of wind or after the towing vehicle changed lanes abruptly. An oscillatory motion that only settles down after a period of time indicates that the trailer is not very stable, and makes one wonder if the trailer could be unstable in slightly different circumstances.

What only a few unlucky people have observed is a trailer that exhibited actual unstable behavior. When this happens, often when a critical towing speed is exceeded, the result is typically a serious accident, which may involve the observer as well as the driver in the towing vehicle. It is difficult for the average person to understand just why most trailers cause no problems, and yet some become unstable for no obvious reason. Even relatively simple models can shed light on this dangerous phenomenon.

In addition to the inherent interest in the stability of trailers (see Bundorf, 1967a; Diboll and Hagen, 1969), these models will illustrate a number of techniques and concepts that will again be useful when studying the stability properties of other vehicles such as automobiles in Chapter 6, and the stability of wheels mounted as casters on motorcycles, automobiles, and aircraft landing gear in Chapter 8.

The chapter will start with the simplest trailer model and gradually add extra effects that finally lead to conclusions about how trailers should be loaded to be stable. It will also be possible to introduce the important idea of a critical speed for trailers that have the possibility of being unstable. Unfortunately, there are many vehicles in addition to trailers that appear to be stable until the first time that they exceed a certain speed, called the critical speed. A surprisingly simple trailer model will be used to illustrate this phenomenon.

II. SINGLE DEGREE-OF-FREEDOM MODEL

In this example, a model of a trailer being towed behind a massive vehicle will be developed. In this case, it is logical to use an inertial coordinate system rather than a body-centered coordinate system because the towing vehicle is assumed to move at constant velocity along a straight path in inertial space no matter what the trailer does. Thus a coordinate system moving with the towing vehicle or on the ground aligned with the path of the towing vehicle can be considered an inertial coordinate system.

As shown in Fig. 5.1, the angle $\theta(t)$ between the centerline of the trailer and the path of the towing vehicle completely describes the motion of the trailer, which is considered to be a rigid body moving in plane motion. When a single position variable such as θ and its derivatives suffice

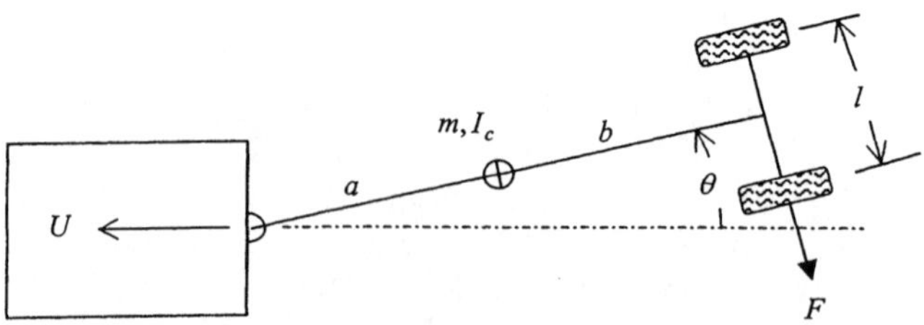

FIG. 5.1 Single degree-of-freedom trailer model.

to describe the dynamics of a system, the system is said to have a single degree-of-freedom.

The trailer is assumed to be a rigid body pivoted without friction at the hitch point. The mass of the wheels is included in the total mass of the trailer. The trailer is assumed to act as though it had no suspension and thus to move in plane motion. Although the height of the center of mass of the trailer above the ground does affect real trailers, particularly those with suspension systems, this simple model essentially assumes that all forces and accelerations lie in the ground plane. The new aspect in this example is the consideration of the interaction of pneumatic tires and the road surface. In Chapter 4, the tire–road interaction was discussed in more detail than is necessary for the present purposes. Here only the elementary concept of a *cornering coefficient* will be used for the analysis.

Fig. 5.1 is a sketch of the simplest model of a trailer to be studied in this chapter. The towing vehicle is assumed to move with constant forward velocity U and with no lateral velocity even if there is a lateral force from the trailer on the body at the hitch. The trailer is considered to be pivoted frictionlessly at the hitch. This means that there is no moment exerted on the trailer at the hitch.

Parameters include a and b, the distances from the center of mass to the hitch point and the axle, respectively; m, the trailer mass; I_c, the centroidal moment of inertia; l, the track of the wheels; and U, the constant towing speed. The single geometric degree-of-freedom variable describing the motion of the trailer is the angle $\theta(t)$, which implies that all velocities and accelerations necessary to write dynamic equations for the trailer can be expressed in terms of θ and its time derivatives.

To find a dynamic equation for θ, one must consider another variable, the force, from the ground acting on the two tires. Fig. 5.2 shows a single tire rolling with velocity U, and being acted upon by a lateral force F. Note that, in Fig. 5.2, F represents a force on the axle of the wheel tending to push the wheel and tire to the right. If one neglects the acceleration of the wheel mass, there is an equal and opposite force on the bottom of the tire from the road that is not shown.

For a given tire at a given normal load, it is found experimentally that the relation between the lateral force, F, and the slip angle, α, has the general form shown in the right side of Fig. 5.2 (see Fig. 4.2 in the previous chapter). For small values of α, the force is nearly proportional to the slip angle and the slope of the relationship is defined as the *cornering coefficient* C_α. The linear range typically applies up to a slip angle of about 5° or 10°. In the linear range, the side velocity $\dot{y}$ is much smaller than the rolling

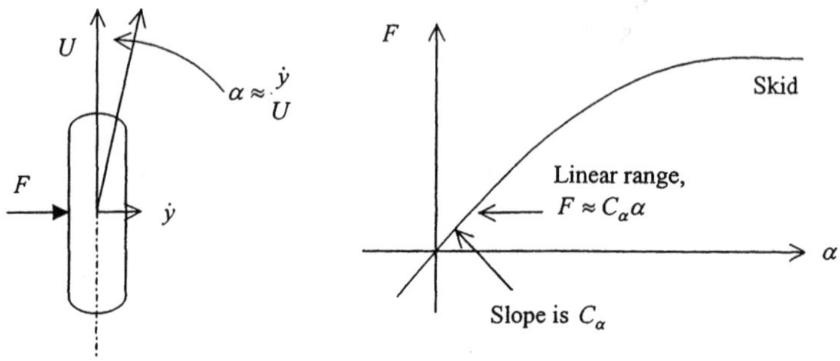

FIG. 5.2 Definition of cornering coefficient.

velocity U, and thus α in radians is nearly $\dot{y}/U$ rather than the tangent of $\dot{y}/U$.

$$\alpha \cong \dot{y}/U \tag{5.1}$$

The force relationship can be written approximately as

$$F = C_\alpha \alpha \tag{5.2}$$

We now consider the *basic motion* to be one in which θ and $\dot{\theta}$ are zero, and the trailer simply follows the towing vehicle. For a stability analysis, we assume a *perturbed motion* in which both θ and $\dot{\theta}$ are small. This means that small angle approximations can be used for trigonometric functions of θ and that the lateral velocity of the tires and thus the slip angles will also turn out to be small.

The major task in developing the equation of motion is to derive an expression for the slip angle of the tires in terms of θ and $\dot{\theta}$. The first step is to make a generic sketch, Fig. 5.3, showing the velocity components at the wheels.

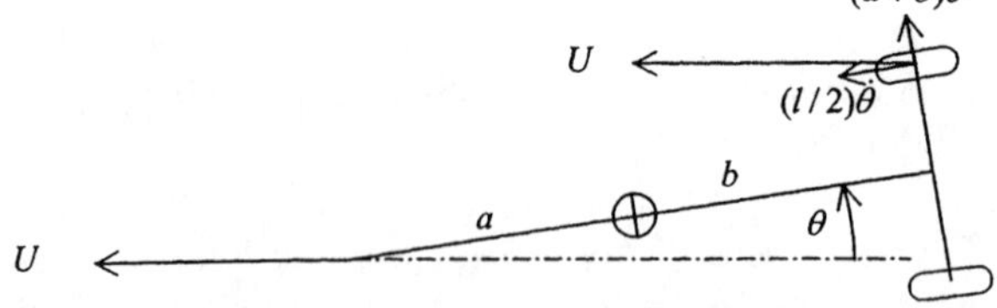

FIG. 5.3 Velocity components at a wheel.

The procedure is to express the velocity at the wheel as a known velocity, U at the hitch point plus $\vec{\omega} \times \vec{r}_{AB}$ where $\vec{r}_{AB}$ is the vector from the hitch point to the wheel. This type of calculation as was previously performed for the shopping cart in Chapter 1, Eqs. (1.15) and (1.16). In this plane motion case, it is easy to write down components of $\vec{\omega} \times \vec{r}_{AB}$ using elementary considerations instead of establishing unit vectors and evaluating the expression formally.

One can consider the vector distance from the hitch point to the upper wheel, $\vec{r}_{AB}$, in Fig. 5.3 as the sum of two vectors. One vector lies along the centerline of the trailer and reaches from the hitch to the center of the axle. The length of this vector is $(a+b)$. A second vector lies between the center of the axle and the upper wheel. It has a length of $l/2$. Because $\dot{\theta}$ plays the role of ω, the result of the velocity components contributed by $\vec{\omega} \times \vec{r}_{AB}$ is the two components proportional to $\dot{\theta}$ shown in Fig. 5.3. The distance $(a+b)$ multiplied by $\dot{\theta}$ gives a lateral velocity to the wheel and the distance $(l/2)$ multiplied by $\dot{\theta}$ gives a (small) forward velocity to the wheel. When the vector velocity with magnitude U is combined with the two components proportional to $\dot{\theta}$, the total velocity of the wheel has been represented in the sketch.

From Fig. 5.3, one can now see how to express the velocity of the wheel as the sum of two other components. The first component is called the *rolling velocity* and is the component of the velocity in the direction the wheel is pointing. The other component is the *lateral velocity* and is the component of the velocity perpendicular to the direction the wheel is pointing. These are the two velocity components shown in Fig. 5.2 as U and $\dot{y}$, respectively.

The rolling velocity in Fig 5.3 is

$$U \cos \theta + (l/2)\dot{\theta} \cong U \tag{5.3}$$

in which the fact that both θ and $\dot{\theta}$ are small have been used ($\cos \theta \cong 1$, $(l/2)\dot{\theta} \ll U$).

The lateral velocity is

$$U \sin \theta + (a+b)\dot{\theta} \cong U\theta + (a+b)\dot{\theta} \tag{5.4}$$

where the approximation $\sin \theta \cong \theta$ has been used.

A combination of Eqs. (5.3) and (5.4) yields the approximate expression for the slip angle as the ratio of the lateral velocity to the rolling velocity as was indicated in Eq. (5.1).

$$\alpha = (U\theta + (a+b)\dot{\theta})/U = \theta + (a+b)\dot{\theta}/U \tag{5.5}$$

Note that since the contribution to the rolling velocity, $(l/2)\dot{\theta}$, was neglected, the expression for slip angle applies equally well to both wheels. (The velocity component associated with the half-track would point in the backward direction on the lower wheel in Fig 5.3, but it would be neglected for the rolling velocity anyway. The lateral velocity components would be identical for both wheels.)

We are now in a position to derive an equation of motion for the trailer. The force F in Fig. 5.1 is the force on the wheels from the ground. When the slip angle is positive as shown in Fig. 5.3, the force on the tires from the ground actually points in the direction shown in Fig. 5.1. Equation (5.5) can now be combined with Eq. (5.2) to yield an expression for the force exerted on the trailer tires

$$F = C_\alpha \alpha = C_\alpha(\theta + (a+b)\dot{\theta}/U) \tag{5.6}$$

Although in Fig. 5.2, the cornering coefficient was for a single tire, in Eq. (5.6) the coefficient is meant to include the net force being generated by both tires. This is possible because the approximate value for the slip angle in Eq. (5.5) applies equally to both tires.

It will be left as an exercise, to use Newton's laws to find the equation of motion. The procedure is straightforward. The sum of the lateral forces must equal the mass times the acceleration of the center of mass, $ma\ddot{\theta}$. The sum of all moments about the center of mass must equal the moment of inertia times the angular acceleration, $I_z\ddot{\theta}$. While the procedure is easily described, it does require the introduction of a side force at the hitch in addition to the total tire force. This hitch force is eventually eliminated to produce the final equation of motion. Here an alternative method for deriving the equation of motion will be introduced.

A. Use of Lagrange's Equations

We will use this opportunity to show how Lagrange's equations can be used to derive the equation of motion (Crandall et al., 1968). In many cases, it is easier to use this method than to use Newton's laws that typically require manipulation to eliminate internal variables such as the hitch force in the case at hand. The algebra required for the elimination of internal variables when using Newton's laws is often subject to human error and the Lagrange equation method eliminates this problem. In any case, it is useful to have available two quite different methods to derive equivalent equations of motion so that one can compare them to check the final result for correctness. It is assumed that the reader has some familiarity with Lagrange's equations or will skip this section and derive the equation using Newton's laws.

Because we have assumed that the hitch has no friction, the connection at the hitch is a so-called *workless constraint*. Because there is assumed to be no friction moment at the hitch, no energy is gained or lost as the trailer rotates. Also, because the hitch has no lateral velocity, the lateral force at the hitch does no work. Generally, forces and moments associated with workless constraints do not have to be considered when using Lagrange's equations. For the present trailer model, the hitch force need not be considered when using Lagrange's equations although it must be considered when using Newton's equations.

For this problem, there is only a single generalized coordinate, θ. The appropriate Lagrange equation is then as follows:

$$\frac{\mathrm{d}}{\mathrm{d}t}\frac{\partial T}{\partial \dot{\theta}} - \frac{\partial T}{\partial \theta} + \frac{\partial V}{\partial \theta} = \Xi_\theta \tag{5.7}$$

In the Eq. (5.7), $T = T(\theta, \dot{\theta})$ is the kinetic energy, $V = V(\theta) = 0$ is the potential energy (which happens to be zero in this case), and Ξ_θ is the generalized force for the θ generalized coordinate. The generalized force can be evaluated by evaluating the work performed by all forces not associated with the kinetic or potential energy during a small virtual displacement of θ,

$$\delta W = \Xi_\theta \delta\theta \tag{5.8}$$

The virtual displacement $\delta\theta$ is an imaginary, infinitesimal change or variation in θ. The work term δW represents the work carried out on the system by all forces not represented by changes in kinetic or potential energy terms included in Eq. (5.7). In this case, the generalized force will be used to account for the tire force of Eq. (5.6).

The general expression for kinetic energy for a rigid body in plane motion is

$$T = (1/2)mv_c^2 + (1/2)I_c\omega^2 \tag{5.9}$$

in which the only square of the velocity of the center of mass and the square of the angular velocity are required. (For Newton's laws, one needs to know the vector linear and angular accelerations, which are often complicated to evaluate.)

To derive linearized perturbation equations for stability analysis using Lagrange equations, the energy expressions should be correct approximations up to second-order terms in small quantities. It is often useful to find the expressions exactly and then later to make the second-order approximation. Fig. 5.4 shows the velocity components for the center of mass when θ is not necessarily small.

Note that although a vector diagram is required to find the vector velocity of the center of mass $\vec{v}_c$, only the square of the magnitude of the velocity is required to compute the kinetic energy. Again, the required velocity is the velocity of the hitch point plus the effect of rotation, which is a vector perpendicular to the line from the hitch point to the center of mass and with a magnitude $a\dot{\theta}$. Using either the law of cosines for one of the velocity triangles in Fig. 5.4, or by evaluating the sum of the squares of two perpendicular components of $\vec{v}_c$, the expression for v_c^2 is found to be the following:

$$v_c^2 = U^2 + (a\dot{\theta})^2 + 2U(a\dot{\theta})\sin\theta \tag{5.10}$$

Now modifying the expression to retain terms in θ and $\dot{\theta}$ up to second order, the results required to evaluate the kinetic energy in Eq. (5.9) are

$$v_c^2 = U^2 + a^2\dot{\theta}^2 + 2Ua\dot{\theta}\theta, \quad \omega^2 = \dot{\theta}^2 \tag{5.11}$$

Note that second-order terms in small quantities include the squares of the quantities as well as product of the quantities. The constant term U^2 could actually be left out of Eq. (5.11) because the Lagrange equation, Eq. (5.7), involves only derivatives of the energy functions.

The generalized force Ξ_θ, is evaluated by computing the work performed by the force, F, exerted by the ground on the two tires when θ is increased by the virtual displacement $\delta\theta$. In this case, the work performed by the tire force on the body is negative when $\delta\theta$ is positive. One can imagine increasing θ by the amount $\delta\theta$ against the force F. In this process, F

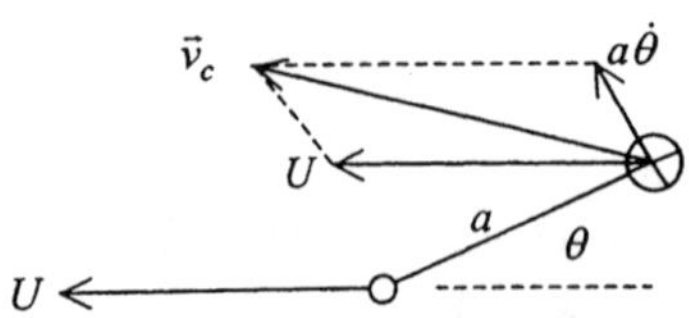

FIG. 5.4 Velocity of the center of mass.

would do negative work on the trailer. This negative work would be given by the force times the distance the force moves,

$$\delta W = -F(a+b)\delta\theta \tag{5.12}$$

Now substituting Eq. (5.6) into Eq. (5.12) and considering Eq. (5.8), we see that the generalized force is the coefficient multiplying $\delta\theta$.

$$\Xi_\theta = -C_\alpha(a+b)\theta - C_\alpha(a+b)^2\dot{\theta}/U \tag{5.13}$$

In this particular case, the generalized force is actually a moment because the generalized coordinate θ is an angle.

Substituting the quantities in Eq. (5.11) into Eq. (5.9), the kinetic energy is

$$T = m(U^2 + a^2\dot{\theta}^2 + 2Ua\dot{\theta}\theta)/2 + I_c\dot{\theta}^2/2 \tag{5.14}$$

Finally, the Lagrange equation, Eq. (5.7), yields the single second-order equation of motion for the variable θ.

$$(I_c + ma^2)\ddot{\theta} + \frac{C_\alpha(a+b)^2}{U}\dot{\theta} + C_\alpha(a+b)\theta = 0 \tag{5.15}$$

This equation is of the form of a single nth order equation as discussed in Chapter 3, Eq. (3.3) with $n = 2$. Alternatively, it is of the form of sets of second-order equations, Eq. (3.7), but in this case of only a single generalized coordinate there, is only one equation and the matrices involved are actually only scalars. For multiple degrees-of-freedom, Lagrange equation formulations will produce equations of the form of Eq. (3.7).

B. Analysis of the Equation of Motion

Equation (5.15) resembles the equation for a torsional oscillator that might be written $J\ddot{\theta} + B\dot{\theta} + K\theta = 0$, where J is a moment of inertia, B is a torsional damping constant, and K is a torsional spring constant. By comparing the two equations, one can give a physical interpretation to stability considerations for the trailer.

First, if the three coefficients of θ, $\dot{\theta}$, and $\ddot{\theta}$ are positive, the results of Chapter 3 indicate that the trailer is stable and that the trailer will oscillate as a damped torsional oscillator would. The three coefficients in Eq. (5.15) are inherently positive with the possible exception of the $\dot{\theta}$ coefficient if a negative value for U were used. It appears that only if U were negative, that is, if the towing vehicle were backing up, would the trailer act like an oscillator with negative damping coefficient and would be therefore be unstable.

In fact, this is a case in which merely using a negative value for U does not correctly yield the equation for a reversed velocity because the slip angle expression of Eq. (5.5) does not correctly represent the slip angle for a trailer backing up when U is considered to be negative. (The use of a negative value for the velocity in the analysis of the shopping cart in Chapter 1 to represent backward motion was legitimate in that case.)

If Fig. 5.3 is redrawn with U shown in the opposite direction, it will be seen that in the expressions for lateral velocity, Eq. (5.4), and the slip angle, Eq. (5.5), a negative sign appears in the term multiplying θ, not in the term multiplying $\dot{\theta}$. What this means is that it is actually the last term in Eq. (5.15) that acquires a negative sign when the trailer is backing up, not the middle term. The conclusion that a backing up trailer is unstable at all speeds is correct but it is the "spring-like" term that is to blame rather than the "damping-like" term.

Secondly, the division by U in the $\dot{\theta}$ term means that what appears to be a torsional damping term becomes very small at very high speed, and the trailer will tend to oscillate a long time if disturbed when U is large. This simplest trailer model never is truly unstable, but the trailer approaches an undamped, neutrally stable system as the speed of travel increases. This gives a hint that a more complete model might well show the possibility of instability at high speed. This is indeed the case as will be demonstrated below using a more complex trailer model.

Finally, one can consider the case at low speeds where one might expect the inertial effects in Eq. (5.15) to be unimportant and that the $\ddot{\theta}$ term might be neglected. Alternatively, one might imagine the case of very rigid tires with high values of C_α, so that any term missing a C_α factor in Eq. (5.15) could be neglected. Either way, if the $\ddot{\theta}$ term in Eq. (5.15) is suppressed the resulting equation is reduced to first order.

$$\dot{\theta} + \frac{U}{(a+b)}\theta = 0 \tag{5.16}$$

This equation, which might be called a "kinematic equation" rather than a dynamic equation, also implies that the slip angle α in Eq. (5.5) vanishes.

Equation (5.16) has the same first-order form as the equation for the shopping cart studied previously in Chapter 1, Eq. (1.19). For a positive value of U, the single coefficient in Eq. (5.16) is positive, indicating that the trailer is stable. This certainly comes as no surprise because the equation only holds for slow motion.

In this case, simply considering a negative value for U does yield the correct equation for reverse motion. The coefficient is then negative and the trailer is unstable just as the backward moving shopping cart was in Chapter 1. The conclusion is that the trailer is stable for slow forward velocities and unstable backing up no matter how slowly. Anyone who has tried to maneuver a boat trailer will agree that this very simple model has some validity at least at low speeds.

III. TWO DEGREE-OF-FREEDOM MODEL

The previous analysis of the stability of a trailer assumed that it was a rigid body attached at the hitch to a towing vehicle that moved in a straight line at a constant velocity. The movement of the trailer could thus be described by a single geometric variable, the angle between the trailer centerline and the direction of motion of the towing vehicle. The result of the analysis was that the trailer never was unstable although it could oscillate with less and less damping as the speed increased. This leads to the suspicion that a more complex model of the trailer might indeed have the possibility of being unstable at high speeds. Real trailers do in fact sometimes exhibit instability at high towing speeds. Therefore another degree-of-freedom will be added to the mathematical model to see whether a slightly more realistic model can shed insight into this phenomenon.

Fig. 5.5 is similar to Fig. 5.1 for the single degree-of-freedom model, except that a flexible connection between the nose of the trailer and the hitch has been introduced. This connection is shown as a spring, with a deflection variable x and a parameter, the spring constant k. The lateral force between the towing vehicle and the nose of the trailer would have the value kx.

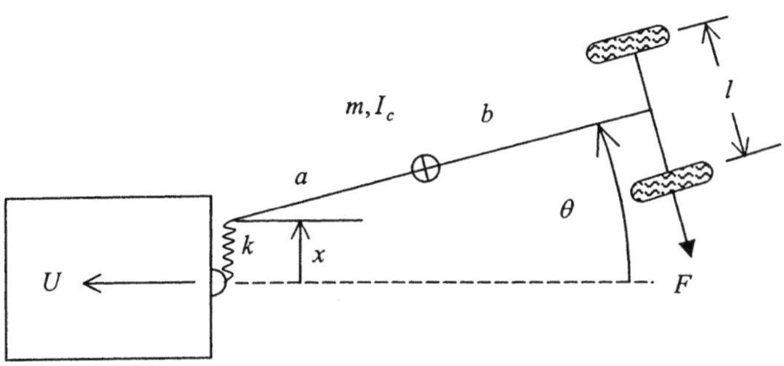

FIG. 5.5 Two degree-of-freedom trailer model.

Actually, the spring can represent several possible effects. It could represent flexibility in the tongue of the trailer or it may represent the lateral flexibility of the towing vehicle's suspension. (If the wheels of the towing vehicle are assumed to move in a straight line along the road surface, the body can be pushed from side to side because of some flexibility in the rear suspension and in the sidewalls of the tires.) Finally, if the connection between the ball at the hitch point and the socket on the tongue of the trailer has excessive play, this would also allow motion between the nose of the trailer and the towing vehicle and this could be represented by a spring with a zero spring constant for a limited range of motion.

In any event, the model of the trailer now has two geometric variables or generalized coordinates, $x(t)$ and $\theta(t)$ that can be used to describe its motion. The analysis of this two degree-of-freedom model proceeds much as the analysis for the single degree-of-freedom case with the addition of the effects of the coordinate x. The basic motion is now represented by $x = \dot{x} = \theta = \dot{\theta} = 0$, and all these variables are assumed to have small values for the perturbed motion.

A. Calculation of the Slip Angle

As in the previous analysis, the force F on the bottom of the two tires will be assumed to be given by a cornering coefficient times a slip angle. The calculation of the slip angle is similar to the previous calculation except that now the entire trailer can move sideways with velocity x, as well as rotate about the hitch point with angular velocity $\dot{\theta}$. The velocity components at one wheel are shown in Fig. 5.6. Fig. 5.6 is identical to Fig. 5.3 with the addition of the velocity component $\dot{x}$.

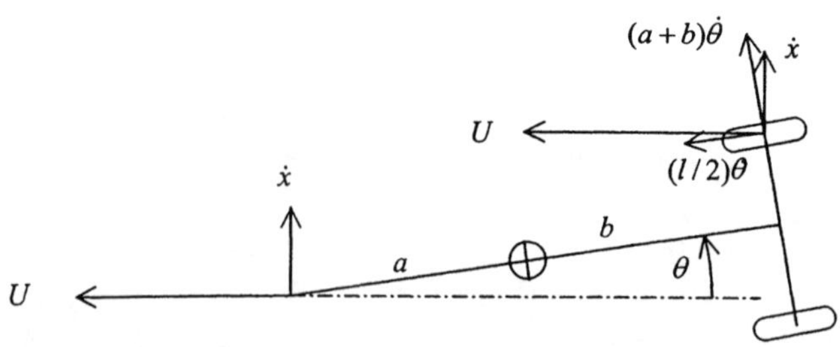

FIG. 5.6 Wheel velocity components for two degree-of-freedom model.

For the perturbed motion, we will assume

$$\theta << 1, \ (l/2)\dot{\theta} << U, \ \dot{x} << U \tag{5.17}$$

The rolling velocity from Fig. 5.6 then is

$$U \cos \theta + (l/2)\dot{\theta} - \dot{x} \sin \theta \cong U \tag{5.18}$$

and as before this expression is approximately true for both wheels because of the approximations allowed by Eq. (5.17).

The lateral velocity is

$$U \sin \theta + (a+b)\dot{\theta} + \dot{x} \cos \theta \cong U\theta + (a+b)\theta + \dot{x} \tag{5.19}$$

Finally, the expression for the slip angle is just the ratio of the lateral velocity to the rolling velocity.

$$\alpha = \theta + (a+b)\dot{\theta}/U + \dot{x}/U \tag{5.20}$$

Again, the force on the wheels, considered positive in the direction shown in Fig. 5.5, will be $F = C_\alpha \alpha$ with the positive cornering coefficient C_α understood to refer to the sum of the two wheel forces or the total force on the axle.

B. Formulation Using Lagrange's Equations

It will again be left as an exercise to find the equations of motion using Newton's laws. Here the use of Lagrange equations for a two degree-of-freedom system will be illustrated. For this system, the Lagrange equation formulation is a straightforward extension of the calculations performed previously for the single degree-of-freedom trailer model. The result will be two second-order differential equations for the two generalized coordinates x and θ.

The kinetic energy has the same general expression used previously, Eq. (5.9). Now, however, the effect of the second generalized velocity $\dot{x}$ must be considered. The three components of the velocity of the center of mass are shown in Fig. 5.7.

The velocity components in Fig 5.7 are identical to those in Fig. 5.4, but now there is the extra component, $\dot{x}$. Because of the extra component, it is not as easy to use the law of cosines, but, by squaring two orthogonal

FIG. 5.7 Velocity components for center of mass.

components and adding, the square of the velocity of the center of mass can be found.

$$\begin{aligned} v_c^2 &= (U + a\dot{\theta}\sin\theta)^2 + (\dot{x} + a\dot{\theta}\cos\theta)^2 \\ &= U^2 + a^2\dot{\theta}^2 + \dot{x}^2 + 2Ua\dot{\theta}\sin\theta + 2\dot{x}a\dot{\theta}\cos\theta \end{aligned} \tag{5.21}$$

This expression is simplified by retaining terms up to second order in the small quantities x, $\dot{x}$, θ, $\dot{\theta}$ so that linearized Lagrange equations will ensue. The final approximate expression for T applying to the perturbed motion is as follows:

$$T = (1/2)m(U^2 + a^2\dot{\theta}^2 + \dot{x}^2 + 2Ua\dot{\theta}\theta + 2\dot{x}a\dot{\theta}) + (1/2)I_c\dot{\theta}^2 \tag{5.22}$$

Because of the spring, there is also potential energy for this model.

$$V = (1/2)kx^2 \tag{5.23}$$

The generalized forces come from considering the increment of work performed on the vehicle by the tire force, F, when the two generalized coordinates are given independent variations δx and $\delta\theta$. In each case, the work performed by F is negative and equal to the force times the distance moved as a result of the variation in the generalized coordinates.

$$\delta W = -F(a+b)\delta\theta - F\cos\theta\delta x \tag{5.24}$$

Using the small angle approximation, the generalized forces are just the coefficients of δx and $\delta\theta$ in Eq. (5.24).

$$\Xi_x = -F, \quad \Xi_\theta = -F(a+b) \tag{5.25}$$

As before, the axle tire force will be the combined cornering coefficient times the slip angle as given by Eq. (5.2), where the slip angle expression is now Eq. (5.20).

The Lagrange equations may now be evaluated. For the generalized coordinate, x, the equation is

$$\frac{\mathrm{d}}{\mathrm{d}t}\frac{\partial T}{\partial \dot{x}} - \frac{\partial T}{\partial x} + \frac{\partial V}{\partial x} = \Xi_x \tag{5.26}$$

After substituting the tire force law, Eq. (5.2), the slip angle, Eq. (5.20), the generalized force, Eq. (5.25) and the energy expressions, Eqs. (5.22) and (5.23) into Eq. (5.26), the result is the first equation of motion.

$$ma\ddot{\theta} + m\ddot{x} + kx = -C_\alpha(\theta + (a+b)\dot{\theta}/U + \dot{x}/U) \tag{5.27}$$

The Lagrange equation for θ is

$$\frac{\mathrm{d}}{\mathrm{d}t}\frac{\partial T}{\partial \dot{\theta}} - \frac{\partial T}{\partial \theta} + \frac{\partial V}{\partial \theta} = \Xi_\theta \tag{5.28}$$

which, after the appropriate substitutions, becomes

$$(I_c + ma^2)\ddot{\theta} + ma\ddot{x} = -(a+b)C_\alpha(\theta + (a+b)\dot{\theta}/U + \dot{x}/U) \tag{5.29}$$

It is instructive to put the two equations in a matrix form as discussed in Chapter 3, Eq. (3.7).

$$\begin{bmatrix} I_c + ma^2 & ma \\ ma & m \end{bmatrix}\begin{bmatrix} \ddot{\theta} \\ \ddot{x} \end{bmatrix} + \begin{bmatrix} C_\alpha(a+b)^2/U & C_\alpha(a+b)/U \\ C_\alpha(a+b)/U & C_\alpha/U \end{bmatrix} \times \begin{bmatrix} \dot{\theta} \\ \dot{x} \end{bmatrix} + \begin{bmatrix} C_\alpha(a+b) & 0 \\ C_\alpha & k \end{bmatrix}\begin{bmatrix} \theta \\ x \end{bmatrix} = \begin{bmatrix} 0 \\ 0 \end{bmatrix} \tag{5.30}$$

C. Analysis of the Equations of Motion

The display of the equations in Eq. (5.30) makes them resemble classical equations for a two degree-of-freedom vibratory system with a mass matrix, a damping matrix and a stiffness matrix. This is the form discussed in Chapter 3, Eq. (3.7). When a comparison is made between Eqs. (3.7) and (5.30), it is seen that the "mass" matrix $[M]$ in Eq. (5.30) does depend on inertia elements and the "stiffness" matrix $[K]$ does contain a spring constant but the "damping" matrix $[B]$ and much of $[K]$ are filled with elements depending on the cornering coefficient. Thus this mathematical model does not represent a normal vibratory system.

It is well known that when vibratory systems are formulated using Lagrange's equations, the matrices are symmetric. Furthermore, a vibratory system with symmetric matrices is not normally unstable. If the matrices are positive definite, which they usually are, one can give an argument to show that the damping in the system can only reduce the potential and kinetic energy. This implies that, in the absence of forcing, the system will tend to return to equilibrium and is thus stable. For vibratory systems that do not have negative spring or damping constants, instability is not really an issue.

In this example, however, while the “mass” and the ”damping” matrices in Eq. (5.30) are symmetric, the “stiffness” matrix is not. Thus one cannot use energy arguments useful for vibratory systems to prove that the trailer system is stable.

To derive the characteristic equation, assume we make assumptions as indicated in Eq. (3.21).

$$\theta(t) = \Theta e^{st}, \quad x(t) = Xe^{st} \tag{5.31}$$

After substituting Eq. (5.31) into Eq. (5.30), and eliminating the common nonzero factor e^{st}, the result is

$$\begin{bmatrix} (I_c + ma^2)s^2 + (C_\alpha(a+b)^2/U)s + C_\alpha(a+b) & mas^2 + (C_\alpha(a+b)/U)s \\ mas^2 + (C_\alpha(a+b)/U)s + C_\alpha & ms^2 + (C_\alpha/U)s + k \end{bmatrix} \times \begin{bmatrix} \Theta \\ X \end{bmatrix} = \begin{bmatrix} 0 \\ 0 \end{bmatrix} \tag{5.32}$$

As indicated in Eq. (3.22), the trivial (zero) solution for Θ and X would be unique unless the determinant of the coefficient array is zero. After quite a bit of algebraic manipulation, the determinant can be written out and the characteristic equation emerges.

$$\begin{aligned} mI_c s^4 + (I_c + mb^2)(C_\alpha/U)s^3 + ((I_c + ma^2)k + mbC_\alpha)s^2 \\ + (C_\alpha(a+b)^2 k/U)s + C_\alpha(a+b)k = 0 \end{aligned} \tag{5.33}$$

This is a complex characteristic equation, but still a few conclusions about stability can be made quite simply. The search for negative coefficients that would indicate instability, shows that, for negative U (backing up), the trailer would be unstable. This is no surprise considering that the same conclusion was reached for the simpler single degree-of-freedom trailer model studied previously. (Strictly speaking, one should recompute the slip angles with the travel velocity reversed to make sure that simply letting U be negative in Eq. (5.33) yields the correct characteristic equation for reverse motion. We will not do this here because it is obvious that the trailer is unstable backing up in any case.)

Furthermore, the coefficients of s^3 and s both approach zero at high speed. This indicates the some eigenvalues have small real parts at high speed. Thus the trailer oscillations would exhibit little damping at high speed. This is also not surprising because the same tendency was found for the simpler trailer model. Finally, the coefficient of s^2 could be negative if b were sufficiently negative, i.e., if the center of mass were behind the rear axle. This is a new result for the two degree-of-freedom model.

In addition to the conditions for instability that can be seen from the characteristic equation coefficients individually, there are almost certainly other conditions that would lead to instability. In principle, one could discover these by filling out a fourth-order Routh array. This can be a formidable task for the present model if the parameters are left in literal form. Instead of attempting this, a partial simplification based on physical reasoning will be made.

IV. A THIRD-ORDER MODEL

It may be that some conditions for instability at low speeds would be present even if the tire slip angles were virtually zero. This could be studied by assuming that the cornering coefficient, C_α, was very large and neglecting any terms in the characteristic equation, Eq. (5.33), not proportional to C_α. This procedure applied to the single degree-of-freedom model reduced the second-order equation of Eq. (5.15) to the first-order equation of Eq. (5.16). After neglecting all terms in Eq. (5.33) not containing C_α and then dividing out C_α, the characteristic equation for the two degree-of-freedom trailer model then reduces to third order.

$$\{(I_c + mb^2)/U\}s^3 + mbs^2 + (a+b)^2(k/U)s + (a+b)k = 0 \qquad (5.34)$$

This third-order characteristic equation appears to have positive coefficients except for $U < 0$. This again suggests that the trailer may be unstable backing up, which is expected. (It is hardly worth checking to see if a negative U correctly represents reverse motion in this case.)

For $b < 0$, the s^2 coefficient is negative—indicating that the trailer would be unstable it the center of gravity were behind the trailer axle. (The fourth-order version did not indicate that the system would be unstable for every case of negative b, but only when a more complicated term involving b was negative.) Trailers normally do not have the center of mass behind the axle. If they did, the weight would tend to make the nose of the trailer lift and the hitch would have to hold the nose down. Trailers are normally loaded in such a way that the weight pushes the nose of the trailer down on the hitch. As b is not usually negative, it is of interest to search for other stability criteria of more practical importance.

A. A Simple Stability Criterion

A new criterion for stability is found by using the Routh criterion for a third order characteristic equation. The result, Eq. (3.24), was derived previously for a characteristic equation in the form of Eq. (3.9). In general, terms a third-order system was found to be stable only if $a_1 a_2 > a_0 a_3$. Using the coefficients in Eq. (5.34), this translates to

$$(mb)(a+b)^2 k/U > ((I_c + mb^2)/U)(a+b)k \tag{5.35}$$

In this inequality, we can multiply both sides by U (assuming U is positive) and divide out the positive quantities k and $(a+b)$. Finally, the positive quantity mb^2 can be subtracted from both sides yielding the elementary stability criterion for stability.

$$mab > I_c \tag{5.36}$$

Another version of this stability criterion is found by defining the radius of gyration, κ, by the relation

$$I_c \equiv m\kappa^2 \tag{5.37}$$

The criterion for stability is then

$$ab > \kappa^2 \tag{5.38}$$

This simple stability criterion is remarkable in several ways. It means that a trailer will not be stable if the center of mass moves too far toward the rear, i.e., if b is positive but has too small a value. Car trailer rental companies try to assure that this does not happen by requiring that the trailer should be loaded such that the "tongue weight" is sufficiently large. The tongue weight is the force down on the nose of the trailer and it is proportional to the weight and the distance the center of gravity is ahead of the axle. Practically, this means that the customer should not load too much weight in the rear of the trailer.

There is also a danger of instability if the moment of inertia of the load is unusually large even when the center of mass is fairly far forward. With the same load location, a compact load with a small moment of inertia in a trailer may result in a stable vehicle while if the load is spread out in space so that the moment of inertia is large, the trailer may be unstable. Several cases of trailer instability have been traced to unusually long loads although the center of gravity was in a normal position. Trailer instability has been the cause on a number of serious accidents and this elementary analysis provides a simple criterion for avoiding some dangerous situations. Certainly more complex models would uncover other possible causes of instability.

It is also unusual that the simple stability criterion of Eqs. (5.36) and (5.38) has nothing to do with speed. Although U appears in the characteristic equation, it cancels out in the Routh criterion—indicating that an unstable trailer will be unstable even at very low speeds. Using experimental trailers, this lack of speed dependence has been verified approximately but common sense indicates that true unstable behavior certainly has to cease as the speed approaches zero. We will now modify the analysis to include a dissipative mechanism for the trailer oscillation and derive an extended stability criterion in which speed plays an important role.

V. A MODEL INCLUDING ROTARY DAMPING

A weakness in the third-order model is observed when the characteristic equation is written out for the case $U=0$. In the characteristic equation, Eq. (5.34), one can multiply by U and then set U to zero. The result is the characteristic equation for zero speed.

$$s\{(I_c + mb^2)s^2 + (a+b)^2k\} = 0 \tag{5.39}$$

This characteristic equation is easy to interpret. Except for the zero eigenvalue, which has no particular significance, the equation has the eigenvalues of an undamped oscillator. A moment's thought will convince one that in fact the trailer can oscillate back and forth around the back axle when it is not moving. There is no sideways motion of the tires and the spring at the hitch provides a restoring moment.

Using the parallel axis theorem, it is clear that the moment of inertia about the trailer axle just $(I_c + mb^2)$. The term $(a+b)^2k$ is a torsional spring constant. When the trailer is not moving, the spring deflection is $x=(a+b)\theta$, the force at the hitch is $F=kx$, and the moment about the axle is $(a+b)F=(a+b)k(a+b)\theta$. Thus the equation for this oscillator can be written $(I_c mb^2)\ddot{\theta} + (a+b)^2k\theta = 0$. This equation has the same finite eigenvalues as Eq. (5.39). The extra zero eigenvalue simply means that the trailer might have a constant angle added to the dynamic angle θ when it is standing still.

It is notable that there is no damping moment for this motion although a real trailer will have various friction elements that would quickly damp out this type of oscillation.

A more realistic assumption is that there is some sort of frictional moment on the trailer. Although the moment is certainly not simply a linear function of the motion, to proceed with an eigenvalue analysis we must assume a linear form of damping moment on the trailer. A more realistic nonlinear friction torque could be used for computer simulation but any nonlinearity in the system equations invalidates the use of eigenvalue analysis. To proceed further, we will simply assume that a friction moment proportional to the angular velocity acts on the trailer. Whether this assumption realistically represents the actual frictional effects that act or not is unclear, but at least this assumption will result in the introduction of a dissipative mechanism into the mathematical model. The friction moment will be assumed to obey the following damping law:

$$M_m = c\dot{\theta} \tag{5.40}$$

where c is an effective rotary damping constant.

In the fourth-order Lagrange formulation, the only modification required is to add a term to the generalized force for the θ coordinate in Eq. (5.25). (The moment has no effect on the generalized force for the x coordinate.)

$$\Xi_\theta = -F(a+b) - c\dot{\theta} \tag{5.41}$$

Ultimately, only a single extra term appears in the "damping matrix" of Eq. (5.30), and this extra term changes the s^3, s^2, and s terms in the characteristic equation. Only one of the new terms is multiplied by C_α, so the reduction to third order by neglecting terms not multiplied by C_α yields a slightly modified version of Eq. (5.34), in which c appears in the s^2 coefficient.

$$\begin{aligned}&\{(I_c + mb^2)/U\}s^3 + (c+mb)s^2 + (a+b)^2(k/U)s \\ &\quad +(a+b)k = 0\end{aligned} \tag{5.42}$$

When the third-order Routh criterion is applied as before, the final criterion for stability changes from Eq. (5.36) to

$$c(a+b)/U + mab > I_c \tag{5.43}$$

The trailer is certainly stable if the previous criterion, Eq. (5.36), is fulfilled because this implies that the inequality of Eq. (5.43) will also be fulfilled. Now, however, at low enough speed, the trailer will be stable no matter what, because of the term $c(a+b)/U$, which becomes large when the speed U becomes small. The conclusion is that the inequality of Eq. (5.43) is always fulfilled if the speed U is small enough. This suggests that all trailers are stable at sufficiently low speeds because all trailers would be expected to have some damping of angular motion. This seems reasonable because no one would expect a trailer to exhibit unstable oscillations at walking speeds.

A. A Critical Speed

If we now consider a trailer that, by the criterion of Eq. (5.36), would be unstable, namely one for which $I_c > mab$, something interesting happens. At low speeds, the criterion of Eq. (5.43) will still be fulfilled and the trailer will be stable. However, there will then be a speed above which the inequality of Eq. (5.43) will not be fulfilled and the system will change from stable to unstable.

This speed is called the *critical speed*, U_{crit}, and such speeds are found in a variety of vehicles. In this case, the critical speed can be found by changing the criterion inequality, Eq. (5.43) to an equality and then solving for the value of U. The resulting value of speed is the speed at which the trailer changes form stable to unstable. The value of U that results form Eq. (5.43) is

$$U_{crit} = c(a+b)/(I_c - mab) \tag{5.44}$$

Note that there is no critical speed if the trailer is inherently stable, so in Eq. (5.44), the denominator is assumed to be positive, i.e., according to the criterion Eq. (5.36), the trailer would be unstable at all speeds. From Eq. (5.44), one can readily see that the critical speed becomes zero when the damping constant c is zero and the critical speed increases as the damping constant increases.

There have been numerous reported cases of accidents involving trailers carrying loads with unusually large moments of inertia that, at a certain speed, begin oscillating in an unstable manner. That the simple third-order mathematical presentation provides an explanation for this possibility is remarkable.

One should, of course, always remember that real trailers exhibit many more complicated modes of motion than those allowed in the simple plane motion model studied in this chapter. No single model can hope to represent all possible types of trailer instabilities. In particular, tall trailers tend to sway back and forth with angular motion around a horizontal axis. This motion is, of course, is not possible for the plane motion model studied here.

Vehicles that having a critical speed are particularly dangerous because they behave normally until the critical speed is exceeded. Thus the operator of a vehicle may assume that there is nothing wrong with his vehicle and yet be surprised by unstable behavior at high speed. In addition to the critical speed discovered using this simple model, there are other possible critical speeds having to do with more complex motions that would require a more complex model to discover.

The fourth-order characteristic equation Eq. (5.33) certainly has the possibility of predicting other parameter combinations that may lead to instability beyond those predicted by the simplified third-order characteristic equation, but it is not clear whether this equation would predict the existence of a critical speed. It could well be that the fourth-order model would simply predict that certain parameter sets would lead to instability at any speed as was the case for the third-order model. Adding a damping moment to the fourth-order model is quite likely to introduce a critical speed because it produced this effect even in the third-order model.

The critical speed discovered for the elementary trailer model is only a specific result. The analysis is not necessarily typical for other vehicles in that the model without dissipation showed no critical speed at all. However, many vehicles are capable of changing their stability properties as a function of speed. We will encounter critical speeds in later chapters for automobiles, rail vehicles, and caster wheels. It is probably fair to say that there are hardly any types of vehicles that have not had some designs exhibit unstable behavior at high speed that was not found at low speeds. Whenever a record speed is attempted, the danger of an unsuspected critical speed is always present.

6
Automobiles

I. INTRODUCTION

Automobiles and trucks have played a dominant role in transportation since the onset of their mass production. Everywhere on earth, people come into contact with these vehicles, often on a daily basis. Typically, a car or a truck is one of the most complex and sophisticated products of the Industrial revolution that individuals encounter and with which they must interact. Although many people have an intuitive understanding of automobile dynamics, relatively few are able to describe, in any quantitative way, just how the physical parameters of these vehicles influence their stability and handling characteristics.

It is the purpose of this chapter to introduce a number of basic features of automobile stability and control by using the simplest mathematical models possible. With the background from this chapter, it should be possible to understand the more sophisticated studies of automobile dynamics that are typically carried out using complex mathematical models and computer simulation. Furthermore, the basic understanding provided by this chapter can help users understand the behavior of real automobiles in ordinary driving circumstances, as well as in extreme conditions, e.g., those that occur in motor sports or accident situations.

II. STABILITY AND DYNAMICS OF AN ELEMENTARY AUTOMOBILE MODEL

In this chapter, we will focus on the lateral dynamics of automobiles. Automobiles do, in fact, move in all three dimensions. Longitudinal motions involve braking and driving forces, but here we will generally assume that the forward speed is substantially constant. Vertical motions, substantially influenced by vehicle suspension and roadway unevenness, will also be neglected. Finally, angular motions in the roll and pitch directions will likewise be neglected. Our stability analysis will deal with lateral acceleration and the yaw rate.

Although all the motions of an automobile are to some extent correlated, it is difficult to understand the fundamentals of lateral motion stability when a complete mathematical model is considered. Models that treat a large number of vehicle motions inevitably contain a large number of parameters, and this means that the models must refer to specific vehicle designs. It is hard to derive any general principles from the study of complex models. By restricting our analysis to plane motion models, we will be able to understand the basics of lateral stability. More complex models are called for when a specific vehicle is under consideration.

Automobiles are obviously stable at very low speeds, so historically there was no particular interest in issues of stability until the propulsion systems had developed enough to allow high speeds to be attained. Even as propulsion systems became more powerful, stability and handling problems were typically addressed more by experiment than by analysis. Other vehicles such as bicycles, motorcycles, and airplanes exhibit stability and control problems at low speeds and, if properly designed, actually become more stable as speeds increase. Particularly in the case of airplanes, there was great interest in developing engineering understanding of stability and control problems well before similar considerations were applied to automobiles. Eventually, however, the analysis of the stability and control aspects of automobile dynamics became a well-studied field.

There are a number of books discussing automobile dynamics in general and stability issues in particular. See, for example, Barstow (1987), Ellis (1969, 1988), Milliken and Milliken (1995), Puhn (1981), Gillespie (1992), Kortuem and Lugner (1994), and Wong (1978). In addition, there are numerous papers dealing with the aspects of automobile dynamics, stability, and control. Some examples are Bundorf (1967b), Olley (1946–1947), Pacejeka (1973a,b), Segal (1956), Garrott et al. (1988), Karnopp (1976), and Whitehead (1988, 1989).

Some of the very first mathematical analyses of elementary mathematical models of automobiles that showed how automobiles could change their stability properties with speed were performed by using an inertial coordinate system. This seemingly obvious choice of coordinates has resulted to equations of motion that are somewhat more complicated than necessary. On the other hand, the use of inertial coordinate systems seemed natural to the pioneers in the field, and this method of analysis brings out some interesting points so it will be presented first. Then, the analysis will be redone using the body-centered coordinates presented in Chapter 2, which are more commonly used in modern studies in the aircraft and automobile fields.

III. STABILITY ANALYSIS USING INERTIAL COORDINATES

Fig. 6.1 shows a plane motion model of an automobile nominally moving along the x-axis, but with a small heading angle $\Psi(t)$. The variables $x(t)$ and $y(t)$ describe the position of the center of mass. In this model, all motion is assumed to occur in the x–y plane, so there is no consideration of vertical (heave) motion or roll, or pitch angular motion.

The parameters of this model are a and b, the distances between the axles and the center of mass; m, the mass; I_z, the polar moment of inertia referred to the center of mass and the z-axis; and the two half-tracks, t_f and t_r, which will subsequently be shown to be unimportant for the following stability analysis.

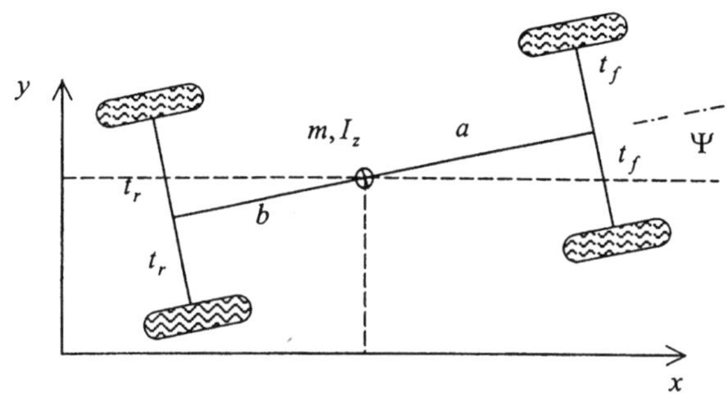

FIG. 6.1 Elementary automobile model described in inertial coordinates.

The basic motion is described by $\dot{x} = U = \text{constant}$, $\dot{y} = \dot{\Psi} = \Psi = 0$. For the perturbed motion, we assume that the x-motion remains approximately the same as for the basic motion but that $\dot{y}$, $\dot{\Psi}$, and Ψ can have small values. This means that no equation of motion for x is required because $x = Ut + \text{constant}$.

The main difficulty in setting up the dynamic equations for this model lies in the computation of the slip angles for the four tires. Fig. 6.2 shows the velocity components of the wheel motion at the left front. As with the vehicle models studied in Chapters 1 and 5, the wheel velocity again involves the summation of the velocity of a particular point (the center of mass) and the velocity components induced by the angular velocity, $\dot{\Psi}$.

The velocity components shown in Fig. 6.2 bear a good deal of resemblance to those shown in Fig. 5.3 for the trailer model. For the trailer, the velocity at the wheel was the velocity at the hitch, plus the velocity induced by the rotation. In Fig. 6.2, the velocity at the wheel is the velocity of the center of mass plus the vector components $a\dot{\Psi}$ and $t_f\dot{\Psi}$. These components can be recognized as components of the general term $\vec{\omega} \times \vec{r}_{AB}$, which was introduced in Eq. (1.15) and applied in Fig. 5.3. In the present case, $\vec{r}_{AB}$ is the vector distance from the center of mass to the center point of the wheel.

By resolving the velocity components shown in Fig. 6.2 in the direction the tire is pointing, the rolling velocity of the tire, using a small angle approximation and the fact that all the perturbed variables are assumed to be small, is given by the expression

$$U \cos \Psi - t_f\dot{\Psi} + \dot{y} \sin \Psi \cong U \tag{6.1}$$

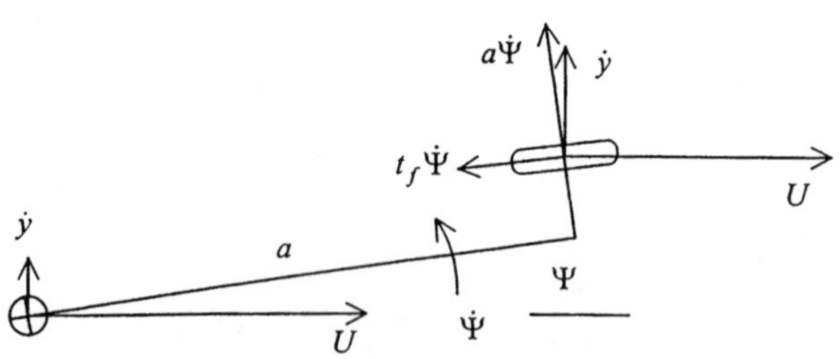

FIG. 6.2 Velocity components for the left front wheel.

Similarly, the lateral velocity perpendicular to the tire pointing direction is found to be

$$\dot{y}\cos\Psi + a\dot{\Psi} - U\sin\Psi \cong \dot{y} + a\dot{\Psi} - U\Psi \tag{6.2}$$

The front slip angle is then,

$$\alpha_f = \frac{\text{Lateral Velocity}}{\text{Rolling Velocity}} = \frac{\dot{y} + a\dot{\Psi}}{U} - \Psi \tag{6.3}$$

Note that after the small perturbed variable assumption has been used, the influence of the front half-track, t_f, disappeared. This means that the slip angle expression applies just as well for the right front wheel. As both rear wheels also have approximately the same slip angles for the perturbed motion, this leads to the common term *bicycle model* for this four-wheel car, because it is possible to consider a single equivalent wheel for the front axle and another for the rear axle.

Fig. 6.3 shows the velocity components at the left rear wheel. The computation of the rear slip angle, α_r, follows closely the procedure for the front wheels.

The rolling velocity is

$$U\cos\Psi - t_r\dot{\Psi} + \dot{y}\sin\Psi \cong U \tag{6.4}$$

The lateral velocity is

$$\dot{y}\cos\Psi - b\dot{\Psi} - U\sin\Psi \cong \dot{y} - b\dot{\Psi} - U\Psi \tag{6.5}$$

The slip angle at the rear is

$$\alpha_r = \frac{\text{Lateral Velocity}}{\text{Rolling Velocity}} = \frac{\dot{y} - b\dot{\Psi}}{U} - \Psi \tag{6.5a}$$

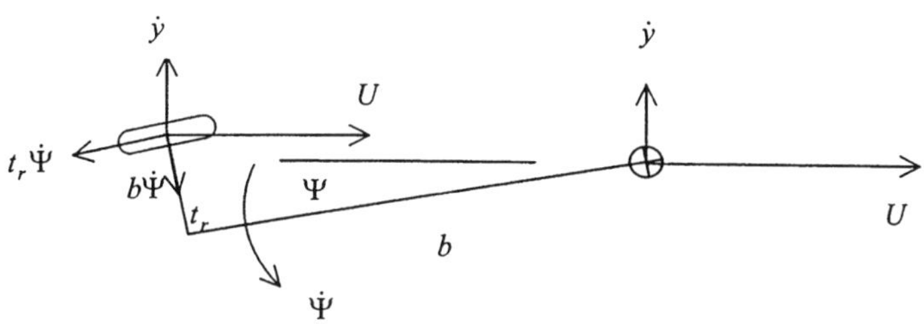

FIG. 6.3 Velocity components for the left rear wheel.

Once again, because of the small angle assumptions, the track has no influence on the slip angle at the rear so the slip angle expression in Eq. (6.5a) applies to both rear wheels.

Fig. 6.4 shows the forces exerted on the automobile by the tires. The forces Y_f and Y_r are shown as positive in the nominal y-direction, and the slip angles are shown as though they were positive (as though the lateral velocities were positive in the nominal y-direction). This is a perfectly logical way to determine the signs of the forces and velocities. However, when the slip angles are positive as shown, the forces exerted on the tires from the ground would actually be in the negative y-direction.

This is a common problem in describing tire–roadway interactions, which leads either to negative cornering coefficients in the linearized case, or to a special way of writing the force laws to have positive coefficients. Later on, when stability criteria are being applied to the equations of motion, it is inconvenient to have some parameters in the equations that are inherently negative. Therefore the laws relating lateral tire forces to slip angles will be written in a special form that will avoid the problem of negative cornering coefficients.

In this chapter, a negative sign will be written in to the force laws so that the cornering coefficients at the front and rear axles, C_f and C_r, are positive. Just as in the case in Chapter 5, the cornering coefficients are assumed to account for the sum of the forces on both of the two tires at each axle.

$$-Y_f = C_f\alpha_f \quad -Y_r = C_r\alpha_r \tag{6.6}$$

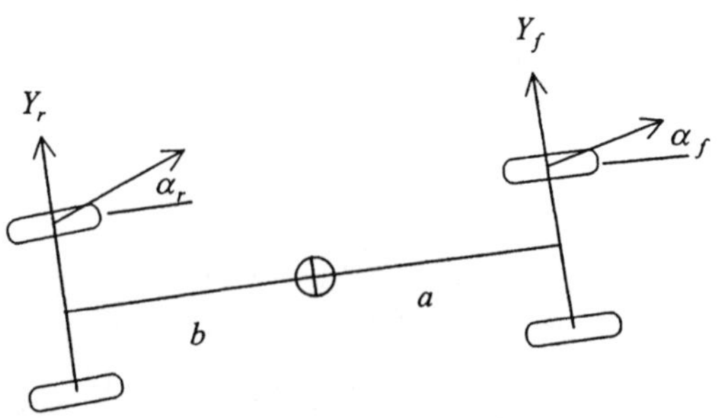

FIG. 6.4 Sketch of axle forces and slip angles.

It should be remembered that the use of linearized force-slip angle relations is only valid for small slip angles. This is fine for a stability analysis that dealing with only small perturbations from the basic motion, but if the vehicle is unstable, the slip angles will build up until the approximation is no longer appropriate.

Because the axle cornering coefficients relate to a summation of the forces on both tires at each axle, they depend partly on the normal forces supported at the individual wheels, the characteristics of the road surface, and of the tires themselves. Although there are many suspension design factors for a vehicle that also contribute to the determination of the axle cornering coefficients beside the characteristics of the tires themselves (Bundorf, 1967b), for now we regard the effective front and rear axle cornering coefficients simply as given quantities.

The equations of motion relate the acceleration of the center of mass to the applied forces and the angular acceleration to the applied moments. Keeping in mind that Ψ is assumed to be a small angle, one can write Newton's law for linear acceleration of the center of mass in the form

$$m\ddot{y} = Y_{\mathrm{f}} + Y_{\mathrm{r}} = -C_{\mathrm{f}}\alpha_{\mathrm{f}} - C_{\mathrm{r}}\alpha_{\mathrm{r}} \tag{6.7}$$

The angular acceleration law is

$$I_z\ddot{\Psi} = aY_{\mathrm{f}} - bY_{\mathrm{r}} = -aC_{\mathrm{f}}\alpha_{\mathrm{f}} + bC_{\mathrm{r}}\alpha_{\mathrm{r}} \tag{6.8}$$

After substituting expressions for the slip angles, Eqs. (6.3) and (6.5), the final equations may be put into the following matrix form:

$$\begin{bmatrix} m & 0 \\ 0 & I_z \end{bmatrix}\begin{bmatrix} \ddot{y} \\ \ddot{\Psi} \end{bmatrix} + \begin{bmatrix} (C_{\mathrm{f}} + C_{\mathrm{r}})/U & (aC_{\mathrm{f}} - bC_{\mathrm{r}})/U \\ (aC_{\mathrm{f}} - bC_{\mathrm{r}})/U & (a^2C_{\mathrm{f}} + b^2C_{\mathrm{r}})/U \end{bmatrix}\begin{bmatrix} \dot{y} \\ \dot{\Psi} \end{bmatrix} + \begin{bmatrix} 0 & -(C_{\mathrm{f}} + C_{\mathrm{r}}) \\ 0 & -(aC_{\mathrm{f}} - bC_{\mathrm{r}}) \end{bmatrix}\begin{bmatrix} y \\ \Psi \end{bmatrix} = \begin{bmatrix} 0 \\ 0 \end{bmatrix} \tag{6.9}$$

Equation (6.9) is in the form of Eq. (3.21), so the characteristic equation is of the form of Eq. (3.23). The same procedure for finding the characteristic equation was used previously for the two degree-of-freedom trailer model. See Eqs. (5.31–5.33). In the case at hand, the expression for the characteristic equation is

$$\det\begin{bmatrix} ms^2+((C_\mathrm{f}+C_\mathrm{r})/U)s & ((aC_\mathrm{f}-bC_\mathrm{r})/U)s-(C_\mathrm{f}+C_\mathrm{r}) \\ ((aC_\mathrm{f}-bC_\mathrm{r})/U)s & I_z s^2+((a^2C_\mathrm{f}+b^2C_\mathrm{r})/U)s-(aC_\mathrm{f}-bC_\mathrm{r}) \end{bmatrix} = \begin{bmatrix} 0 \\ 0 \end{bmatrix} \tag{6.10}$$

When this expression is evaluated, the fourth-order characteristic expression emerges explicitly.

$$\begin{aligned}(mI_z)s^4 &+ \left[m(a^2C_\mathrm{f}b^2C_\mathrm{r})/U+I_z(C_\mathrm{f}+C_\mathrm{r})/U\right]s^3 \\ &+[(C_\mathrm{f}+C_\mathrm{r})(a^2C_\mathrm{f}+b^2C_\mathrm{r})/U^2-(aC_\mathrm{f}-bC_\mathrm{r})^2/U^2 \\ &- m(aC_\mathrm{f}-bC_\mathrm{r})]s^2+0s^1+0s^0=0\end{aligned} \tag{6.11}$$

Because the last two coefficients vanish, the equation can be written in the following form:

$$\begin{aligned}\{(mI_z)s^2 &+ \left[m(a^2C_\mathrm{f}b^2C_\mathrm{r})/U+I_z(C_\mathrm{f}+C_\mathrm{r})/U\right]s \\ &+[(C_\mathrm{f}+C_\mathrm{r})(a^2C_\mathrm{f}+b^2C_\mathrm{r})/U^2-(aC_\mathrm{f}-bC_\mathrm{r})^2/U \\ &- m(aC_\mathrm{f}-bC_\mathrm{r})]\}s^2=0\end{aligned} \tag{6.12}$$

This form of the characteristic equation makes it clear that two of the eigenvalues are zero. The stability of the automobile will depend on the eigenvalues associated with the remaining quadratic part of the equation when the factor s^2 is removed. The two zero eigenvalues are associated with solution components that have the time behavior of $e^{0t}=\text{constant}$. These solutions have no particular significance for the analysis of stability. Obviously, adding a constant distance to the y-coordinate or adding a constant heading angle to Ψ merely means that location and direction of the basic motion path have been changed. The stability of the automobile actually has nothing to do with where in the x–y plane the x-axis is located or with the direction of the x-axis. Thus the use of inertial coordinates has produced a fourth-order system that is more complicated then necessary for stability analysis.

If we now consider the three coefficients of the quadratic part of the characteristic equation when s^2 in Eq. (6.12) has been eliminated, we see that the coefficients of s^2 and s are inherently positive, but that the complicated last term is not necessarily positive.

As we have seen in Chapter 3, for a second-order characteristic equation, the eigenvalues will represent a stable system only if all coefficients are positive. Therefore the criterion for stability in this case has to do with the last coefficient in the quadratic part of Eq. (6.12) that must be positive if the car model is to be stable.

$$(C_f + C_r)(a^2C_f + b^2C_r)/U^2 - (aC_f - bC_r)^2/U^2 \\ -m(aC_f - bC_r) > 0 \tag{6.13}$$

After some algebraic manipulation, this criterion for stability can be simplified to

$$(a+b)^2C_fC_r/U^2 + m(bC_r - aC_f) > 0 \tag{6.14}$$

If the inequality in Eq. (6.14) is true, the car is stable; if not, it is unstable.

A. Stability, Critical Speed, Understeer, and Oversteer

Several interesting facts can be easily observed from the simplified stability criterion, Eq. (6.14). First, if the speed is sufficiently low, the first term involving $1/U^2$ will be a large enough positive number that the criterion will be satisfied for any parameter set. This confirms the obvious idea that all automobiles are stable at very low speeds.

Second, if the second term in Eq. (6.14) is positive, the car will surely be stable at any speed. The second term is positive when

$$bC_r > aC_f \tag{6.15}$$

This condition is described by the term *understeer*. Clearly, this term relates to steering properties of the car. For now, all we know is that an understeer vehicle is stable even at high speeds, where the $1/U^2$ term in the stability criterion becomes small, but later we will discuss how understeer affects the steady state cornering behavior.

Finally, if

$$bC_r < aC_f \tag{6.16}$$

we see that the second term in the stability criterion, Eq. (6.14), is negative. This means that if the speed U is gradually increased from zero, the positive first term will decrease to a point at which the first term will just balance the negative second term. This speed is called the *critical speed*, U_{crit}, and above this speed the stability criterion will no longer be satisfied and the car will be unstable. This speed is determined by equating the positive and negative terms in Eq. (6.14) and solving for the speed.

$$U^2_{\text{crit}} = -(a+b)^2 C_f C_r / m(bC_r - aC_f) \tag{6.17}$$

Note that this expression yields a positive number for ${U_{\text{crit}}}^2$ because we assume that $bC_r < aC_f$.

This situation is described by the term *oversteer*, and it does not only imply that the car is unstable for speeds greater than the critical speed, but also that the speed drastically affects the steady state cornering behavior of the vehicle. The steady state cornering of automobiles with understeer and oversteer characteristics will be discussed in a later section.

B. Body-Fixed Coordinate Formulation

Most modern vehicle dynamic studies use a coordinate system attached to the vehicle as described in Chapter 2 to describe the vehicle motion. This indicates that the coordinate system rotates with the vehicle, which somewhat complicates the acceleration terms in the basic dynamic equations. On the other hand, the use of body-centered coordinates often results in simpler final equations. In the case at hand, this formulation will directly result in a second-order characteristic equation, and will eliminate the bother of the two zero eigenvalues, which were found in the previous analysis to result from fourth-order equation of motion.

Fig. 6.5 shows the so-called *bicycle model* of an automobile, in which only a single equivalent front wheel and a single equivalent rear wheel are shown together with the body-centered coordinate system discussed in Chapter 2. A new feature is the steer angle, δ, shown at the front. This feature is not needed for stability analysis at present, but is included for later use when steering behavior is discussed. For the stability analysis, the steer angle δ will simply be assumed to be zero.

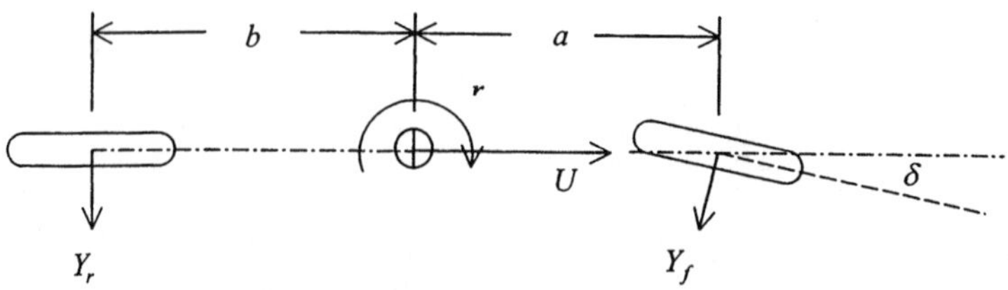

FIG. 6.5 Body-centered coordinate system for an automobile.

Now U and V are velocity components of the center of mass, but with respect to the body coordinates x and y that are attached to the vehicle body and moving with it. Once again, it will be assumed that the velocity in the forward x-direction, U, is strictly constant for the basic motion and is essentially constant for the perturbed motion, so it is a parameter rather than a variable. The two variables needed to describe the motion are V, the velocity of the center of mass in the y-direction, and r, the angular velocity or the yaw rate about the z-axis.

The lateral tire forces are denoted Y_r and Y_f, although the front lateral tire force does not point strictly in the y-direction if the front wheel is steered as shown. These definitions are in conformity with the general notation introduced in Chapter 2. The parameters of the vehicle are the same as those used previously in the inertial coordinate system analysis.

The equations of motion are specialized from the general equations of Chapter 2. It is assumed that the steer angle δ is a small angle, so that $\cos\delta \cong 1$. The equation

$$m(\dot{V} + rU) = Y_f + Y_r \tag{6.18}$$

determines the lateral acceleration of the center of mass and the equation

$$I_z\dot{r} = aY_f - bY_r \tag{6.19}$$

determines the angular acceleration.

Calculation of the slip angles is accomplished with the help of the sketches shown in Fig. 6.6. Note that β is a sort of slip angle for the center of gravity of the vehicle itself. We will later see that when cornering, the car does not generally go exactly in the direction it is pointed, and thus $\beta \cong V/U$ is not zero. Again, the velocities at the wheels are found by adding the velocity of the center of mass to components induced by the angular velocity, r. In this coordinate system, the extra components, ar and br are easily observed to add in the fashion shown in Fig. 6.6.

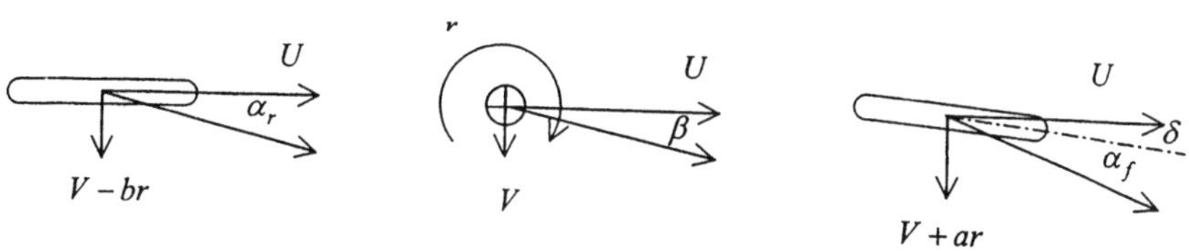

FIG. 6.6 Velocity components for slip angle calculations.

Using the sketches and assuming small angles, the slip angles are readily expressed as ratios of the lateral velocities (with respect to the wheel pointing direction) to the rolling velocities.

$$\alpha_r = (V - br)/U, \quad \alpha_f = (V + ar)/U - \delta \tag{6.20}$$

Note how the steer angle δ affects the front slip angle.

As in the previous analysis, it is once again found that positive slip angles as shown result in forces on the tires in the negative Y direction. Because it is preferable to use positive cornering coefficients rather than negative ones, the linearized force laws will be again be written thus:

$$-Y_f = C_f\alpha_f, \quad -Y_r = C_r\alpha_r \tag{6.21}$$

Combining Eqs. (6.18–6.21) the equations of motion then become

$$m(\dot{V} + rU) = -(C_f + C_r)V/U - (aC_f - bC_r)r/U + C_f\delta \tag{6.22}$$

$$I_z\dot{r} = -(aC_f - bC_r)V/U - (a^2C_f + b^2C_r)r/U + aC_f\delta \tag{6.23}$$

For the basic motion, many variables vanish, $\delta = V = r = Y_f = Y_r = \alpha_f = \alpha_r = 0$, and U is constant. For the perturbed motion, we assume $\delta = 0$, U is still considered to be constant, and that variables $V(t)$ and $r(t)$ take on small enough values so that the slip angles remain small. The dynamic equations for the perturbed motion then assume the following vector-matrix form:

$$\begin{bmatrix} m & 0 \\ 0 & I_z \end{bmatrix}\begin{bmatrix} \dot{V} \\ \dot{r} \end{bmatrix} + \begin{bmatrix} (C_f + C_r)/U & (aC_f - bC_r)/U + mU \\ (aC_f - bC_r)/U & (a^2C_f + b^2C_r)/U \end{bmatrix} \times \begin{bmatrix} V \\ r \end{bmatrix} = \begin{bmatrix} 0 \\ 0 \end{bmatrix} \tag{6.24}$$

This equation set is essentially in the form of Eq. (3.5), and the characteristic equation is found following the general procedure in Eq. (3.15).

$$\det\begin{bmatrix} ms + (C_f + C_r)/U & (aC_f - bC_r)/U + mU \\ (aC_f - bC_r)/U & I_zs + (a^2C_f + b^2C_r)/U \end{bmatrix} = 0 \tag{6.25}$$

When the determinant is written out, the characteristic equation emerges explicitly.

$$\begin{aligned} &mI_zs^2 + \left[m(a^2C_f + b^2C_r)/U + I_z(C_f + C_r)/U\right]s \\ &\quad + (C_f + C_r)(a^2C_f + b^2C_r)/U^2 - (aC_f - bC_r)^2/U^2 \\ &\quad - m(aC_f - bC_r) = 0 \end{aligned} \tag{6.26}$$

If the last term is algebraically simplified, the characteristic equation becomes

$$\begin{aligned} mI_z s^2 &+ \big(m(a^2 C_f + b^2 C_r)/U + I_z(C_f + C_r)/U\big)s \\ &+ (a+b)^2 C_f C_r/U^2 + m(bC_r - aC_f) = 0 \end{aligned} \tag{6.27}$$

This characteristic equation is exactly the same as the quadratic part of the fourth-order characteristic equation that was previously obtained from the model described in inertial coordinates when the zero eigenvalues were eliminated. Here, we see an example of the fact mentioned in Chapter 3 that all correct dynamic equations describing a system will yield the same eigenvalues, apart from possible zero eigenvalues, which are of no particular significance for stability analysis.

The body-centered coordinate formulation yields a second-order characteristic equation directly and avoids the more complex fourth-order characteristic equation that, in the end, has two zero eigenvalues of no particular interest for stability analysis. It should be clear that the formulation using the body-fixed coordinate system is, in some ways, simpler for this problem than the analysis using an inertial coordinate system.

The discussion of understeer, oversteer, and critical speed are, of course, independent of the particular differential equations used to describe the vehicle as long as they are correct, so they need not be repeated here.

The "bicycle model" has a long history of use in automobile steering and stability studies. For those interested in bond graph representations, the Appendix develops a bond graph for the bicycle model of an automobile using the rigid body bond graphs in Chapter 2. This bond graph could be used to construct more complex system models. For example, it would be possible to append an actuator bond graph model for active steering studies.

IV. TRANSFER FUNCTIONS FOR FRONT AND REAR WHEEL STEERING

In the preceding section, the stability of an automobile traveling in the straight path with zero steering angle was studied. We now extend our study to include steering dynamics using the same vehicle model. The body-centered coordinate system approach is particularly useful for this purpose. When linear tire characteristics are assumed, transfer functions relating steering inputs to various response quantities are a convenient way to represent the steering dynamics. Transfer functions relate input and output variables for linear systems and are described in automatic control texts such as Ogata (1970).

For generality, we will extend our considerations to both front wheel and rear wheel steering. This will create an opportunity to discuss the significant differences in the dynamic behavior between vehicles steered from the front and from the rear.

It may seem odd to consider rear wheel steering systems because only low-speed vehicles such as lawn mowers, forklift trucks, and street sweepers are commonly found to steer the rear wheels. At low enough speeds, stability considerations are not normally considered to be important for ground vehicles. On the other hand, there have been a number of studies dealing with a combination of front and rear steering (Furukawa, 1989; Sharp and Crolla, 1988; Tran, 1991).

In principle at least, there are several good reasons for considering the steering the wheels both at the front and at the back of a vehicle. In a conventional front-steered automobile, there is a coupling of angular motion in yaw and lateral motion, and only a single control element—the steering wheel—simultaneously influences both aspects of motion.

For example, a driver wishing to change lanes on a straight road needs to generate lateral forces at the front and rear axles. With front wheel steering, the slip angle at the front can be changed directly through the steering system, and thus the driver has direct control of the lateral force at the front. At the rear, however, the slip angle only changes after the car has rotated in yaw. This means that the car must change its angular attitude in order to move sideways. There is an inevitable lag in the lateral motion response because the car cannot generate a lateral force at the rear until it has rotated in yaw.

For the lane change maneuver, if the rear wheels are steered in the same sense as the front, lateral forces can be simultaneously generated at the front and rear. This direct control of the lateral forces at the front and rear could speed up the lateral motion response, which might be useful in an emergency lane change maneuver.

In contrast, if the driver wishes to turn a tight corner, then the desire is specifically to change the yaw angle, and this would be aided by steering the rear wheel in the opposite sense to the front wheels. Some production steering systems use this scheme to shorten the turning radius of long vehicles for low-speed maneuvering.

Finally, the angle β shown in Fig. 6.6 will be shown later to be nonzero when a conventional car negotiates a corner. The reason for this has to do with the necessity of having a slip angle at the rear axle. This means that a passenger sitting near the center of mass would have the impression that the car is not traveling exactly in the direction it was pointed. With front and rear steering, one can make β as zero. This possibility has been used as an argument in favor of front and rear steering, although in normal driving situations, it is hard to imagine β to be large enough to raise concern.

In this section, we will not deal with the many proposals for relating the front and rear steering angles, but rather we will concentrate on the differences in the dynamics associated with front and rear steering when the tires are in their linear range.

Fig. 6.7 shows the model of Fig. 6.6 extended to include front and rear steer angles, δ_f and δ_r.

When there is a rear steer angle, the rear slip angle becomes

$$\alpha_r = (V - br)/U - \delta_r \tag{6.28}$$

which is an extension of the relation given in Eq. (6.20), but the remaining equations derived previously still apply. The equations of motion are slightly modified versions of Eqs. (6.22) and (6.23).

$$\begin{aligned} m(\dot{V} + rU) &= -(C_f + C_r)V/U - (aC_f - bC_r)r/U + C_f\delta_f + C_r\delta_r \\ I_z\dot{r} &= -(aC_f - bC_r)V/U - (a^2C_f + b^2C_r)r/U + aC_f\delta_f - bC_r\delta_r \end{aligned} \tag{6.29}$$

Now we take the Laplace transform of Eq. (6.29). The Laplace transforms of time-dependent variables are, in principle, new variables which are functions of the transform variable, s. In conformance with common practice, we now let the variable names V, r, δ_f, and δ_r also stand for the Laplace transforms of the time-dependent functions. The context in which these variables appear will clarify whether the variable stands for a function of time or of the Laplace variable.

In deriving transfer functions, zero initial conditions are assumed. The only difference between the time-dependent Eq. (6.29) and the trans-

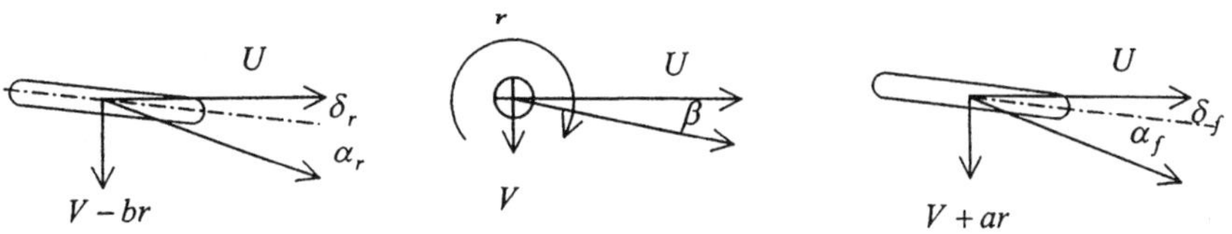

FIG. 6.7 Velocity components for bicycle model with front and rear wheel steering.

formed version is that $\dot{V}$ and $\dot{r}$ become sV and sr, where s is the Laplace variable. In matrix form, Eq. (6.29) then transforms into a set of algebraic rather than differential equations.

$$\begin{bmatrix} ms + (C_f + C_r)/U & mU + (aC_f - bC_r)/U \\ (aC_f - bC_r)/U & I_z s + (a^2 C_f + b^2 C_r)/U \end{bmatrix} \begin{bmatrix} V \\ r \end{bmatrix} = \begin{bmatrix} C_f \\ aC_f \end{bmatrix} \delta_f + \begin{bmatrix} C_r \\ -bC_r \end{bmatrix} \delta_r \tag{6.30}$$

The transfer functions are found by solving these algebraic equations for variables such as V and r given either as δ_f or δ_r. This can be carried out by using Cramer's rule from linear algebra.

Cramer's rule states that every variable in a set of linear algebraic equations can be expressed as the ratio of two determinants. The denominator in the ratio is always the determinant of the array at the left side of Eq. (6.30). When written out, this determinant is exactly the *characteristic polynomial*. (When the characteristic polynomial is set to zero, the result is just the characteristic equation, which was used to determine the eigenvalues and the stability of the vehicle.) Because this polynomial in s will be used frequently, we will give it the symbol Δ. The last term in the polynomial is given in the simplified form as it was given in the characteristic equation in Eq. (6.27).

$$\begin{aligned} \Delta = {} & mI_z s^2 + \left(m(a^2 C_f + b^2 C_r)/U + I_z(C_f + C_r)/U\right)s \\ & + (a^2 + b^2)C_f C_r/U^2 + m(bC_r - aC_f) \end{aligned} \tag{6.31}$$

The numerator determinant in Cramer's rule has to do with whatever forcing variable is involved, as well as with whatever variable of the system is of interest. The numerator determinant is the determinant of the array to the left in Eq. (6.30), but with one of the column vectors at the right in Eq. (6.30) substituted for a column in the array. The transfer functions are then found by dividing the numerator determinant by Δ. The procedure is best understood by example.

The forcing column to be used in the numerator determinant will involve one of the two input variables, δ_f or δ_r. If this column is substituted in the first column of the array, the result will be the first variable, V. If the forcing column is substituted in the second column, the result will be the second variable, r.

As an example of Cramer's rule applied to Eq. (6.30), when the first column in the coefficient array is substituted by the forcing column for δ_f

and the determinant is divided by Δ, the result is a relationship between the first variable V and the first forcing quantity δ_f.

$$V = \det \begin{vmatrix} C_f\delta_f & mU + (aC_f - bC_r)/U \\ aC_f\delta_f & I_z s + (a^2C_f + b^2C_r)/U \end{vmatrix} / \Delta$$

When this result is expressed as a ratio of V to δ_f, it becomes the transfer function relating the lateral velocity to the front steering angle.

$$\frac{V}{\delta_f} = \frac{I_z C_f s - maC_f U + (a+b)bC_f C_r/U}{\Delta} \tag{6.32}$$

Other transfer functions are derived in a similar manner.

$$\frac{r}{\delta_f} = \frac{maC_f s + (a+b)C_f C_r/U}{\Delta} \tag{6.33}$$

$$\frac{V}{\delta_r} = \frac{I_z C_r s + mbC_r U + (a+b)aC_f C_r/U}{\Delta} \tag{6.34}$$

$$\frac{r}{\delta_r} = \frac{-mbC_r s - (a+b)C_f C_r/U}{\Delta} \tag{6.35}$$

One should remember that all transfer functions have the same denominator (Eq. (6.31)), and that this polynomial also appears in the characteristic equation used to determine the eigenvalues and the stability of straight line running. This denominator is second order in the Laplace variable s. The four transfer functions above all have numerators that are first order in s. The minus signs in r/δ_r are simply attributable to the positive direction assumed for the rear steer angle.

Another set of transfer functions of interest deals with the lateral acceleration of the center of mass. They are derived by combining the transfer functions given above. In these expressions, the derivative of the lateral velocity is expressed in the Laplace transform variable, $\dot{V} = sV$.

$$\frac{\dot{V} + rU}{\delta_f} = \frac{I_z C_f s^2 + (a+b)bC_f C_r s/U + (a+b)C_f C_r}{\Delta} \tag{6.36}$$

$$\frac{\dot{V} + rU}{\delta_r} = \frac{I_z C_r s^2 + (a+b)aC_f C_r s/U - (a+b)C_f C_r}{\Delta} \tag{6.37}$$

These acceleration transfer functions are notable in several ways. First, they have second-order numerators and denominators. This is significant because transfer functions become complex exponential frequency response functions when the variable s is replaced by $j\omega$, where j is the imaginary number and ω stands for frequency in radians per second. A transfer function with the same order in s for the numerator and the denominator has a response ratio that approaches a constant as the forcing frequency becomes very large. In contrast, a transfer function with a numerator of lower order than the denominator such as those in Eqs. (6.32)–(6.35) will have a frequency response ratio that becomes small at very high frequencies.

This implies that the acceleration response to steering inputs does not fall off at high frequencies, as is the case for the lateral velocity and the yaw response. This would have implications, e.g., if one were to use accelerometers in a feedback steering control system instead of yaw rate sensors.

Second, there is an important difference between the acceleration response to front wheel steering and to rear wheel steering. The minus sign in the rear wheel steering transfer function numerator means the transfer function has a zero in the right half of the s-plane. [The variable s is generally complex, but for the transfer function of Eq. (6.37), it is easily shown that there is a real, positive value of s for which the numerator is zero.] This implies that a rear wheel steered automobile is a "nonminimum phase" system, and will have an unusual type of response to sudden rear steering inputs (Ogata, 1970).

If the rear wheel is steered quickly in one direction, the acceleration will have a "reverse response." That is, the acceleration will first be in one direction, and will later be in the opposite direction. A person sitting near the center of gravity of a rear-steering car would feel this back-and-forth acceleration whenever the rear steering angle was suddenly changed. This effect makes rear wheel steering control more difficult than front wheel steering. This is one reason why most people have trouble backing up a car at high speed and why most vehicles capable of high speed are steered from the front.

Reverse action acceleration response is an inherent property of rear wheel steering and has nothing to do with stability or instability. Stability has to do with whether the denominator of the transfer functions can be made to vanish for a value of s lying in the right half of the s-plane. If this were to be the case, there would be an eigenvalue with a positive real part, which would imply that the system is unstable. The reverse action has to do with right half plane zeros of the numerators, rather than the denominator of the transfer functions, and occurs for rear steering cars whether they are stable or unstable.

Another case of a nonminimum phase system will be encountered for steering controlled tilting vehicles in Chapter 7.

V. YAW RATE AND LATERAL ACCELERATION GAINS

When the Laplace variable s is set equal to zero in transfer functions, the steady state relation of output to input is given. For example, the steady yaw rate for a constant front steer angle is found from the r/δ_f transfer function by setting $s=0$. In this case, the result is

$$\left.\frac{r}{\delta_f}\right|_{s=0} = \frac{U}{(a+b)+m(bC_r-aC_f)U^2/(a+b)C_fC_r} \tag{6.38}$$

This expression is simplified by defining the so-called *understeer coefficient*, K, which appears in the transfer functions and plays a prominent role in the discussion of steady cornering, as will be seen in the next section. The understeer coefficient is defined in Eq. (6.39).

$$K = \frac{m(bC_r-aC_f)}{(a+b)C_fC_r} \tag{6.39}$$

The sign of K is the same as the sign of the quantity $(bC_r - aC_f)$. Thus positive values of K correspond to what we have previously labeled as understeer [Eq. (6.15)], negative values correspond to oversteer [Eq. (6.16)], and if K is zero, the car is called *neutral steer*.

It is common to call the transfer functions evaluated at $s=0$ the *zero frequency gains*. Here, we will designate these gains by G with subscript "r" for yaw and "a" for acceleration. We will also use subscripts "f" and "r" for front and rear steer, respectively. This notation will be used again in the discussion of cornering although only front wheel steering will be considered. Using the understeer coefficient, the zero frequency gains are as follows:

$$\begin{aligned} G_{rf} &= \frac{U}{(a+b)+KU^2}, \quad & G_{rr} &= \frac{-U}{(a+b)+KU^2} \\ G_{af} &= \frac{U^2}{(a+b)+KU^2}, \quad & G_{ar} &= \frac{-U^2}{(a+b)+KU^2} \end{aligned} \tag{6.40}$$

The terms understeer, neutral steer, and oversteer have significance with respect to stability, but the use of this nomenclature may be more obvious in the discussion of cornering behavior below.

A. The Special Case of the Neutral Steer Vehicle

The neutral steer vehicle with $K=0$ is sometimes thought of as ideal from the point of view of handling. Indeed, when good handling is a main priority, designers often strive for equal weight distribution on the front and rear axles, which, with the same tires and tire pressures front and rear, favors a neutral or nearly neutral steer characteristic. A number of other factors besides weight distribution enter into the determination of the understeer coefficient and handling behavior, and they allow a designer to adjust the dynamics of an automobile (see Bundorf, 1967b). Here, we examine some special features of neutral steer vehicles.

In the linearized model, a neutral vehicle is one for which $aC_f = bC_r$, so $C_r = aC_f/b$ can be eliminated in favor of C_f in all previous results. When this is carried out, a number of terms disappear. The denominator of the transfer functions simplifies from Eq. (6.31) to

$$\Delta = (ms + (a+b)C_f/bU)(I_z s + a(a+b)C_f/U) \tag{6.41}$$

The factored form of Eq. (6.41) indicates that if this characteristic polynomial were set equal to zero to find the eigenvalues, there would be two real, negative eigenvalues. The car would not only be stable, but also would have no tendency to oscillate at all in this case. In addition, for a neutral steer car, the yaw rate transfer functions have an exact pole-zero cancellation that reduces them to first order.

$$\frac{r}{\delta_f} = \frac{aC_f}{I_z s + a(a+b)C_f/U} \tag{6.42}$$

$$\frac{r}{\delta_r} = \frac{-aC_f}{I_z s + a(a+b)C_f/U} \tag{6.43}$$

The acceleration transfer functions for a neutral steer car remain second order because there is no pole-zero cancellation.

$$\frac{\dot{V} + rU}{\delta_f} = \frac{I_z C_f s^2 a(a+b)C_f s/U + a(a+b)C_f^2/b}{\Delta} \tag{6.44}$$

$$\frac{\dot{V} + rU}{\delta_r} = \frac{(aC_f/b)I_z s^2 + (a^2/b)(a+b)C_f s/U - a(a+b)C_f/b}{\Delta} \tag{6.45}$$

A final interesting simplification occurs for a neutral steer vehicle that has the special value of its polar moment of inertia,

$$I_z = mab \tag{6.46}$$

If this is the case, the moment of inertia can be eliminated as an independent parameter. This may seem to be a very special case, but this relation is approximately true for many cars. Under this assumption, the two real negative eigenvalues become equal and the denominator of the transfer functions becomes

$$\Delta = ab(ms + (a+b)C_f/bU)^2 \tag{6.47}$$

The transfer functions simplify quite a bit for a neutral steer car when in addition Eq. (6.46) is true.

$$\frac{r}{\delta_f} = \frac{C_f}{b(ms + (a+b)C_f/bU)} \tag{6.48}$$

$$\frac{r}{\delta_r} = \frac{-C_f}{b(ms + (a+b)C_f/bU)} \tag{6.49}$$

$$\frac{\dot{V} + rU}{\delta_f} = \frac{abC_f(ms^2 + (a+b)C_f s/bU + (a+b)C_f/b^2)}{\Delta} \tag{6.50}$$

$$\frac{\dot{V} + rU}{\delta_r} = \frac{a^2C_f(ms^2 + (a+b)C_f s/bU - (a+b)C_f/ab)}{\Delta} \tag{6.51}$$

Although these highly simplified transfer functions for neutral cars with a special value for the moment of inertia certainly do not apply in general, they have been used in general studies of handling dynamics and steering control because they contain a reduced number of parameters. Furthermore, neutral steering cars are often thought of as representing a sort of optimum with respect to handling qualities, and these specialized transfer functions can be used to compare the steering response of real vehicles with the response of one version of an ideal vehicle.

VI. STEADY CORNERING

In a previous section, the stability of a simple model of an automobile was studied. The model assumed that the car moved only in plane motion and that the relation between the tire forces and the (small) slip angles was linear. Under these assumptions, it was shown that single equivalent wheels could represent the front and rear axles and cornering coefficients could represent the total forces acting on the two axles.

Although stability was analyzed only for motion in a straight line, the equations of motion in body-centered coordinates were derived, including a small steer angle δ for the front wheels. By allowing δ to have nonzero values, the model can also be used to study cornering behavior. It will be shown that there is an interesting link between the stability properties of a car and the manner in which the steer angle required to negotiate a turn varies with speed.

A. Description of Steady Turns

To simplify the discussion, we will consider *steady turns*, i.e., turns taken at a constant speed and having a constant turn radius. Naturally, such turns do not strictly occur in normal driving, but often there is a period of time in actual turns when the car nearly is in a steady state. Furthermore, it is easy to construct a circular skid pad for testing cars and thus steady turns at various speeds are often used to characterize a car's steering behavior. The terms understeer and oversteer, which were already encountered in the previous section on stability analysis, are most easily related to steady turn behavior.

As a first step, consider the general expression for the front and rear slip angles derived for the simple model in body-centered coordinates including the front steer angle δ, see Eq. (6.20).

By eliminating V from these two equations, one can find a relation for the steer angle in terms of the slip angles and the yaw rate.

$$\delta = -\alpha_f + \alpha_r + (a+b)r/U \tag{6.52}$$

In Fig. 6.8, we consider a steady right-hand turn of radius R. To keep the slip angles small, we assume that the turn radius is large compared to the wheelbase, $R \gg (a+b)$. This is normally the case for high-speed turns, in which skidding is not involved. The angles are exaggerated somewhat in the sketch in Fig. 6.8 for clarity.

The yaw rate (for the body and the entire figure) $r \cong U/R$ and the lateral acceleration of the center of mass is approximately U^2/R. Notice that in the sketch of Fig. 6.8, the tire forces are in the positive direction but the slip angles are negative, as is the lateral velocity, V. Using the relation $r = U/R$, we can rewrite the steer angle expression Eq. (6.52) as

$$\delta = -\alpha_f + \alpha_r + (a+b)/R \tag{6.53}$$

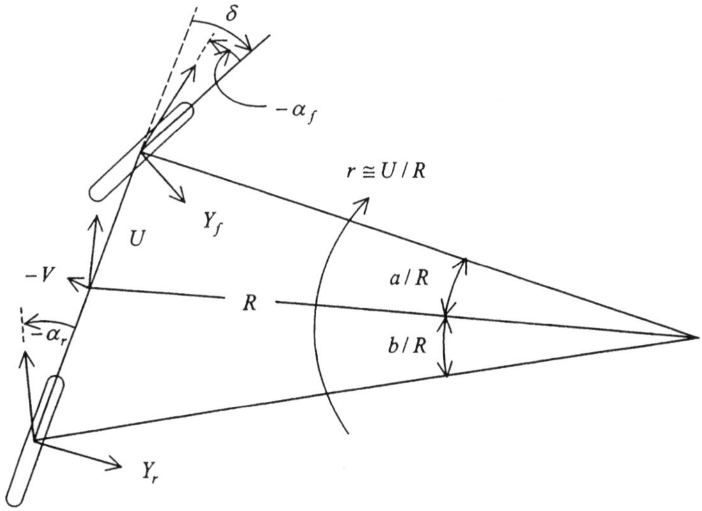

FIG. 6.8 Automobile in a steady turn.

The dynamic equations of motion in the body-centered coordinate system have been written before, Eqs. (6.18) and (6.19), but now for the steady turn, $\dot{r} = \dot{V} = 0$. When $r = U/R$ is substituted in these dynamic equations, the acceleration of the center of mass appears as U^2/R. The equations then simplify to

$$mU^2/R = Y_f + Y_r \tag{6.54}$$

$$0 = aY_f - bY_r \tag{6.55}$$

This allows us to solve for the tire forces.

$$Y_f = \frac{mU^2 b}{R(a+b)} \tag{6.56}$$

$$Y_r = \frac{mU^2 a}{R(a+b)} \tag{6.57}$$

It is a peculiar fact that these lateral forces at the front and rear axles are distributed in a similar fashion to the normal or weight forces supported by the axles when the automobile is in a steady turn. These weight forces, W_f and W_r, are proportional to the total weight mg, and can be derived by considering a free body diagram showing the vertical forces on the car neglecting any possible aerodynamic forces.

$$W_f = \frac{mgb}{(a+b)} \tag{6.58}$$

$$W_r = \frac{mga}{(a+b)} \tag{6.59}$$

This implies that

$$Y_f = W_f \frac{U^2}{Rg} \tag{6.60}$$

$$Y_r = W_r \frac{U^2}{Rg} \tag{6.61}$$

Now the slip angles can be found using the linearized relations, Eq. (6.21), which allow the cornering coefficients to be positive quantities. Using these expressions for the slip angles and substituting into the formula for the steer angle, Eq. (6.53), the result is

$$\delta = \left(\frac{Y_f}{C_f} - \frac{Y_r}{C_r}\right) + \frac{(a+b)}{R} \tag{6.62}$$

Then, using the expressions for the lateral forces derived from the dynamic equations, Eqs. (6.56) and (6.57), the steer angle relation becomes

$$\delta = \left(\frac{m(bC_r - aC_f)}{(a+b)C_f C_r}\right)\frac{U^2}{R} + \frac{(a+b)}{R} \tag{6.63}$$

Another version using the weight forces from Eqs. (6.60) and (6.61) is

$$\delta = \left(\frac{W_f}{C_f} - \frac{W_r}{C_r}\right)\frac{U^2}{Rg} + \frac{(a+b)}{R} \tag{6.64}$$

Both these expressions relate the steer angle to the speed via *understeer coefficients*. The coefficients actually appear in two slightly different forms in Eqs. (6.63) and (6.64).

$$\delta = K_1 U^2/R + (a+b)/R \tag{6.65}$$

and

$$\delta = K_2 U^2/Rg + (a+b)/R \tag{6.66}$$

The first version,

$$K_1 = \frac{m(bC_r - aC_f)}{(a+b)C_f C_r} \tag{6.67}$$

has the dimensions rad/(m/s^2), and is exactly the same understeer coefficient K seen previously in the study of transfer functions, Eq. (6.39). The second version

$$K_2 = \frac{W_f}{C_f} - \frac{W_r}{C_r} \tag{6.68}$$

has the dimensions of *rad*/"*g*", where the acceleration is expressed in dimensionless "g" units. This later version is often preferred in presenting experimental results.

The sign of the understeer coefficient determines how the steer angle changes as the speed (and hence the lateral acceleration) changes in a steady turn. However, the numerator of K_1 has the same factor that we found to be important in the analysis of stability for straight-line running. If either version of K is positive, the car is *understeer*, and we know that the car is stable for all speeds. If K is zero, the car is called *neutral steer* and is also stable. If K is negative, the car is *oversteer*, and will be unstable in straight running but only for speeds above the critical speed.

B. Significance of the Understeer Coefficient

It turns out that the critical speed is also the speed at which the steer angle is zero in a steady turn for an oversteer car. This can be seen by setting $\delta = 0$ in the expression for steer angle given above.

$$0 = \left(\frac{m(bC_r - aC_f)}{(a+b)C_f C_r}\right)\frac{U_{crit}^2}{R} + \frac{(a+b)}{R} \tag{6.69}$$

This yields

$$U_{crit}^2 = \frac{-(a+b)^2 C_f C_r}{m(bC_r - aC_f)}$$

which is exactly the formula for critical speed for an oversteering vehicle found in the stability analysis [Eq. (6.17)].

Fig. 6.9 shows how the steer angle varies as a function of lateral acceleration according to either Eq. (6.65) or Eq. (6.66). In Fig. 6.9, K is used to represent either K_1 or K_2.

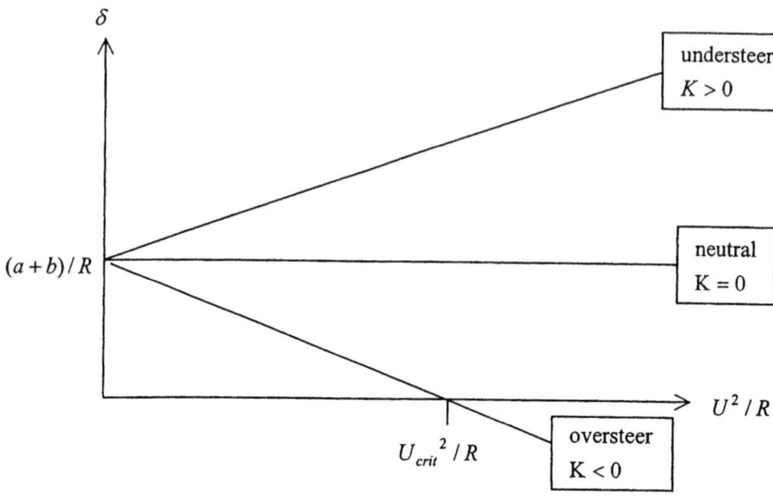

FIG. 6.9 Steer angle as a function of lateral acceleration.

Because most cars are designed to understeer, the understeer line in Fig. 6.9 shows the common experience that if a constant radius turn is taken at increasing speeds, the driver has to increase the steer angle. Another way to say this is that, for an understeering car, the magnitude of the front slip angle increases faster than the magnitude of the rear slip angle as the lateral acceleration increases.

From Fig. 6.8, it can be appreciated that as a car increases its lateral acceleration in a steady turn, the rear slip angle must increase in magnitude so the rear tires can generate a larger force. This can only happen if the centerline of the car rotates with respect to the turn radius. If the steer angle is maintained constant, this change in attitude of the car with respect to the turn radius will increase the magnitude of the front slip angle as much as the rear slip angle.

For a neutral steer car, the forces at the front and rear can be adjusted to give a larger lateral acceleration by simply changing the car attitude without any change in steer angle, as indicated in Fig. 6.9. In contrast, an understeer car requires a larger change in slip angle at the front than the rear to maintain equilibrium for a higher lateral acceleration. This can be accomplished by increasing the steer angle. The opposite is true for an oversteering car for which the steer angle must be reduced to reduce the front slip angle change, so that it is less than the rear slip angle change necessary to accommodate to the increased lateral acceleration.

One should keep in mind that the definition of understeer coefficient given here strictly applies only to the linear range of the force-slip angle relation. However, automobile racers often speak of "terminal understeer," which means that the limiting speed for a corner occurs when the front wheels reach the maximum lateral force that can be generated at very large slip angles, although the rear wheels could generate larger forces. A more accurate description for such a car is that it "pushes" at the limit. As will be shown later, a car that is designed to understeer for normal driving with small slip angles may not necessarily push at the limit because of the nonlinearity of the tire force relation to the slip angle.

As Fig. 6.9 shows, at very low speeds, all cars require a steer angle of $(a+b)/R$, which depends only on the wheelbase and the turn radius. This is the case when the lateral forces and the slip angles are vanishingly small. For a neutral steer car, this steer angle remains constant as the speed in the turn increases. For increasing lateral acceleration, the front and rear slip angles increase equally. This means that the car rotates relative to the turn radius to increase slip angles at both axles, but the steer angle does not need to change as the speed is increased. (See the sketch of the turn geometry in Fig. 6.8.) Car racers sometimes extend this idea to the large slip angle case to speak of a neutral car as one in which the front and rear axles reach their maximum lateral forces simultaneously. Again, a car that is neutral steer for small slip angles may not remain neutral in the extended sense at the limit.

We already know that an oversteer car becomes unstable for straight running at speeds greater than the critical speed. In steady cornering, it is now clear that the critical speed is the speed at which an oversteering car can negotiate a turn with no steer angle at all. For speeds above the critical speed, there is a *control reversal*, which means that the driver must *countersteer*. This implies that the driver must steer left to turn right.

The reason for this is that, for an oversteer car, the rear slip angle magnitude increases faster than the front slip angle magnitude as the lateral acceleration increases. Thus the driver must reduce the front slip angle by reducing the steer angle as the car rotates with respect to the turn radius to increase the rear slip angle when increasing speed above the critical speed. Above the critical speed, the actual steer angle is in the reverse direction to the turn direction.

Automobile racers who refer to "terminal oversteer" really mean that the limit speed in a turn occurs when the rear axle reaches its limiting lateral force before the front axle. Strictly speaking, this terminology is not related to oversteer as it exists for small slip angles. They may also describe such a car as "loose," which has some justification. Even in the linear case for small slip angles, when an oversteering car exceeds the critical speed, not only is countersteering necessary, but the car also becomes unstable. This indicates that it is up to the driver to continually correct the car's tendency to deviate from the desired path. The subjective feeling is that the rear of the car is only loosely connected to the road.

VII. ACCELERATION AND YAW RATE GAINS

The results derived above can be manipulated to reveal some interesting aspects of the problem facing a driver acting as a controller of an automobile. In short, the response of the vehicle to steering inputs varies drastically with speed, and is fundamentally different depending on whether the car is understeer, neutral, or oversteer. Two particular aspects of steering response will be considered: the yaw rate and the lateral acceleration of the center of mass.

When a driver moves the steering wheel and thus changes the angle of the road wheels, δ, two easily sensed effects are produced. One is a change in the lateral acceleration and another is a change in the yaw rate. Although the two effects are coupled in a conventionally steered automobile, frequently one is more important to the driver than the other. For example, if the task is to change lanes on a straight freeway when traveling at high speed, the desire is to accelerate laterally without much change in heading or yaw rate. On the other hand, when rounding a right angle corner, one must establish a yaw rate to change the heading angle by 90°, and the lateral acceleration is simply a necessary by-product during the maneuver. Thus the driver must use only one control input, the steering angle, to accomplish two quite different types of tasks. Furthermore, the input–output relationships are not constant but vary depending on how fast the vehicle is traveling. It is no wonder that it takes some training and practice to become a good driver.

Consider the fundamental steady turn relation defining the understeer coefficient, Eq. (6.65). In this equation, K represents the understeer coefficient in the form of Eq. (6.39), or as denoted by K_1 in Eq. (6.67). The *acceleration gain*, G_a, is defined to be the ratio of the lateral acceleration to the steer angle considering only the case of a steady turn. (For sharp changes in steer angle, there is a transient response but here we only consider the steady state response.) The results apply most directly to relatively slow changes in steer angle. The result is

$$G_a = \frac{U^2/R}{\delta} = \frac{1}{K + (a+b)/U^2} = \frac{U^2}{KU^2 + (a+b)} \tag{6.70}$$

(This corresponds exactly to the zero frequency gain G_{af}, defined in Eq. (6.40), from the acceleration transfer function for front wheel steering, although it is derived in a completely different manner from steady turn considerations.)

It is useful to separately discuss the three cases: $K > 0$, $K = 0$, and $K < 0$, which correspond to understeer, neutral, and oversteer, respectively.

For the neutral case, $K = 0$ and

$$G_a = \frac{U^2}{(a+b)} \tag{6.71}$$

This means that the acceleration the driver feels per unit steer angle increases as U^2. In this sense, driving a neutral steer car involves controlling a variable gain system. The acceleration the driver experiences as a result of a change in steer angle varies as the square of the car speed.

For the understeer case, $K > 0$, and it proves to be useful to define a *characteristic speed*, U_{ch}, such that at the characteristic speed, the acceleration gain is one-half that of a neutral car. Considering Eq. (6.70), this requires that

$$KU_{ch}^2 = (a+b) \tag{6.72}$$

With this definition, the (positive) understeer coefficient K can be algebraically eliminated in favor of the characteristic speed in Eq. (6.70).

$$G_a = \frac{U_{ch}^2}{(a+b)} \left[\frac{U^2/U_{ch}^2}{1 + U^2/U_{ch}^2} \right] \tag{6.73}$$

For the oversteer case, $K < 0$, the *critical speed*, U_{crit}, defined in Eq. (6.17) or Eq. (6.69), is found to obey the relation

$$KU_{crit}^2 = -(a+b) \tag{6.74}$$

Using this relation the understeer coefficient K can be replaced with the critical in Eq. (6.70).

$$G_a = \frac{U_{crit}^2}{(a+b)} \left[\frac{U^2/U_{crit}^2}{1 - U^2/U_{crit}^2} \right] \tag{6.75}$$

Fig. 6.10 shows how, in general, acceleration gains vary with speed. One should remember that the three cases shown have no particular relation to each other. They are for three separate automobiles. There is no obvious connection between the characteristic speed for an understeering car and the critical speed for an oversteering car in Fig. 6.10. On the other hand, it is possible to change a single car from understeer to neutral to oversteer by changing the loading, the tires, or even the tire pressure. Most cars are designed to be as insensitive to these changes as possible, so it usually takes a major change in one of these factors to change the understeer coefficient significantly.

At very low speeds, the acceleration gains for all three cases, Eqs. (6.71), (6.73), and (6.75), approach identical expressions. The gains simply become the ratio of the square of the speed divided by the wheelbase. Note, however, that the lateral acceleration becomes very sensitive to steer angle for a neutral car at high speed. For an oversteer car, this sensitivity even

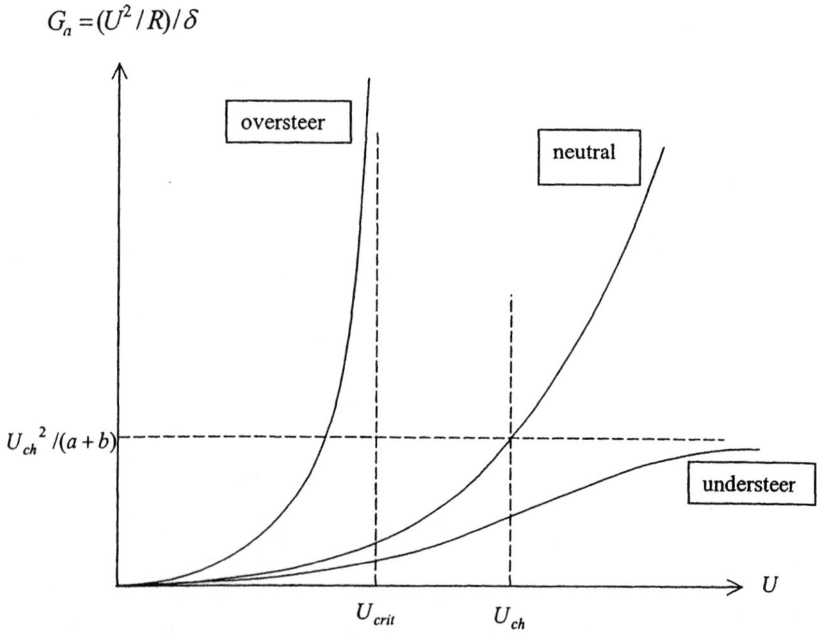

FIG. 6.10 The acceleration gain as a function of speed for oversteer, neutral, and understeer automobiles.

becomes infinite as the finite critical speed is approached. On the other hand, the acceleration gain approaches a finite limit for an understeer car for speeds that are high compared to the characteristic speed. Thus the three types of cars behave quite differently from one another as the speed varies.

Another quantity of interest is the yaw rate gain, *i.e.*, the steady yaw rate per unit steer angle, G_r. This can be expressed as

$$G_r = r/\delta = (U/R)/\delta = G_a/U = \frac{U}{(a+b) + KU^2} \tag{6.76}$$

(This is exactly the same gain that was defined in Eq. (6.40) as the zero frequency yaw rate gain for front wheel steering, G_{rf}.)

Again, there are three cases in which the understeer coefficient can be eliminated in favor the characteristic speed and the critical speed.

Neutral steer:

$$K = 0, \quad G_r = \frac{U}{(a+b)} \tag{6.77}$$

Understeer:

$$K > 0, \quad G_r = \frac{U}{(a+b)}\left[\frac{1}{1 + U^2/U_{ch}^2}\right] \tag{6.78}$$

Oversteer:

$$K < 0, \quad G_r = \frac{U}{(a+b)}\left[\frac{1}{1 - U^2/U_{crit}^2}\right] \tag{6.79}$$

Fig. 6.11 shows the general form of the yaw rate gains for the three cases. Once again, one must be careful not to assume that the three cases shown are necessarily related. It is true that the equality of the initial slopes of the three cases plotted in Fig. 6.11 implies that the three cars have the same wheelbase, $(a+b)$. Equations (6.78) and (6.79) approach Eq. (6.77) at low speeds, and in the figure the three curves approach each other at low speeds, but there is no connection between the values of the critical speed for the oversteering car and the characteristic speed for the understeering car.

The oversteering car is on the verge of instability at the critical speed, and at that speed the yaw rate gain becomes infinite. This means that the yaw rate is infinitely sensitive to steering inputs at the critical speed. Above the critical speed, control reversal occurs and the driver must countersteer as well as to attempt to stabilize an unstable vehicle. This effect is not

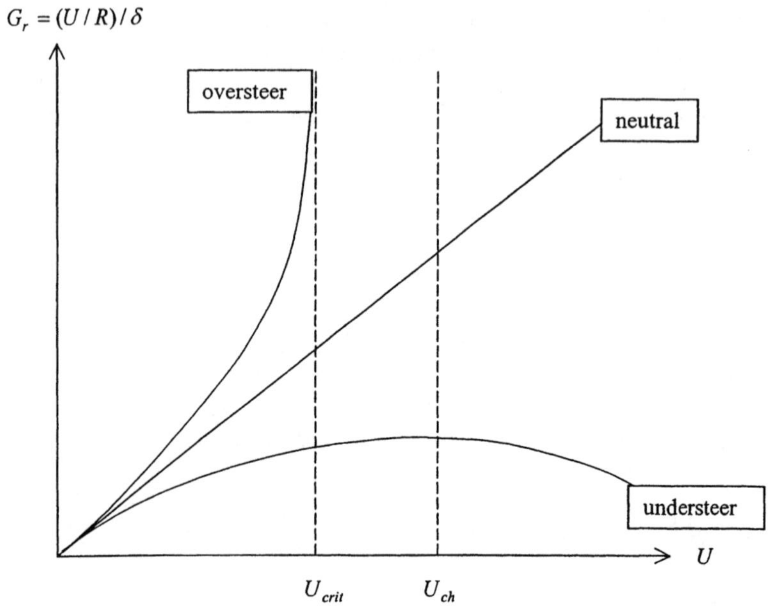

FIG. 6.11 The yaw rate gain as a function of speed for oversteer, neutral, and understeer automobiles.

shown in Fig. 6.11. When the critical speed is exceeded, the driver not only has to countersteer, but also has to try to stabilize the car. The driver feels that the car continually seems to want to slew around and ultimately to travel in reverse.

The yaw rate gain for a neutral car increases linearly with speed, so that if a specific yaw rate is desired, smaller steering inputs are required at high speed than at low speed.

For the understeering car, the yaw rate gain initially rises with speed, but the gain reaches a maximum at the characteristic speed and declines at even higher speeds. The understeering car is stable at high speed, but requires large steer inputs to achieve a given yaw rate at very high speeds. This correlates intuitively with the notion that an understeering car is not only stable at high speeds, but also becomes less responsive to steering inputs at very high speeds. This is an example of the common property of many vehicles that, a high degree of stability, is often purchased at the expense of maneuverability.

As Figs. 6.10 and 6.11 indicate, the sign of the understeer coefficient has a large effect on the compromise between stability and controllability. A very stable car may not respond to steering inputs extremely well in the sense of generating lateral acceleration or yaw rate particularly at high speeds. On the other hand, a car that is very responsive to steering inputs may become unstable at high speed. Neither extreme situation is desirable. An extremely stable car may not be able to swerve to avoid an obstacle, and a car that is unstable at high speed may well be hard for the driver to control. These plots make plausible the idea that a neutral or near-neutral steer automobile represents a sound compromise in steering response.

Automobile designers have a number of means at their disposal to adjust the handling of a car (Bundorf, 1967b). Among these are the location of the center of gravity, the spring rates at front and rear, antisway bar rates at the front and rear, the tire sizes and pressures at front and rear, and even the shock absorber characteristics. These effects are not strictly accounted for in the simple bicycle model. Qualitatively, it is possible to consider that these effects change the effective axle cornering coefficients, and thus modify the understeer coefficient. Another complicating feature not taken into account yet has to do with driving and braking. This not only introduces transient effects, but also requires the tires to generate longitudinal forces as well as lateral forces. As was discussed in Chapter 4, this significantly affects a tire's ability to provide lateral forces. In a rough sense, one can also see that braking and driving forces on a tire also can be thought of as affecting the front and rear cornering coefficients.

Steady braking results in an increase in normal force at the front axle and a decrease in normal force at the rear. From the discussion in Chapter 4, one might conclude that this would increase the front cornering coefficient and decrease the cornering coefficient at the rear. This would seem to lead to a tendency toward oversteer, which is a frequent occurrence. The complication is that the requirement on the tires to produce longitudinal braking force can be seen as a factor reducing the effective cornering coefficient. This is clear from the graph in Fig. 4.6, which illustrates the effect when a tire simultaneously produces longitudinal and lateral forces. Thus, depending on how the braking force is distributed between the front and rear axles, braking could result in steering characteristics toward either the understeer or oversteer direction.

The situation under acceleration is even more complicated. In addition to the normal force shift associated with acceleration, there is the question of how the power is delivered. Cars with front-wheel drive, rear-wheel drive, or all-wheel drive behave quite differently upon application of power. When only modest power is applied, the change in steering behavior is small. A large amount of power applied at the front wheels promotes understeer, while a large amount of power at the rear can result in what is commonly called "power induces oversteer." For all-wheel drive power trains, the situation depends on the ratio of torque applied at the front axle to that applied at the rear axle. The understeer coefficients defined in Eqs. (6.67) and (6.68) depend on the difference of two quantities, so they are quite sensitive to rather small changes in the two cornering coefficients. This means that in transient conditions, cars that are designed to understeer can at least temporarily appear to change steering characteristics quite drastically.

Without getting into the details of just how designers can adjust understeer and oversteer characteristics, it may be worthwhile to discuss some general tendencies. For example, for practical reasons, most cars use the same type of tire at all four wheels, and often the tire pressures are nearly identical as well. When this is the case, a forward center of gravity tends to promote understeer and a rearward center of gravity tends to result in oversteer. Thus front-engined cars (and particularly front-wheel drive cars) have a strong tendency to be understeer. Rear-engined cars tend toward oversteer. High-performance cars typically strive to have the center of gravity near the center of the wheelbase. Most sports cars are intended to be very maneuverable and strive for a neutral steering characteristic, but most passenger cars are designed to be understeer for safety reasons.

A problem for many automobiles and trucks has to do with the effects of payload. Often, loads shift the center of gravity toward to rear and thus tend to reduce the understeer and, in the extreme case, to shift the steering behavior to oversteer. An overloaded pickup truck with low tire pressure at the rear can be almost undriveable even at moderate speeds. To reduce the possibility of this happening, the designers may provide more understeer for an empty pickup truck than would be desirable from a handling point of view.

One should recognize that it is not possible to absolutely characterize understeer, oversteer, or even neutral steer as optimal. One must consider the type of vehicle involved and the normal operating speeds. There is not much to fear from an oversteering vehicle if the critical speed is significantly higher than the vehicle's top speed. Similarly, an understeer car will remain responsive to steering inputs as long as the characteristic speed is not too low relative to the car's normal operation speed range. In fact, the recent trend to install electronic stability enhancement systems on automobiles makes some handling problems less significant than they once were from a safety point of view. Electronic means to enhance stability will be discussed in Chapter 11. On the other hand, it is certainly better to start with a car that handles well without electronic aids than to use modern electronics to try to remedy handling problems.

VIII. DYNAMIC STABILITY IN A STEADY TURN

The considerations of the stability of an automobile as well as the study of the steady turn behavior have so far considered only linearized relations between lateral tire forces and slip angles. This is logical for the consideration of the stability of straight-line motion because the basic motion required no lateral forces and the perturbed motion required only small forces and slip angles. This assumption is more restrictive for steady turns, because it means that the steady lateral forces must remain fairly small if the linear assumption is to remain valid. This restriction implies that the lateral acceleration must also be limited if the conclusions are to be valid. In normal driving, most maneuvers are actually accomplished at rather low values of lateral acceleration. Most turns are taken at speeds resulting in accelerations of no more than 10% or 20% of the acceleration of gravity. Under these conditions, for most cars and tires, the linear tire force assumption is a fairly good approximation.

On the other hand, if we consider emergency maneuvers or limit speeds in turns, the tire force-slip angle becomes significantly nonlinear. As we have seen in Chapter 4, at high slip angles, the tire force reaches a maximum value and further increases in slip angle typically cause to tire force to decline as the tire begins to slide along the roadway. Under these conditions, the understeer and oversteer properties of the automobile model discussed in the previous sections are no longer constant properties of the car itself but rather change with the severity of the maneuver.

The characterization of the dynamics of automobiles in the nonlinear range can be a complex task. To gain some insight into this area and yet to keep the discussion reasonably simple, we will restrict the discussion to the stability analysis of cars while they are in a steady turn. The new feature is that we will no longer assume that the lateral acceleration is sufficiently small such that a linear tire force model is adequate. The bicycle model will be retained, which means that each axle will be represented by a single force-slip angle law including the effects of both tires.

Fig. 6.8, which was used previously to define terms for a negotiating a steady turn, can still be used, but now the slip angles will not be assumed to be small enough that a nearly constant cornering coefficient can be used. For typical tires on dry pavement, the linear range extends to slip angles of about 5° or so, and the lateral forces at the top end of this range correspond to accelerations of more than one-half the acceleration of gravity. This means that for normal driving, the linear approximation has validity. We now consider larger slip angles and steer angles, but even as the tires begin to behave in a nonlinear way, the angles are assumed to be small enough that the approximations used for the trigonometric functions in the analysis remain reasonable.

A. Analysis of the Basic Motion

The basic motion for a stability analysis is now a steady turn and the variable values necessary to accomplish the turn will be denoted with the subscript "s." We still assume that the forward speed U = constant. The remaining variables for the basic motion are defined as follows:

$$\begin{aligned} &V = V_s, \quad r = r_s = U/R, \quad \dot{V}_s = \dot{r}_s = 0, \quad Y_f = Y_{fs} \\ &Y_r = Y_{rs}, \quad \alpha_f = \alpha_{fs}, \quad \alpha_r = \alpha_{rs}, \quad \delta = \delta_s \end{aligned} \tag{6.80}$$

The steady steer angle is determined by a version of Eq. (6.52), because we still assume that the angles are small enough to use small angle approximations for trigonometric functions. Note that in the tire force plots in Fig. 4.1, significant nonlinear behavior is exhibited for tire slip angles of only 10° or 15°. From Eq. (6.52), the steer angle for the basic motion is approximately

$$\delta_s = -\alpha_{fs} + \alpha_{rs} + (a+b)/R \tag{6.81}$$

The dynamic equations remain as before [see Eqs. (6.18) and (6.19)]

Now, however, there are the so-called *trim conditions*, which constrain the steady values of variables of the basic motion so that the car can execute a steady turn.

$$\begin{aligned} &mr_sU = Y_{fs} + Y_{rs} \\ &0 = aY_{fs} - bY_{rs} \quad r_s = U/R \end{aligned} \tag{6.82}$$

Fig. 6.12 shows generally how the lateral tire forces are related to slip angles when the slip angles are no longer assumed to be very small. As discussed previously, positive slip angles yield negative forces when the notation of Chapter 2 is used, so it is convenient to plot negative forces vs. slip angles.

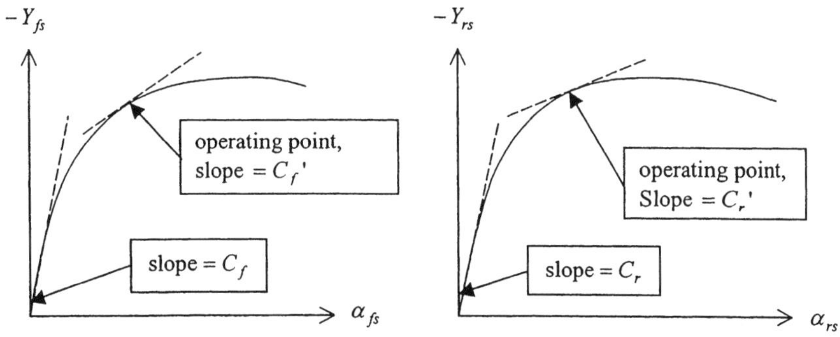

FIG. 6.12 Lateral forces as functions of slip angles in the nonlinear region.

The previous stability analysis treated the case of straight-line motion. Then the basic motion corresponded to zero values for both lateral force and slip angle. The cornering coefficients used, C_f and C_r, are the slopes of the force curves at the origin. In the present case, there are steady values for the forces and slip angles at the *operating points* shown in Fig. 6.12. These steady values are determined by the trim conditions, Eq. (6.82). The slopes of the force curves at these operating points will be designated C_f' and C_r'. As Fig. 6.12 shows, these slopes can be significantly different form the cornering coefficients when the steady forces in the turn become large.

B. Analysis of the Perturbed Motion

The perturbed motion is described by variables representing small deviations from the constant steady values associated with the basic motion. These time-varying variables will be denoted using the symbol Δ. In every case, the actual variables will be written as the steady variables plus the deviation variables. For example, the yaw rate, $r(t)$, will be expressed as $r = r_s + \Delta r$, and the front lateral force, $Y_f(t)$, is written as $Y_f = Y_{fs} + \Delta Y_f$.

Fig. 6.13 shows how the change of the front lateral force is related to the change of the slip angle away from the steady value by the local slope of the force law C_f'.

The force-slip angle is now linearized about the operating point determined by the basic motion.

$$\Delta(-Y_f) \cong \frac{d}{d\alpha_f}[-Y_f(a_f)]\Delta\alpha_f = C_f'\Delta\alpha_f \tag{6.83}$$

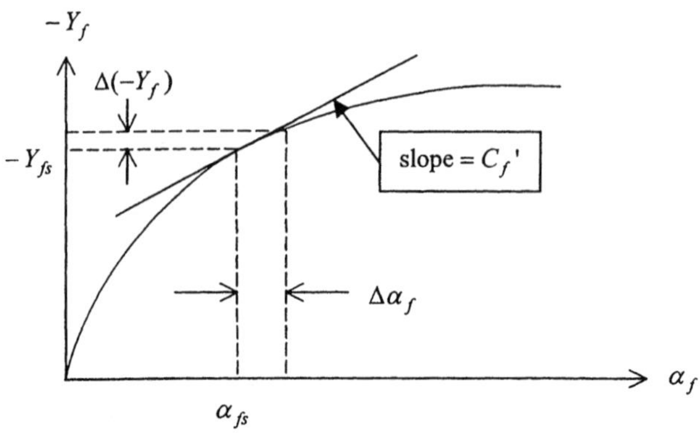

FIG. 6.13 The change in lateral force due to a change in slip angle.

A similar expression is used at the rear.

The dynamic equations for the perturbed motion can be now written by substitution into the general dynamic equations, Eqs. (6.18) and (6.19). Because the steady, basic motion variables are constant in time,

$$\dot{V} = \Delta\dot{V}, \quad \dot{r} = \Delta\dot{r} \tag{6.84}$$

The dynamic equations are then

$$m\Delta\dot{V} + m(r_s U + \Delta r U) = Y_{fs} + \Delta Y_f + Y_{rs} + \Delta Y_r \tag{6.85}$$

$$I_z\Delta\dot{r} = a(Y_{fs} + \Delta Y_f) - b(Y_{rs} + \Delta Y_r) \tag{6.86}$$

Returning to the trim conditions, Eq. (6.82), we find that all the steady variables cancel out of Eqs. (6.85) and (6.86), leaving dynamic equations that resemble the equations which apply to perturbations from straight-line motion [Eqs. (6.18) and (6.19)]. Now, however, the dynamic equations apply to perturbations from a steady cornering situation.

$$m\Delta\dot{V} + m\Delta r U = \Delta Y_f + \Delta Y_r \tag{6.87}$$

$$I_z\Delta\dot{r} = a\Delta Y_f - b\Delta Y_r \tag{6.88}$$

The slip angles for the basic motion are related to the motion variables as derived previously, Eq. (6.20).

$$\alpha_{fs} = (V_s + ar_s)/U - \delta_s, \quad \alpha_{rs} = (V_s - br_s)/U \tag{6.89}$$

The perturbed variables obey similar equations.

$$\Delta\alpha_f = (\Delta V + a\Delta r)/U - \Delta\delta, \quad \Delta\alpha_r = (\Delta V - b\Delta r)/U \tag{6.90}$$

Now, using linearized lateral force-slip angle laws such as Eq. (6.83), the final equations for the perturbed variables may be written.

$$m(\Delta\dot{V}+\Delta rU) = -(C_f' + C_r')\Delta V/U - (aC_f - bC_r) \times\Delta r/U + C_f'\Delta\delta \tag{6.91}$$

$$I_z\Delta\dot{r} = -(aC_f' - bC_r')\Delta V/U - (a^2C_f' + b^2C_r')\Delta r/U + aC_f'\Delta\delta \tag{6.92}$$

These equations closely resemble the equations previously derived using body-centered coordinates for the stability analysis of straight-ahead running [Eqs. (6.22) and (6.23)]. The lateral velocity and the raw rate are now replaced by the corresponding perturbation variables and local slopes of the force-slip angle curves replace the cornering coefficients.

Because of the similarity between the dynamic perturbation equations for straight line running and for steady turns, the previous stability criterion, Eq. (6.14), can be used with the new variables. The car in a steady turn will be stable if the expression

$$m(bC_r' - aC_f') + C_f'C_r'(a+b)^2/U^2 > 0 \tag{6.93}$$

is satisfied.

For straight-ahead running, it was possible to classify cars as understeer, oversteer, or neutral steer depending on the sign of $bC_r - aC_f$ independent of speed. It was also clear that the criterion [Eq. (6.14)] would always be satisfied for small enough U, and therefore all cars are stable at low speeds. The situation is less clear in the case of a steady turn because C_f' and C_r' are the local slopes of the force-slip angle curves that change if either the turn radius or the speed is changed.

It is clear that the cornering coefficients C_f and C_r are positive as defined for straight line running, but C_f' and C_r' can approach zero or even become negative at very large slip angles. Thus, although the criterion for stability [Eq. (6.93)] will always be satisfied for low speeds, because then C_f' and C_r' approach C_f and C_r, the situation in a turn at high speeds is not clear. The term involving $1/U^2$ diminishes with increasing speed, but the quantity $bC_r' - aC_f'$ may change sign as the parameters U and R of the basic motion change. The car can therefore behave quite differently in curves than it does in a straight line, and it is not possible to define critical speeds or characteristic speeds for cars in steady turns unless the tire forces remain in the linear region.

C. Relating Stability to a Change in Curvature

If the tire characteristics of an automobile are assumed to be strictly linear, then there is no difference in the stability analysis between a basic motion in a straight line and in a steady curve. In the discussion of steady turns for a linear car model, it was noted that instability occurred at a steady lateral acceleration level for an oversteer vehicle at which the steer angle in the turn became zero. When nonlinear tire characteristics are assumed, the situation is more complex, but it turns out to be possible to relate dynamic stability to the change in steady steer angle with a change in turn radius at constant speed. This allows one to extend the definition of understeer and oversteer to the nonlinear case. The terms do not apply just to the car but to the car and the particular turn it is negotiating.

Using the trim conditions, Eq. (6.82), the steady forces are

$$Y_{fs} = (mU^2/R)b/(a+b) = \frac{(mU^2 b)}{(a+b)^2}\left(\frac{a+b}{R}\right) \tag{6.94}$$

$$Y_{rs} = (mU^2/R)a/(a+b) = \frac{(mU^2 a)}{(a+b)^2}\left(\frac{a+b}{R}\right) \tag{6.95}$$

and the local slopes of the force laws are

$$C_f' = -\frac{dY_{fs}}{d\alpha_{fs}}, \quad C_r' = -\frac{dY_{rs}}{d\alpha_{rs}} \tag{6.96}$$

The steady steer angle obeys Eq. (6.81). Now, using a simplified notation for the wheelbase,

$$l \equiv (a+b) \tag{6.97}$$

we can calculate the change in steer angle when the turn radius is changed or, more precisely, when l/R is changed by differentiating Eq. (6.81).

$$\begin{aligned}\frac{\partial \delta_s}{\partial(l/R)} &= -\frac{\partial \alpha_{fs}}{\partial(l/R)} + \frac{\partial \alpha_{rs}}{\partial(l/R)} + 1 \\ &= -\frac{d\alpha_{fs}}{dY_{fs}}\frac{\partial Y_{fs}}{\partial(l/R)} + \frac{d\alpha_{rs}}{dY_{rs}}\frac{\partial Y_{rs}}{\partial(l/R)} + 1 \\ &= \frac{1}{C_f'}\frac{mU^2 b}{l^2} - \frac{1}{C_r'}\frac{mU^2 a}{l^2} + 1\end{aligned} \tag{6.98}$$

If Eq. (6.98) is rearranged into Eq. (6.99)

$$\frac{l^2 C_f' C_r'}{U^2}\frac{\partial \delta_s}{\partial(l/R)} = m(bC_r' - aC_f') + l^2 C_f' C_r'/U^2 \tag{6.99}$$

one sees that the stability expression, Eq. (6.93), appears on the right-hand side of the equation. If the expressions on the two sides of Eq. (6.99) are positive, the car is stable. If they are negative, the car is unstable.

Let us suppose that the two slopes C_f' and C_r' are positive. (This is generally the case, except for very large slip angles.) Then the stability of the vehicle depends on the sign of $\partial\delta_s/\partial(l/R)$, because this determines the sign of the stability expression on the left-hand side of Eq. (6.99). If $\partial\delta_s/\partial(l/R)$ is positive, the car is stable; if $\partial\delta_s/\partial(l/R)$ is negative, the car is unstable for the particular speed and curve radius.

To interpret this fact, it is worthwhile to return for a moment to the strictly linear case in which a constant understeer coefficient, K, can be defined. In the section on steady turns, an expression was derived for the steer angle required given a turn radius, R, and a constant speed, U [Eq. (6.65)]. Using the notation of this section, it is

$$\begin{aligned}\delta_s &= KU^2/R + l/R = (KU^2/l)(l/R) + l/R \\ &= \left[(KU^2/l) + 1\right](l/R)\end{aligned} \tag{6.100}$$

In the linear case then,

$$\frac{\partial\delta_s}{\partial(l/R)} = \frac{KU^2}{l} + 1 \tag{6.101}$$

This result confirms what we already know. All cars are stable when U is sufficiently small. No matter what the sign of K happens to be, both sides of Eq. (6.101) will be positive if U is small enough. Understeer cars, for which K is positive, or neutral steer cars with $K=0$, are stable at all speeds. Oversteer cars for which K is negative become unstable only when U exceeds the critical speed and the terms on both sides of Eq. (6.101) become negative.

In the section on steady turns, the steer angle was plotted vs. speed for a constant radius turn. It is useful now to plot steer angle vs. l/R for constant speed. The plot is shown in Fig. 6.14.

For the linear tire force assumption, the slopes of the plot of δ_s as a function of (l/R) are constant because K and U are both constant. Note that an oversteer situation means that the slope of δ vs. l/R is less than unity, but only if the slope is negative is the car unstable. This only happens for the oversteer case and at a speed higher than the critical speed.

For nonlinear tire characteristics, the slopes are not constant and vary with the turn radius when the speed is constant. The reason, of course, is that sharper turns require greater lateral acceleration and thus

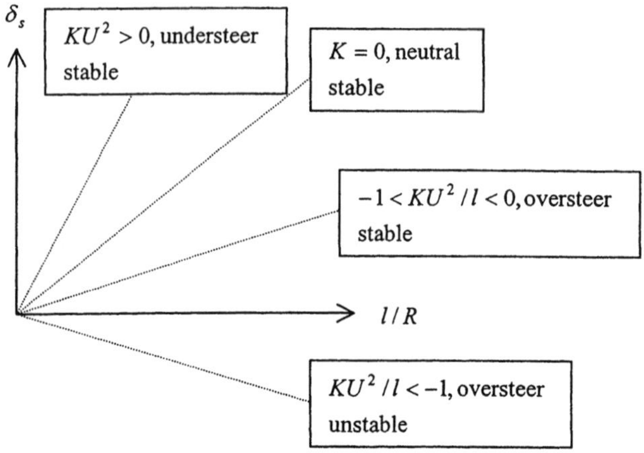

FIG. 6.14 Steer angle as a function of wheelbase divided by turn radius.

the tires operate at different points on their tire characteristics when the radius changes. In the nonlinear case, the terms understeer and oversteer can only be defined for particular turn radii and speeds, but one can show regions of oversteer, understeer, stability and instability on a plot of δ_s as a function of (l/R), as shown in Fig. 6.15.

As Fig. 6.15 shows, understeer and oversteer relate to the local slopes of the steer angle curve. Instability sets in only when the slope becomes negative. Under the assumption that C_f' and C_r' remain positive, the vehicle becomes unstable only from an oversteer condition.

Fig. 6.15 is somewhat difficult to understand, and one should remember that it is just one possible example of how the steer angle could vary with turn radius at a constant speed. However, it does illustrate the fact that the nonlinearity of tire (or, more correctly, axle) characteristics mean that cars cannot be classified as understeer, neutral, or oversteer except for maneuvers during which the axle forces and slip angles remain essentially in their linear range.

IX. LIMIT CORNERING

In a previous section on steady cornering, the assumption was made that the relation between tire side force and the slip angle remained in the linear

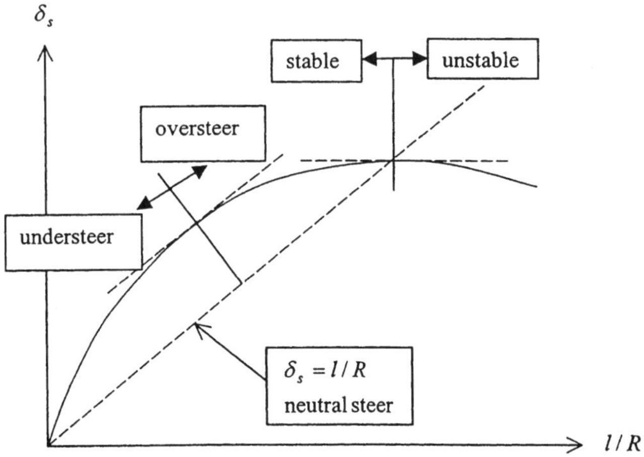

FIG. 6.15 Steer angle as a function of l/R for a nonlinear case.

region. Now, steady cornering in the nonlinear range will be studied. We will describe, at least qualitatively, what happens to a car as it approaches the limiting speed in a corner. By limit speed, we mean the highest speed that the tires will allow the car to have while maintaining a given curve radius.

It is almost obvious that a tire with a fixed normal force can only produce a limited lateral force before it begins to slide. Roughly speaking, the maximum lateral force corresponds to the normal force times a coefficient of friction for rubber on pavement. The Coulomb friction model suggests that once slipping starts, the friction force does not change. As was pointed out in Chapter 4, it is not quite that simple because tests on real tires show that the lateral force vs. slip angle curve typically reaches a maximum lateral force at moderate slip angles, and then begins to decrease at very large slip angles. At any rate, the maximum side forces possible from the tires limit the maximum speed in a given corner because they limit the maximum centripetal acceleration of the center of mass, U^2/R.

It may be less obvious that the limitation may come either at the front or at the rear. In order for the car to remain in a steady turn, the moments about the center of mass from the front and rear axles must sum to zero. (Otherwise, constant angular momentum and angular velocity cannot be maintained.) Thus, if either end of the car has reached its maximum force, the extra force capability at the other end cannot be used because the moment would not remain at zero. The result is that attempts to increase the cornering speed beyond the limit will result in either the front end or the rear end beginning to slide out.

The analysis to be presented below will still use some small angle approximations. This is reasonable because the slip angle at which the maximum lateral tire force occurs is often only about 10° or 15°. Even racecars approaching limit speeds in the middle of a corner usually do not exhibit large attitude angles. If the speed limit in a corner is exceeded, the car will either slide out of the corner or spin, at which time the small angle approximation will no longer apply.

Finally, the "bicycle model" will continue to be used. When we talk about forces and slip angles at the front and rear, we will mean the total force at the front and rear axles from both wheels. We also assume that the slip angles are essentially equal for the two wheels at the front and also for the two wheels at the rear. This means that the force-slip angle relationships are not attributable only to the tires themselves, but to the design of the suspensions at the front and rear.

There are many factors that can influence the axle characteristics (Bundorf, 1967b). For example, consider the camber angle, i.e., the angle the tires make with the normal to the road surface. When a car is cornering, it tends to roll toward the outside of the turn. Different suspension designs result in different camber angle changes when the car body rolls and a camber change affects lateral tire force generation.

Also, in a steady turn, the normal forces for the two outside tires must be larger than the normal forces on the two inside tires. However, the distribution of the changes in normal forces at the between the tires at the front and rear is affected by the suspension stiffness and the presence or absence of antisway bars at the front or at the rear. Because of the nonlinear dependence of lateral tire forces on the normal forces, the distribution of total roll stiffness between the front and rear axles affects how the front and rear axle forces are related to the corresponding slip angles. This important means of adjusting the handling characteristics of vehicles was discussed briefly in Chapter 4 with respect to the tire characteristics plotted in Fig. 4.2 (see Problem 6.8).

Finally, it should be mentioned that even with identical tires on all wheels, handling engineers can affect the axle characteristics by specifying different tire pressures at the front and rear.

The number of factors affecting understeer and oversteer (or handling properties in general) is too large to discuss in any detail here, but one should just keep in mind that the axle force-slip angle curves are determined largely, but not entirely, by the tire characteristics. As will be seen, what are important are always *differences* between front and rear forces that are often similar in magnitude. Thus sometimes small changes have large effects, particularly in the limit.

A. Steady Cornering with Linear Tire Models

Before discussing cornering with a nonlinear tire model, which is necessary when considering limit cornering behavior, it is worthwhile to review some results for the case in which it is assumed that the tires remain in the linear region. In the section on steady cornering, the relation between steer angle and the slip angle at the front and rear was derived, see Eq. (6.52). This relation remains true regardless of the relationship between lateral force and slip angle. If a linear relationship is assumed, the slip angles are proportional to the lateral acceleration, U^2/R, and a constant understeer coefficient K_2 was defined [see Eq. (6.69)].

The steer angle relationship then takes the form of Eq. (6.66). In this form, K_2 indicates how δ varies when the lateral acceleration in "g's" changes. If $K_2 > 0$, δ increases as the lateral acceleration increases and this is an understeer situation. As can be seen in Fig. 6.8 in the section on steady cornering, when δ is positive, the slip angles are negative. Understeer really means that $(-\alpha_f)$ is larger in magnitude than $(-\alpha_r)$. For this reason, it is sometimes convenient to redefine the slip angles so that only positive quantities appear in the steer angle equation.

$$\alpha_1 \equiv -\alpha_f, \quad \alpha_2 \equiv -\alpha_r \tag{6.102}$$

With these definitions, Eq. (6.52) becomes

$$\delta = (\alpha_1 - \alpha_2) + (a + b)/R \tag{6.103}$$

With these new definitions of the slip angles, for the linear case,

$$K_2 = (\alpha_1 - \alpha_2)/(U^2/Rg) \tag{6.104}$$

and K_2 is positive (understeer) when $\alpha_1 > \alpha_2$ and negative (oversteer) when $\alpha_1 < \alpha_2$.

B. Steady Cornering with Nonlinear Tire Models

For the nonlinear case, there is no constant understeer coefficient; however, one can define a variable coefficient that expresses how the slip angle difference or the steer angle changes as the lateral acceleration changes.

$$K_2 \equiv \frac{d(\alpha_1 - \alpha_2)}{d(U^2/Rg)} = \frac{d\delta}{d(U^2/Rg)} \tag{6.105}$$

The variable understeer coefficient can be found graphically from plots of the axle force characteristics $Y_f(\alpha_1)$ and $Y_r(\alpha_2)$. It proves to be convenient to normalize the axle lateral forces by the weight forces. As was shown in the previous section on steady cornering, in the absence of aerodynamic forces, the lateral forces are proportional to the weight forces, Eqs. (6.60) and (6.61), and in fact, the ratios of the side forces to the weight forces is always equal to the lateral acceleration in g's.

$$\frac{Y_f}{W_f} = \frac{Y_r}{W_r} = \frac{U^2}{Rg} \tag{6.106}$$

From plots of the axle forces normalized by the weight forces, one can determine the slip angles given the lateral acceleration, as shown in Fig. 6.16.

Combining the two plots in Fig. 6.16, one can see how $(\alpha_1 - \alpha_2)$ changes as U^2/Rg varies, as shown in Fig. 6.17.

In the case shown in Fig. 6.17, α_1 is always greater than α_2, and the difference $(\alpha_1 - \alpha_2)$ increases as U^2/Rg is increased, until a value is reached at which the front slides out and a steady turn cannot be achieved.

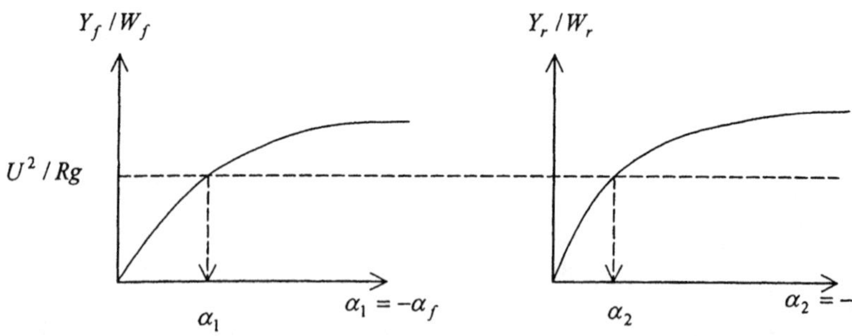

FIG. 6.16 Front and rear slip angles determined from normalized force vs. slip angle curves.

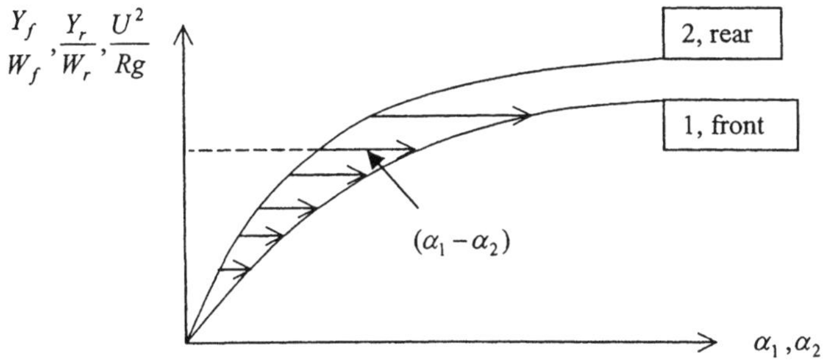

FIG. 6.17 Determination of slip angle difference from lateral acceleration.

Finally, a single plot relating the acceleration to the slip angle difference can be constructed from the combined plot of Fig. 6.17, as shown in Fig. 6.18.

In the example of Fig. 6.18, the variable understeer coefficient is always positive and increases continuously as the lateral acceleration is increased from zero to the maximum value possible before the front axle slides and a steady turn is no longer possible. However, there are many other possibilities, as shown in Fig. 6.19. Because the understeer coefficient has to do with the difference in front and rear slip angles, occasionally small effects that come into play at certain values of lateral acceleration can make significant changes in the steering behavior of the vehicle.

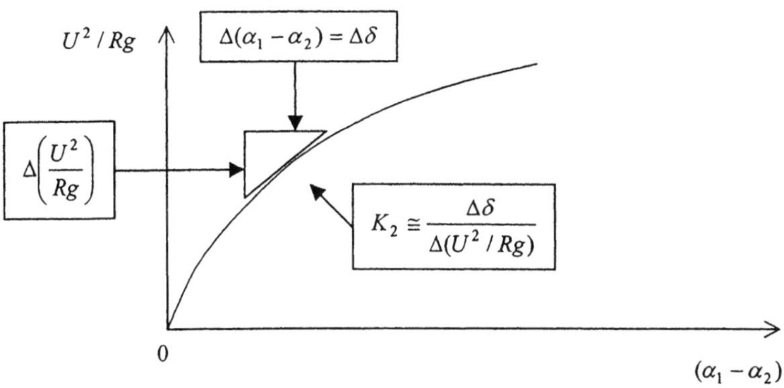

FIG. 6.18 Lateral acceleration related to slip angle difference.

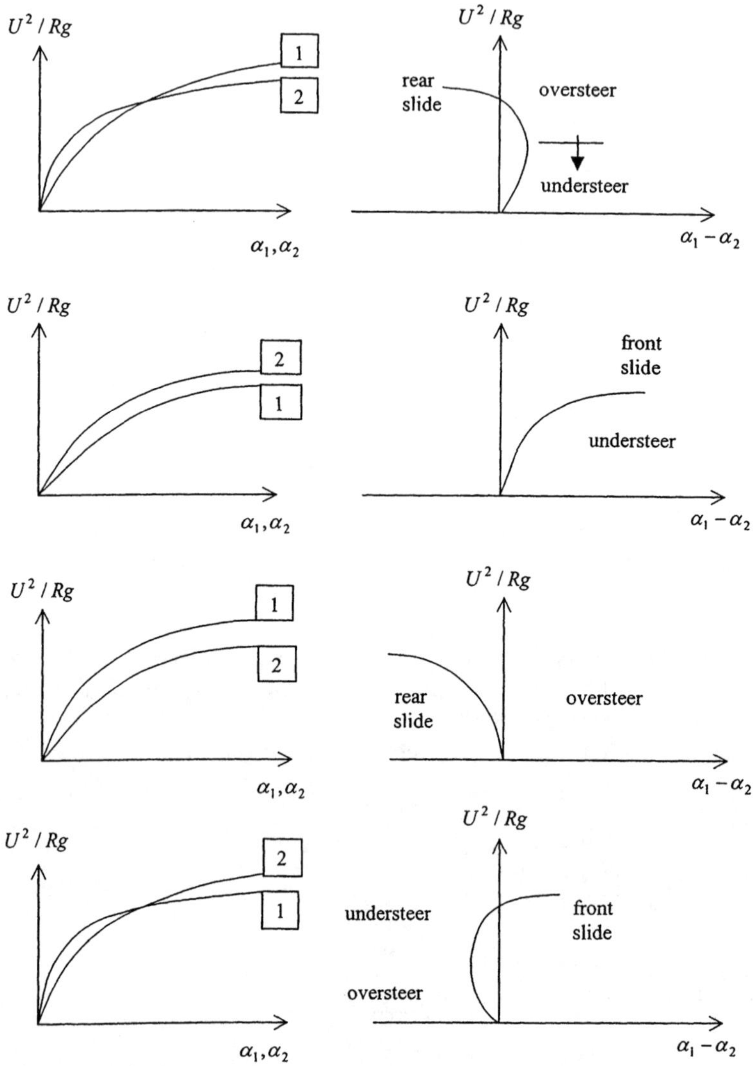

FIG. 6.19 Plots equivalent to Figs. 6.17 and 6.18 for a number of particular cases.

From the plots in Fig. 6.19, one can see that the understeer coefficient can vary with the lateral acceleration in a steady turn. A car that understeers at low acceleration can begin to oversteer at higher acceleration for example. Furthermore, the limiting lateral acceleration may be achieved when either the front axle or the rear axle finally slides out.

It is also possible to imagine a car for which both slip angle curves are identical. The understeer coefficient would then always be zero and both ends of the car would reach their maximum lateral for simultaneously. One can think of this as the situation sometimes described as a "four-wheel drift." This is the origin of the notion that a neutral steer car is a sort of optimum even when the concept is extended into the nonlinear regime.

7
Two-Wheeled and Tilting Vehicles

I. INTRODUCTION

In this chapter, the problem of stabilizing two-wheeled vehicles such as bicycles and motorcycles through active control of steering will be studied. The same ideas can be applied to vehicles with three or four wheels if they are free to tilt or bank in curves.

While it is obvious that two-wheeled vehicles must bank when negotiating a curve, there are good reasons for allowing other vehicles to tilt toward the inside of curves. The general tendency is for a conventional automobile or truck to tilt toward the outside of a curve. In extreme cases, if the curve is sharp enough and the speed is high enough, a vehicle with a high center of gravity can even overturn. The practical effect of this is that high-performance ground vehicles must be fairly wide compared to the height of their centers of gravity. If a vehicle was to tilt toward the inside of curves, much as a motorcycle does, the overturning tendency would be reduced or eliminated and, for the same center of gravity height, the vehicle width could be reduced.

These concepts have been suggested for narrow but tall commuter vehicles, the compact dimensions of which might allow extra lanes to be constructed on existing freeway rights-of-way and could save space in parking structures. The advantage for a tilting vehicle having more than two wheels is that such a vehicle could have some passive stability in lean, which is absent for a two-wheeled vehicle. A motorcycle that encounters a slippery spot in the road while in a turn will immediately fall over while a multiwheeled tilting vehicle need not. There are various ways in which vehicles can be made to lean in curves, but in this chapter, we will concentrate on studying the use of steering control to stabilize the lean angle and to follow a curved path. This will provide a simple explanation of how a human operator can control a bicycle or motorcycle.

The pertinent literature for this chapter deals with bicycles, motorcycles, and tilting vehicles in general. See, for example, Booth (1983), Karnopp and Fang (1992), Jones (1970), Li et al. (1968), Schwarz (1979), and Sharp (1971).

II. STEERING CONTROL OF BANKING VEHICLES

There are a number of vehicles that tilt or bank toward the inside of turns in the manner of an airplane when it is executing a coordinated turn. In a steady turn, these vehicles tilt to an angle such that the vector combination of the acceleration of gravity and the centrifugal acceleration lies along a symmetry axis of the vehicle. Passengers in such vehicles have the sensation that there is no lateral acceleration relative to their own bodies, but they are pushed down in their seats as if the acceleration of gravity had increased somewhat. Coffee in a cup tilted with the vehicle has no tendency to spill even when the centripetal acceleration in a steady turn is high.

Cars and trucks with conventional suspensions, of course, tend to tilt toward the outside of turns. This direction of tilt is undesirable from the point of view of passenger comfort, and the tilt also shifts the center of mass position toward the outside wheels thus increasing the possibility of overturning. The suspension springs and devices called antisway bars resist this tilting tendency, but their effectiveness is limited by the necessity providing a certain amount of suspension compliance for the sake of passenger comfort and to allow the wheels to follow roadway unevenness. Recently developed partially active suspensions are able to reduce the amount of tilt in conventional automobiles, but generally, these active suspension systems do not attempt to tilt the car toward the inside of a turn.

Among ground vehicles that tilt toward the inside of turns, bicycles and motorcycles are obvious examples. In addition, a number of advanced trains tilt for passenger comfort in high-speed turns. Tall thin commuter vehicles have also been developed which tilt toward the inside of turns to reduce the chance of overturning in sharp turns.

Some tilting rains and commuter vehicles are tilted by direct action of an actuator and an automatic control system that forces the vehicle body to tilt to the inside of a turn against its natural tendency to tilt toward the outside. This requires an active system with an energy supply, but the design of this type of tilting servomechanism is fairly straightforward.

Single-track vehicles such as motorcycles are tilted in a completely different way by action of the steering system. In this chapter, the dynamics and control of steering-controlled banking vehicles will be discussed. These vehicles are unstable in the absence of active control of the steering either by a human operator or an automatic control system.

Obviously, steering loses effectiveness as a means for balancing such vehicles at very low speeds or at rest, so another means of achieving balance must be provided. (Bicycle and motorcycle riders use their feet for this purpose.) Once a certain speed has been reached, the steering system not only is used to cause the vehicle to follow a desired path, but also is used to stabilize the vehicle and to cause it to bank at the proper angle in a steady turn.

The present analysis applies not only to single-track two-wheeled vehicles, but also to three- or four-wheeled vehicles with a roll axis near the ground, a high center of gravity, and a suspension with very low roll stiffness. For such vehicles, direct tilt control using an actuator would have to be used to supplement steering control at low speeds.

There have been numerous mathematical studies of the balancing problem for bicycles and motorcycles over a period of many years. Many of the more recent studies include several degrees of freedom, geometric nonlinearities, and nonlinear tire-force models. Such mathematical models are often so complex that insight into the essential dynamics and control of the vehicles is nearly impossible. Here we will introduce a number of simplifications in order to achieve a low-order linear model that is easily understandable and yet illustrates the essential dynamics and control features of steering-controlled banking vehicles.

A. Development of the Mathematical Model

Fig. 7.1 shows a number of the dimensions and variables associated with the mathematical model of a tilting vehicle. For a single-track vehicle, the sketch in Fig. 7.1 is to be imagined as existing in the ground plane.

For a multiwheeled vehicle, the two wheels represent equivalent single wheels for the front and rear axles much as was done for the "bicycle" model for automobiles. The "ground" plane would then pass through the roll center of the suspension. The dimensions a and b relate to the distances from the projection of the center of mass to the front and rear axles in the ground plane. The velocity components U and V describe the

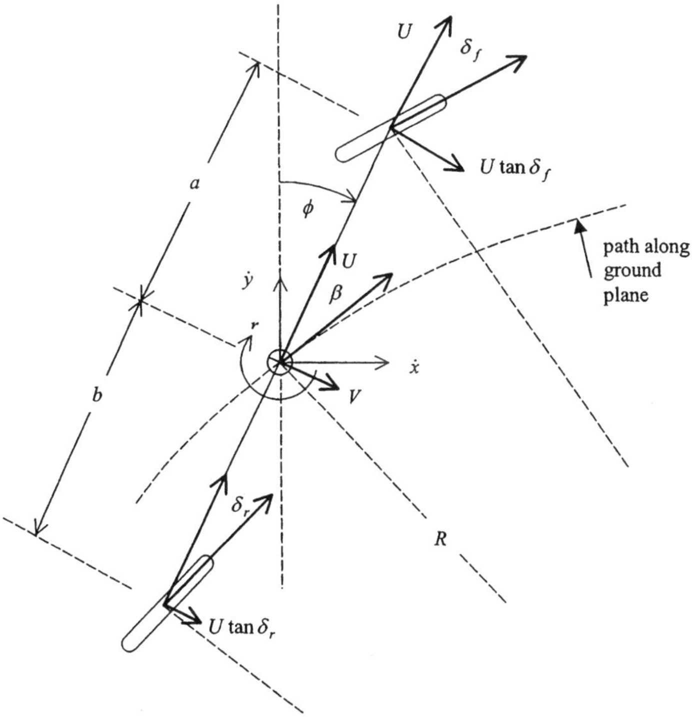

FIG. 7.1 Ground plane geometry for a two-wheeled vehicle in a turn.

velocity of the center of mass projection on the ground plane. (Because of the time-varying tilt angle, the center of mass has other velocity components besides these components of the projection in the ground plane.)

The coordinates x and y locate the ground plane projection of the center of mass in inertial space, and the angle ϕ, which may be large, represents the orientation of the vehicle with respect to the y-axis. Although most vehicles use only front-wheel steering, both front and rear-steering angles, δ_f and δ_r, will be included. It will be shown that rear-wheel steering alone poses difficult control problems, but experimental vehicles have been constructed using a combination of front- and rear-wheel steering. Both steer angles will be assumed small since the model is intended for use at relatively high speeds when the turn radius, R, is large with respect to the wheelbase, $(a+b)$.

A major simplifying assumption is that the slip angles are negligible. This may seem to be an odd assumption since in the analysis of automobiles in Chapter 6, slip angles played a major role in the stability analysis. In this chapter, however, the dynamics of tilting is prominent and the slip angle effects are not important as long as the tires do not skid. This assumption certainly breaks down at high lateral acceleration and it precludes the use of nonlinear tire characteristics, but it has the great advantage that no tire characteristics at all are involved in the model. It is certainly common experience, when riding a bicycle, that it is hard to discern any slip angle when riding at moderate lean angles. The wheels appear to roll almost exactly in the direction that they are pointed with no noticeable slip angles.

We also assume that the forward velocity U is constant. With these assumptions, motion in the ground plane is determined purely kinematically. Using the small angle assumption,

$$\tan\delta_{\mathrm{f}} \cong \delta_{\mathrm{f}}, \quad \tan\delta_{\mathrm{r}} \cong \delta_{\mathrm{r}} \tag{7.1}$$

simple geometric considerations result an expression for the turn radius. (Problem 7.1 involves the derivation of this relationship.)

$$R = (a + b)/(\delta_{\mathrm{f}} - \delta_{\mathrm{r}}) \tag{7.2}$$

It may be useful to note here that Eq. (7.2) is a generalization of the steer angle relationships encountered previously in the "bicycle model" of an automobile. If there is no rear steer angle and if the magnitudes of the front and rear slip angels are equal, then Eqs. (6.53) and (7.2) are identical. In addition, if the understeer coefficient should vanish, Eqs. (6.65) and (6.66) would match Eq. (7.2) in the absence of a rear steer angle. This means that if the tilting vehicle was a strictly neutral steer vehicle, the relationships would be the same as if the slip angles were actually zero (see Fig. 6.8).

The yaw rate r is given by the expression

$$r \cong U/R = U(\delta_{\mathrm{f}} - \delta_{\mathrm{r}})/(a + b) \tag{7.3}$$

The lateral velocity in the ground plane is found again purely kinematically by considering the lateral velocities at the front and the rear.

$$V \cong U(b\delta_{\mathrm{f}} + a\delta_{\mathrm{r}})/(a + b) \tag{7.4}$$

(Problem 7.2. deals with this relationship.)

Finally the "slip angle" for the center of mass projection in the ground plane, β, is

$$\beta = V/U = (b\delta_{\mathrm{f}} + a\delta_{\mathrm{r}})/(a + b) \tag{7.5}$$

If it is desired to track the location of the center of mass projection in the ground plane during computer simulation, for example, the following equations can be used:

$$\dot{\phi} = r, \quad \dot{y} = U\cos\phi - V\sin\phi, \quad \dot{x} = U\sin\phi + V\cos\phi \tag{7.6}$$

With the assumption of zero slip angles for the front and rear wheels, the motion in the ground plane is completely determined by the time histories of the front and rear steer angles. Now the dynamics of the tilting and acceleration of the vehicle body will be modeled.

B. Derivation of the Dynamic Equations

Fig. 7.2 shows the vehicle body with its center of mass a distance h above its ground plane projection and tilted at the lean angle θ. The angle θ functions as the single geometric degree of freedom. The principal axes of the body are assumed to be parallel to the 1-, 2-, and 3-axis system shown in the figure. The 1-axis lies along the longitudinal axis of the vehicle in the x–y plane, the 3-axis is aligned with the vehicle axis that is vertical when the lean angle is zero, and the 2-axis is perpendicular the 1- and 3-axes. The principal moments of inertia relative to the mass center are denoted I_1, I_2, and I_3. When writing the expression for kinetic energy, one can imagine

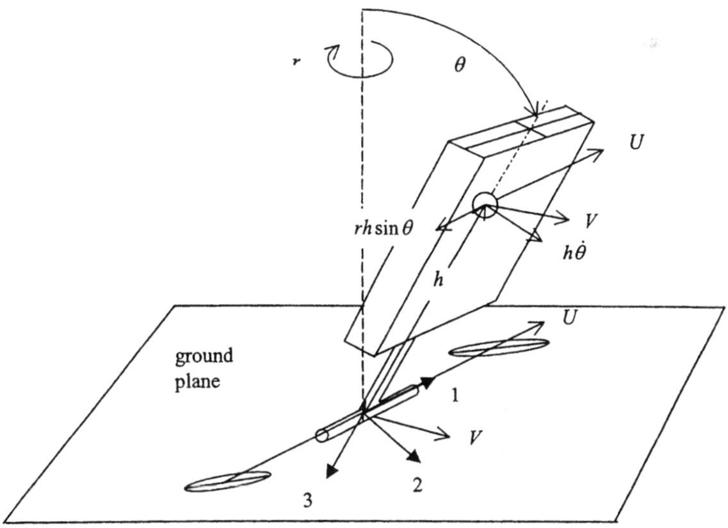

FIG. 7.2 The vehicle body tilting around the ground plane axis.

the 1-, 2-, and 3-axes translated to the center of mass and actually being then the principal axes for the body.

The equation of motion will be derived using Lagrange's equation much as was described in Chapter 5, Eqs. (5.7)–(5.9). In this case of three-dimensional motion, the kinetic energy expression is more complicated than for plane motion and there is a potential energy term having to do with the height of the center of mass in the gravity field. Because of the zero slip angle assumption, the tire lateral forces are perpendicular to the tire velocities and thus do no work. This means that there is no need for a generalized force to represent these forces. The steer angles merely provide a prescribed kinematic motion of the ground plane axis about which the vehicle tilts. The motion variables U, which is assumed to be constant, and V, which is determined by the steer angles through Eq. (7.4), will enter the expression for the kinetic energy.

From Fig. 7.2, one can find the square of the velocity of the center of mass, noting that the U and V velocity components lie in the ground plane. The velocity of the center of mass is composed of the velocity of its projection in the ground plane with components U and V added to the components induced by the angular rotation rates r and $\dot{\theta}$. All the components of the center of mass velocity are shown translated to the center of mass location in Fig. 7.2. The square of the center of mass velocity can be written as the sum of the squares of two orthogonal components. The first component is the component in the 1-direction, $(U - rh\sin\theta)$. The second component has to do with the vector sum of V and $h\dot{\theta}$. These two vectors are separated by the angle θ. The square of the vector sum can be found using the law of cosines to be $(V^2 + h^2\dot{\theta}^2 + 2Vh\dot{\theta}\cos\theta)$. The final expression for the square of the velocity of the center of mass is

$$v_c^2 = (U - rh\sin\theta)^2 + (V^2 + h^2\dot{\theta}^2 + 2Vh\dot{\theta}\cos\dot{\theta}) \tag{7.7}$$

The angular velocities along the 1-, 2-, and 3-directions are seen to be

$$\omega_1 = \dot{\theta}, \quad \omega_2 = r\sin\theta, \quad \omega_3 = r\cos\theta \tag{7.8}$$

Then, the kinetic energy expression appropriate for three-dimensional motion (Crandall et al., 1968) is

$$T = mv_c^2/2 + (I_1\omega_1^2 + I_2\omega_2^2 + I_3\omega_3^2)/2 \tag{7.9}$$

in which Eqs. (7.7) and (7.8) will be substituted.

The potential energy expression is just mg times the height in the gravity field,

$$V = mgh\cos\theta \tag{7.10}$$

where in this case, V represents the potential energy rather than a lateral velocity.

Finally, substituting Eqs. (7.7)–(7.10) into Lagrange's equation for $\theta(t)$, which happens to be exactly Eq. (5.7), the resulting equation of motion is found to be

$$\begin{aligned}(I_1 + mh^2)\ddot{\theta} + (I_3 - I_2 - mh^2)r^2\cos\theta\sin\theta - mgh\sin\theta \\ = -mh\cos\theta(\dot{V} + rU)\end{aligned} \tag{7.11}$$

This equation is certainly a more complicated equation of motion than necessary for present purposes since no small-angle approximations have yet been made for the lean angle θ. Only in extreme cases do bicycles or motorcycles achieve large lean angles so θ will be assumed to be small enough to allow the use of the usual trigonometric small-angle approximations. The middle term on the left-hand side of Eq. (7.11) will be neglected since it involves the product of the small lean angle and the square of the yaw rate, which is also small for cases of practical interest. After applying the small-angle approximations for functions of θ, a linearized equation results.

$$(I_1 + mh^2)\ddot{\theta} - mgh\theta = -mh(\dot{V} + rU) \tag{7.12}$$

(This version of the equation can also be derived using Newton's laws and the small-angle approximations from the beginning, but it requires the introduction of the tire lateral forces and then their subsequent elimination.)

The terms on the right side of Eq. (7.12) are determined entirely by the time-varying steer angles using the kinematic equations derived above [Eqs. (7.3) and (7.4)]. After using these relations, the final form of the linearized equation of motion is

$$(I_1 + mh^2)\ddot{\theta} - mgh\theta = -\frac{mh}{(a+b)}(b\dot{\delta}_f U + \delta_f U^2 + a\dot{\delta}_r U - \delta_r U^2) \tag{7.13}$$

This equation has several interesting features. For example, if the steer angles and their rates were all zero, the equation would describe an upside-down pendulum (for small angles). The term $(I_1 + mh^2)$ can be recognized as the moment of inertia of the pendulum about a pivot point in the ground plane. This should be no surprise since a motorcycle with locked steering would surely tend to fall over just as an upside-down pendulum would.

The right side of Eq. (7.13) indicates that as $U \to 0$, steering action becomes ineffective in influencing the lean angle as would be expected. What may come as a surprise is that not only do the steer angles influence the lean angle, but also the steer angle rates have an effect. In fact, for low speeds, the steer angle rates are more important than the angles themselves since as the speed decreases, the effectiveness of the rates declines only with U while the effectiveness of the angles declines with U^2.

III. STEERING CONTROL OF LEAN ANGLE

In order to study the control aspects of banking vehicles, it is useful to define some combined parameters that allow the structure of the dynamic equation to be seen clearly. By dividing Eq. (7.13) by mgh and defining three time constants and a gain, the equation appears in the form

$$\tau_1^2\ddot{\theta} - \theta = -K(\tau_2\dot{\delta}_f + \delta_f + \tau_3\dot{\delta}_r - \delta_r) \tag{7.14}$$

where

$$\tau_1^2 \equiv (I_1 + mh^2)/mgh, \quad \tau_2 \equiv b/U, \quad \tau_3 \equiv a/U, \quad \text{and} \quad K \equiv U^2/g(a+b) \tag{7.15}$$

Note that the time constant τ_1 is determined solely by the physical parameters of the vehicle but that the remaining two time constants, τ_2 and τ_3, as well as the gain, K, change with the speed. As the speed increases, the gain K increases but the variable time constants τ_2 and τ_3 become shorter.

For the basic motion of a stability analysis, the lean angle, both steer angles, and their derivatives all vanish. This assumption obviously satisfies Eq. (7.14). For the perturbed motion, the right-hand side of Eq. (7.14) is still zero, but θ no longer is assumed to vanish. The characteristic equation for the system is then

$$\tau_1^2 s^2 - 1 = 0 \tag{7.16}$$

which, when solved, results in two real eigenvalues,

$$s = \pm 1/\tau_1 \tag{7.17}$$

The existence of a positive real eigenvalue confirms that the vehicle is unstable in lean when it is not controlled by a manipulation of the steering angles.

A. Front-Wheel Steering

Suppose we consider first a lean control system using only front-wheel steering. The transfer function relating θ to δ_f can be found in the standard way from Eq. (7.14),

$$\frac{\theta}{\delta_f} = -\frac{K(\tau_2 s + 1)}{(\tau_1^2 s^2 - 1)} \tag{7.18}$$

To stabilize the lean angle, let us first try the simplest possible proportional control system,

$$\delta_f = -G(\theta_d - \theta) \tag{7.19}$$

in which G is the proportional gain, θ_d is a desired lean angle, and the minus sign compensates for the minus sign inherent in the transfer function, Eq. (7.18). [The minus sign on the gain in Eq. (7.19) is not particularly important and is simply the result of the choice of signs for the direction of the steer angles in the derivation of the dynamic equation. The control system is a negative feedback system when all the negative signs are considered.] A block diagram for the controlled system is shown in Fig. 7.3.

The transfer function for the closed-loop system can be found either by algebraically combining the control law Eq. (7.19), with the open-loop transfer function, Eq. (7.18), or by block diagram algebra to be

$$\frac{\theta}{\theta_d} = \frac{GK(\tau_2 s + 1)}{\tau_1^2 s^2 + GK\tau_2 s + (GK - 1)} \tag{7.20}$$

The denominator of this transfer function (which is the characteristic polynomial for the closed-loop system as discussed in Chapter 6) indicates that the system will be stable only if

$$GK = GU^2/g(a + b) > 1 \tag{7.21}$$

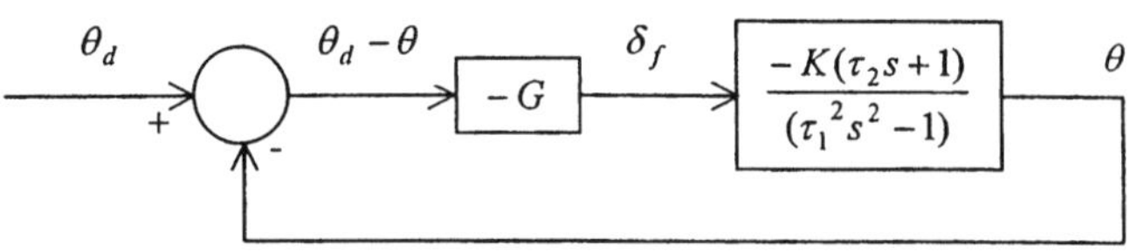

FIG. 7.3 Block diagram for front-wheel steering lean angle control system.

since otherwise, the constant term in the second-order characteristic equation would be negative. At high speeds, this is no problem, but as U decreases, G must increase drastically to maintain stability. This is ultimately not possible as $U \to 0$.

A bicycle rider slowing to a stop typically steers in ever-wider excursions trying to maintain balance but is ultimately forced to put down a foot to remain upright. In terms of the present model and the proportional control assumed, the rider appears to be trying to increase G as U decreases in order to maintain stability. A large gain means that the steer angle becomes large even for small deviations of the lean angle from the desired lean angle. If a rider is stopping, the desired lean angle is zero. The rider can react to small lean angles with large steer angles in an attempt to keep the bicycle upright, but eventually, the attempt fails as the speed approaches zero. (It is true that because of the geometry of actual bicycle front forks, skilled riders can actually balance at standstill, but this effect is not in the present model. In any event, balancing at zero speed is a difficult trick that most riders can hardly perform.)

By setting $s = 0$ in the closed-loop transfer function, Eq. (7.20), one can see how the steady-state lean angle θ_{ss} is related to the desired lean angle θ_d. Using Eq. (7.15), an expression for the relation as a function of speed can be found.

$$\frac{\theta_{ss}}{\theta_d} = \frac{GK}{GK - 1} = \frac{GU^2/g(a+b)}{(GU^2/g(a+b)) - 1} \tag{7.22}$$

If the system is stable, $GK > 1$ and the steady lean angle is at least proportional to the desired lean angle. For a given gain G, as the speed increases, the steady lean angle approaches the desired lean angle. Thus this simple proportional control system not only stabilizes the system (except at low speeds), but also forces the lean angle to approach any given desired lean angle at high speeds or if the gain G is large.

Now by considering the transfer function relating the front steer angle and the lean angle, Eq. (7.18), when $s = 0$, it can be seen that in the steady state, the lean angle is directly related to the steering angle δ_f,

$$\theta_{ss} = K\delta_{fss} = (U^2/g(a+b))\delta_{fss} \tag{7.23}$$

The steering angle is kinematically related to the turn radius and to the yaw rate Eqs. (7.2) and (7.3) with no rear steer angle,

$$R = (a+b)/\delta_f, \quad r = U\delta_f/(a+b) \tag{7.24}$$

These relations provide a philosophy for determining the desired lean angle. First, the rider (in the case of a bicycle or motorcycle) or an automatic control system (in the case of a controlled tilting vehicle) determines a turn radius or a yaw rate required to follow a desired path. This determines a steer angle from Eq. (7.24)

$$\delta_{\mathrm{f}} = (a + b)/R \quad \text{or} \quad \delta_{\mathrm{f}} = (a + b)r/U \tag{7.25}$$

Then, assuming that $\theta_{\mathrm{d}} \cong \theta_{\mathrm{ss}}$, the steady-state relation, Eq. (7.23), can be used to derive a desired lean angle given the desired turn radius or desired yaw rate.

$$\theta_{\mathrm{d}} \cong \theta_{\mathrm{ss}} = U^2/gR \quad \text{or} \quad \theta_{\mathrm{d}} \cong \theta_{\mathrm{ss}} = Ur/g \tag{7.26}$$

The block diagram in Fig. 7.4 indicates the process.

Beginning motorcycle riders are often taught to think about leaning the motorcycle in the direction of a desired turn rather than steering directly in the desired direction as one would in a tricycle or a car. The block diagram makes this advice seem reasonable.

B. Countersteering or Reverse Action

There is nothing particularly unusual about the response of the lean angle of the controlled system to a change in desired lean angle as long as the gain G in combination with the speed is large enough to make $GK > 1$ so that the system is stable. The closed-loop transfer function in Eq. (7.20) has a first-order numerator and a second-order denominator just like the transfer function relating yaw rate to front-steering angle for the automobile model studied in Chapter 6, Eq. (6.33). What may come as a surprise is that the steer angle itself does have unusual dynamics for the closed-loop system.

Using the dynamic and controller equations directly, Eqs. (7.14) and (7.19), or using block diagram algebra on Fig. 7.3, one can derive the transfer function relating the steer angle and the desired lean angle.

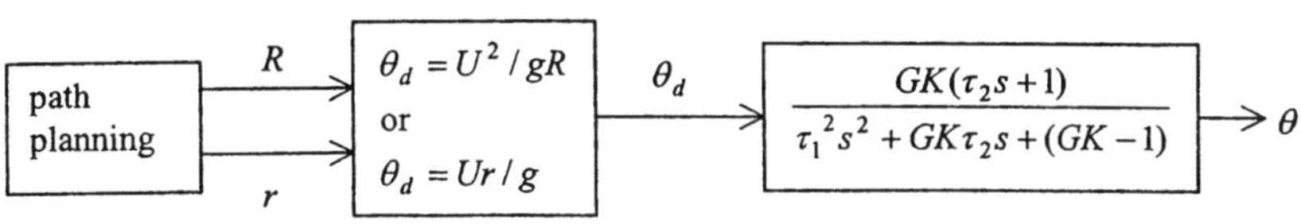

FIG. 7.4 Diagram showing how a desired turn radius or yaw rate determines the desired lean angle for the lean control system.

$$\frac{\delta_{\mathrm{f}}}{\theta_{\mathrm{d}}} = \frac{-G(\tau_1^2 s^2 - 1)}{\tau_1^2 s^2 + GK\tau_2 s + (GK - 1)} \tag{7.27}$$

For a constant desired lean angle, the steady-state steer angle produces essentially the turn radius specified if Eq. (7.26) is used. The steady steer angle for a constant desired lean angle is found by setting $s=0$.

$$\delta_{\mathrm{fss}} = \left(\frac{G}{GK - 1}\right)\theta_{\mathrm{d}} = \left(\frac{G}{GU^2/g(a+b) - 1}\right)\theta_{\mathrm{d}} \tag{7.28}$$

At high speeds, when

$$GK \gg 1, \quad \delta_{\mathrm{fss}} \to (g(a+b)/U^2)\theta_{\mathrm{d}} \tag{7.29}$$

If the desired lean angle is given in terms of the speed and the turn radius

$$\theta_{\mathrm{d}} = U^2/gR \qquad \text{(7.26) repeated}$$

the result is

$$\delta_{\mathrm{fss}} \to (a+b)/R \tag{7.30}$$

as expected considering Eq. (7.24). At lower speeds, or for lower values of G, the steady-state steer angle would not exactly result in the desired turn radius, but a rider or a controller could correct this by adjusting θ_{d} somewhat if a larger or smaller turn radius would be necessary to follow a desired path.

The initial value of the steer angle to a step change in desired lean angle is found by letting $s \to \infty$ in the transfer function (see Ogata, 1970). When s approaches infinity in Eq. (7.27), the result is

$$\delta_f(0) = -G\theta_d \tag{7.31}$$

Equations (7.28) and (7.31) mean that for a positive step in desired lean angle, the steer angle is initially negative but ultimately becomes positive. See the sketch of the steer angle step response shown in Fig. 7.5.

The reason for this peculiar behavior lies in the numerator of the transfer function that has one "zero" in the right half of the s-plane. This means that the transfer function between the desired lean angle and the steering angle is nonminimum phase and has a reverse action steering angle response. The same type of nonminimum phase response was encountered previously in the acceleration transfer function for rear-wheel steering of an automobile, Eq. (6.37).

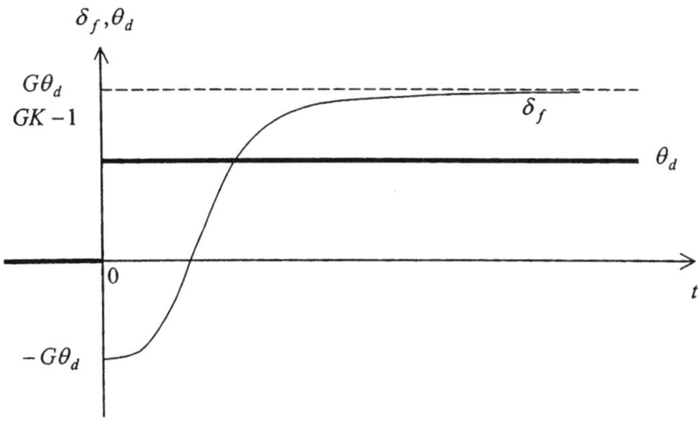

FIG. 7.5 Response of the steer angle to a step change in desired lean angle.

Motorcycle and bicycle riders recognize this reverse action phenomenon that is required to initiate a sudden turn. The initial phase of steering, which is in the opposite direction to the final steer direction, is called "countersteering." In order to make a sharp turn in one direction, one has to initially turn the handlebars in the opposite direction. Some motorcycle-riding instructors actually encourage their students to practice this counterintuitive countersteering in order to better handle rapid avoidance maneuvers.

The reason for countersteering is often explained in less technical terms as follows. To make a right turn, the vehicle must establish a lean to the right. Starting from an upright position, the vehicle is made to lean right only by initially steering left. By steering left, the wheels move to the left relative to the center of mass thus creating a lean to the right. As the lean develops, the steer angle must change from left to right to stop the lean angle from increasing too much. In the end, the vehicle turns to the right at a constant lean angle with gravity tending to make the vehicle fall to the inside of the turn and the so-called centrifugal force tending to make the vehicle lean to the outside of the turn. In a steady turn, these two tendencies just balance each other.

Riders of bicycles and motorcycles are often not aware of the need to countersteer if they negotiate only gentle turns. In normal riding, they rarely are perfectly upright and instead are continually making small steering corrections to keep the lean angle near zero. When the time comes to initiate a turn, the rider simply allows a lean angle in the correct direction to develop without correction and then steers to stabilize the lean angle at a value that corresponds to the speed and the turn radius desired.

C. Rear-Wheel Steering

The idea that to avoid an obstacle on the left by turning right one must first turn left toward the obstacle has lead some to speculate that steering the rear wheel might be preferable to steering the front. A prototype "safety" motorcycle was once constructed based on this concept, but it proved to be virtually impossible to ride (Schwarz, 1979). It has proven possible for some people to learn to ride a rear-steering bicycle, but it is a remarkably difficult task for a human being.

On the other hand, there may be advantages to be found from a coordinated front and rear-wheel steering scheme. It is known that at least one motorcycle manufacturer has experimented with a two-wheel steering prototype. For instance, a long wheelbase vehicle could have a shorter turning radius at slow speeds or better stability at high speeds with such a system particularly if it was an active system (Karnopp and Fang, 1992). Here we will simply point out the unusual dynamics of a purely rear-wheel steering system.

Returning to the linearized dynamic equation for the lean angle, Eq. (7.14), the transfer function relating the lean angle to the rear-wheel steer angle is

$$\frac{\theta}{\delta_r} = \frac{-K(\tau_3 s - 1)}{(\tau_1^2 s^2 - 1)} \tag{7.32}$$

This transfer function closely resembles the transfer function for front-wheel steering, Eq. (7.18). The denominator is, of course, identical indicating that the open-loop system is unstable. The numerator, however, has an extra minus sign when compared to the numerator of the corresponding transfer function for front-wheel steering. Clearly, Eq. (7.32) has a zero in the right half of the s-plane. This indicates a non-minimum phase system and reverse action in the open-loop system. (Our previous analysis found a similar effect only for the steer-angle response of the closed-loop system.)

The steady-state lean angle for a constant steer angle is found by stetting $s = 0$.

$$\theta_{ss} = -K\delta_{rss} \tag{7.33}$$

Now a positive rear steer angle corresponds to a negative lean angle, whereas a positive front steer angle corresponds to a positive lean angle

[Eq. (7.23)]. This is not particularly significant and really only has to do with the positive directions chosen for the steer and lean angles.

Now let us once again try a simple proportional control system for stabilizing the lean angle as was done for the front-steering case.

$$\delta_r = G(\theta_d - \theta) \tag{7.34}$$

In this case, the plus sign seems logical instead of the minus sign used for the front-wheel steer case because of the different signs for the two steady-state relations, Eqs. (7.23) and (7.33).

Combining this with the open-loop transfer function, Eq. (7.32), as was done before in deriving Eq. (7.20), the result is

$$\frac{\theta}{\theta_d} = \frac{-GK(\tau_3 s - 1)}{\tau_1^2 s^2 - GK\tau_3 s + (GK - 1)} \tag{7.35}$$

Now it is clear that this simple proportional control absolutely will not work. If G is positive, the middle term multiplying s in the denominator will be negative indicating that the system will be unstable. If G is negative, the last term in the denominator will be negative also indicating that the system will be unstable. Thus there is no possible proportional gain that will stabilize the system.

This does not mean that no control system can be developed to stabilize the system, but rather that it would take a more sophisticated control scheme to do the job for rear-wheel steering than for front-wheel steering. A more sophisticated model for the vehicle would reveal even more dynamic problems with pure rear-wheel steering.

It was mentioned previously that an experimental motorcycle was constructed based on the idea that steering the rear wheel rather than the front would eliminate the need for counter steering (Schwarz, 1979). The thought was that this might be a safety feature because in attempting to avoid an obstacle, a rider would not have to steer briefly toward the object before being able to steer away from it. These thoughts were correct as far as they went, but the more complicated dynamics of rear-steering two wheelers proved to make the concept impractical. Test riders found the prototype motorcycle very hard to ride without crashing so it hardly qualified as a safety advance.

However, just because the simple control scheme of Eq. (7.34) leads to a result, Eq. (7.35), that is clearly unsatisfactory does not mean that a rear-steering two-wheeler cannot be controlled. Rear-steering bicycles have been constructed, and it has been proved that talented human beings can learn to ride them. There is no doubt that a sufficiently complicated automatic control scheme could be devised for a rear-steering motorcycle, but it is hard to invent a practical reason to prefer rear-wheel steering to front-wheel steering.

On the other hand, a combination of front- and rear-wheel steering could conceivably have advantages particularly for racing motorcycles (Karnopp and Fang, 1992). A short wheelbase makes a motorcycle responsive to steering inputs but limits longitudinal acceleration and deceleration because of the high center of gravity compared with the wheelbase. Long wheelbased motorcycles are known to be less maneuverable but have fewer problems under extreme accelerating or braking conditions. Combined front- and rear-wheel steering, perhaps with the combination speed-dependent, might result in a better comprise than is possible for front-wheel steering alone. It is known that some manufacturers have experimented with rear-wheel steering, but apparently, the benefits found so far did not outweigh the disadvantages of the extra complication and expense of introducing a steering mechanism at the rear.

8
Stability of Casters

I. INTRODUCTION

Casters are wheels attached to a pivoting arrangement that allows them to line up with the direction of motion. They are found typically on the front of shopping carts and on other pieces of moveable equipment. Furthermore, the caster effect is found on many steered wheels on vehicles such as cars, airplanes, bicycles, and motorcycles.

The caster effect is intended to provide a sort of self-steering tendency that prevents the wheel from generating side forces or that keeps the vehicle moving in a straight line in the absence of steering torques. If a steering torque is applied to a wheel that has some caster effect, the steering torque required to keep the wheel turned is related to the lateral force being generated by the wheel. In many vehicles, the steering torque provided by the caster effect provides useful feedback to a driver about the magnitude of the lateral force or acceleration that the steering input is causing. At the dawn of the age of powered vehicles, a few were constructed with no caster effect at the steered wheels. Soon it was recognized that caster not only stabilized the wheel itself, giving the wheel the ability to straighten out by itself in the absence of steering torque, but it also made the steering feel more natural to drivers and helped them to better control their vehicles.

Casters, however, often exhibit unstable oscillatory motion that is merely annoying in the case of a shopping cart wheel but can become dangerous for other types of vehicles. This type of unstable motion is called shimmy (Gillespie, 1992), in the case of automobile or aircraft landing gear wheels, and wobble (Sharp, 1971) in the case of motorcycle front wheels. The causes of this type of instability are many, including excess flexibility in the structure holding the wheel, play in the pivot bearings, and the dynamics of tire force generation. In this chapter some relatively simple models of casters will be analyzed. Some of the analysis in this chapter follows closely the previous analysis of trailers in Chapter 5, but a number of new concepts will be introduced that have general application to steered wheels.

II. A VERTICAL AXIS CASTER

Fig. 8.1 shows an idealized caster with a vertical pivot axis. As all the bearings are considered to be ideal, there is only a single degree-of-freedom and gyroscopic moments play no role in the motion. (Moments associated with gyroscopic effects are counteracted by the rigid structure.) Note that the center of gravity applies to the combination of the wheel and the pivot bracket holding it.

The parameters of the caster are the distances from the pivot axis to the center of mass, a, and from the center of mass to the wheel contact point, b, the mass, m, the moment of inertia about the center of mass,

$$I_c \equiv m\kappa^2 \tag{8.1}$$

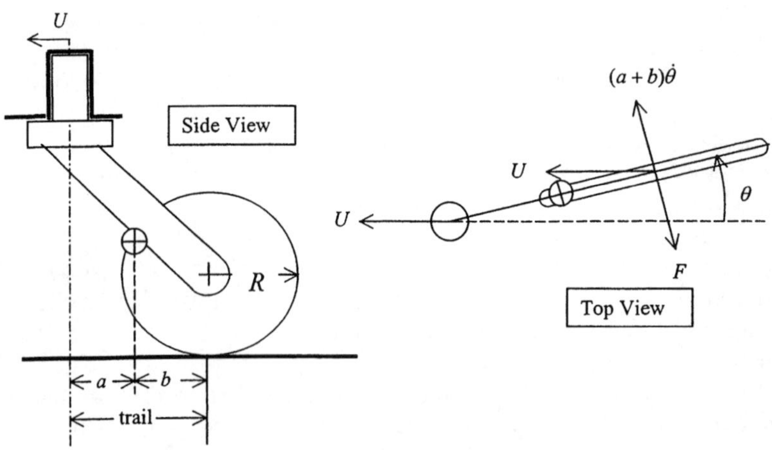

FIG. 8.1 Dimensions of a vertical axis caster.

where κ is the radius of gyration and, finally, the cornering coefficient for the tire, C_α. The distance from the pivot axis to the wheel contact point with the ground is often called the *trail* and, for this caster, is just $(a+b)$.

For a single degree-of-freedom analysis, the only motion variable needed is the angle θ. For the basic motion θ and $\dot{\theta}$ are zero. Assuming that $\dot{\theta}$ is small for the perturbed motion, the rolling velocity is

$$U\cos\theta \approx U \tag{8.2}$$

and the lateral velocity is

$$U\sin\theta + (a+b)\dot{\theta} \cong U\theta + (a+b)\dot{\theta} \tag{8.3}$$

Then the slip angle α for the tire is the lateral velocity divided by the rolling velocity assuming as has been done previously that small angle assumption can be made.

$$\alpha = \theta + (a+b)\dot{\theta}/U \tag{8.4}$$

The lateral force F shown in the sketch will be expressed as

$$F = C_\alpha \alpha \tag{8.5}$$

where, with the force direction and the slip angle direction as they are shown in Fig. 8.1, the cornering coefficient C_α is positive.

The equation of motion is easily written by equating the moment about the pivot point to the rate of change of the angular momentum about the pivot point. The moment of inertia about the pivot is the moment of inertia about the center of mass plus ma^2 by the parallel axis theorem. Using Eq. (8.1), the moment of inertia about the pivot axis is given by two equivalent expressions.

$$I_c + ma^2 = m(\kappa^2 + a^2) \tag{8.6}$$

The equation of motion is found by expressing the rate of change of the angular momentum as equal to the applied moment. Using Eqs. (8.4) and (8.5) the result is

$$m(\kappa^2 + a^2)\ddot{\theta} = -F(a+b) = -C_\alpha(\theta + (a+b)\dot{\theta}/U)(a+b) \tag{8.7}$$

Written in the standard form of Eq. (3.3), the equation of motion becomes

$$m(\kappa^2 + a^2)\ddot{\theta} + (C_\alpha(a+b)^2/U)\dot{\theta} + C_\alpha(a+b)\theta = 0 \tag{8.8}$$

In deriving Eq. (8.8), both U and $(a+b)$ are assumed to be positive and thus the system is stable because all the coefficients in this second-order equation are positive. The coefficient of $\dot{\theta}$ does become small at high speed indicating the caster could exhibit lightly damped oscillations as the speed is increased.

There are two situations in which one might simplify the analysis. First, one might imagine that for low speeds the inertia term in Eq. (8.8) could simply be neglected because the angular acceleration would be small. Another case might be if the wheel and particularly its tire were assumed to be made of a fairly rigid material such as hard rubber. The cornering coefficient has to do with distortion of the tire under load and if the tire has little distortion, the cornering coefficient is large. If $C_\alpha \to \infty$ and the inertia term, which is not proportional to C_α, is suppressed, the equation reduces first order. Following either argument, the resulting simplified equation of motion amounts to a statement that the slip angle vanishes.

$$\dot{\theta} + (U/(a+b))\theta \tag{8.9}$$

As $U = \mathrm{d}x/\mathrm{d}t$, where x is the distance traveled, this equation can be transformed by writing $\dot{\theta}/U = (\mathrm{d}\theta/\mathrm{d}t)(\mathrm{d}t/\mathrm{d}x) = \mathrm{d}\theta/\mathrm{d}x$ to yield

$$\mathrm{d}\theta/\mathrm{d}x + \theta/(a+b) \tag{8.10}$$

Equation (8.10) has, as a solution, an exponential trajectory in space,

$$\theta(x) = \theta_0 e^{-x/(a+b)} \tag{8.11}$$

This represents the kinematic behavior of a caster at low speeds or with a very large cornering coefficient.

III. AN INCLINED AXIS CASTER

The front wheels of motorcycles and bicycles are casters with inclined pivot axes. This introduces a number of geometric complications although the basic dynamic behavior of inclined axis casters is similar to the dynamics of those with vertical axes. Fig. 8.2 shows a generic inclined axis caster. Although the sketch in Fig. 8.2 does not resemble the front wheels of actual bicycles or motorcycles, it does serve to define parameters. It can even represent the front wheels of an automobile in an approximate way because these wheels also have a caster effect.

For large rake angle β and large steering angle θ, the geometry of an inclined axis caster is quite complicated, but typical rake angles for bicycles and motorcycles are only about 25° to 30° and, for stability analysis, the steering angle is assumed small so some approximations are justified. It should also be noted that the sketch in Fig. 8.2 has an exaggerated amount

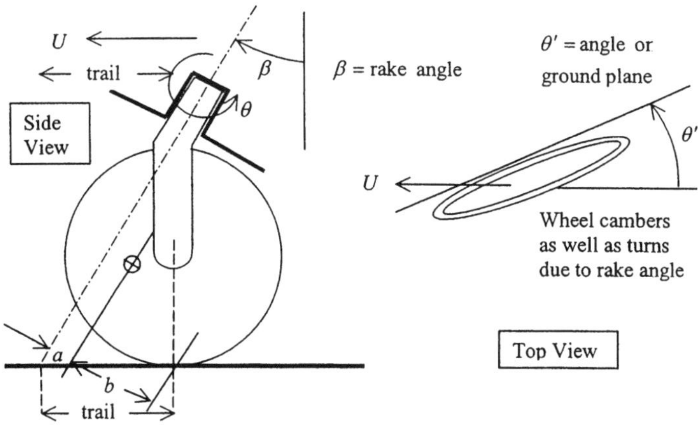

FIG. 8.2 Caster with an inclined axis.

of trail for clarity in the definition of parameters. In this case the trail is defined as the distance along the ground between the intersection of the steering axis with the ground and the contact point of the wheel with the ground. In realistic cases, the center of mass may be ahead of the steering axis so that the distance a is actually negative and the trail is typically much smaller than the radius of the wheel. In any case the formula

$$\text{trail} = (a + b)/\cos\beta \tag{8.12}$$

remains correct even if a is negative.

Among the effects that will be neglected here is the vertical motion associated with the steering of the caster wheel. The frame in which the caster is mounted does move vertically as a function of θ and this effect does have some significance, particularly for a light-weight vehicle such as a bicycle (Jones, 1970). However, for small steer angles and for vehicles that are not leaned over, this effect is of little significance so it will not be considered.

The wheel of an inclined axis caster does not remain in a vertical plane but rather the wheel will lean over at a so-called camber angle when the wheel is steered. The top view in Fig. 8.2 attempts to indicate this. The camber angle is not very large if the rake angle is small but the camber angle does have an effect on the lateral force generation at the tire so this effect will be considered when the tire force is considered below.

Another effect present even for small steer angles is that the angle that the tire makes with a line on the ground plane in the travel direction is not the same as the steer angle if the steer axis is inclined. That is, the steer angle θ and the angle at which the tire touches the ground θ' in Fig. 8.2 are not equal as they are for a vertical axis caster but are related by the rake angle β. The relationship is essentially the same as for a Hooke or Cardan universal joint. Fig. 8.3 may help make this clear. The two angles θ and θ' are only equal when the rake angle β is zero.

The exact relationship between θ' and θ is derived in Martin (1982).

$$\theta' = \tan^{-1}(\cos\beta \tan\theta) \tag{8.13}$$

This relationship means that θ' and θ are only equal for values of θ of 0 and $\pi/2$ (or 90°) and otherwise the ground plane angle is less than the steer angle. For steer angles up to about 30°, a useful approximation is

$$\theta' \cong (\cos\beta)\theta \tag{8.14}$$

The effect of the rake angle in practical cases is simply to slow the steering as the angle on the ground is less than the steer angle.

The other effect of the rake angle to be considered is that the wheel tilts with respect to the ground when the wheel has a steer angle. The tilt angle is called the *camber angle* and given the symbol γ. For reasonably small steer angles the camber angle is approximately

$$\gamma = (\sin\beta)\theta \tag{8.15}$$

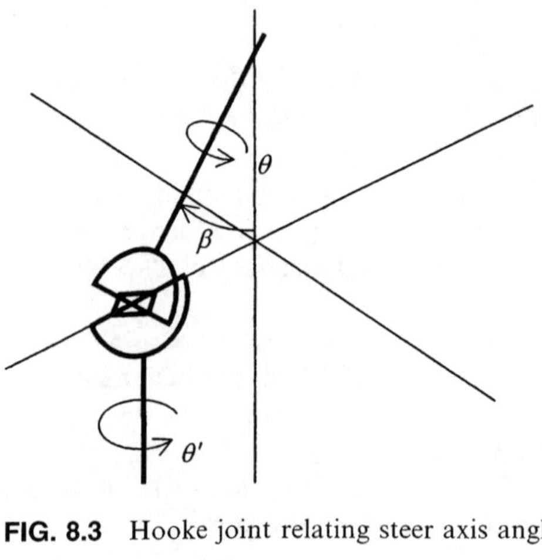

FIG. 8.3 Hooke joint relating steer axis angle θ to ground plane angle θ'.

The camber angle produces a *camber thrust* that is separate from the force due to the slip angle. The total lateral force then can be written in the linearized case, using a camber thrust coefficient C_γ,

$$F = C_\alpha \alpha + C_\gamma \gamma \tag{8.16}$$

In practical cases, the camber effect is not large as the camber thrust coefficient is typically about 1/10 of the cornering coefficient and the camber angle is typically a fairly small fraction of the steer angle.

Considering the geometry of the inclined axis caster in Fig. 8.2, one can see that the rolling velocity of the wheel is again

$$U \cos \theta' \cong U \tag{8.17}$$

and the lateral velocity is, using Eq. (8.14),

$$U \sin \theta' + (a+b)\dot{\theta} \cong U\theta' + (a+b)\dot{\theta} \cong U(\cos\beta)\theta + (a+b)\dot{\theta} \tag{8.18}$$

Thus the approximate expression for the slip angle is

$$\alpha = (\cos \beta)\theta + (a+b)\dot{\theta}/U \tag{8.19}$$

and the final expression for the lateral force is

$$F = C_\alpha((\cos \beta)\theta + (a+b)\dot{\theta}/U) + C_\gamma(\sin \beta)\theta \tag{8.20}$$

Again using $m(\kappa^2 + a^2)$ to represent the moment of inertia about the steering axis, the equation of motion is

$$\begin{aligned} &m(\kappa^2 + a^2)\ddot{\theta} + C_\alpha((a+b)^2/U)\dot{\theta} + (C_\alpha \cos\beta + C_\gamma \sin \beta) \\ &\quad \times (a+b)\theta = 0 \end{aligned} \tag{8.21}$$

As is readily apparent, this equation is very similar to the equation for the vertical axis caster, Eq. (8.8). The camber angle effect is quite small and probably can be neglected in most practical cases. The main effect of the rake angle is to reduce the coefficient of θ by multiplying by the cosine of the rake angle.

The conclusions about the stability of the single degree-of-freedom model of the vertical axis caster also apply with minor modifications to the inclined axis caster. Both caster models are stable but the damping of oscillations declines as the speed increases. It is logical then to assume that a more complex model might exhibit the instabilities that are sometimes observed in practice.

IV. A VERTICAL AXIS CASTER WITH PIVOT FLEXIBILITY

A two degree-of-freedom model of a caster produces a fourth-order system that almost certainly is capable of being unstable under certain conditions. The analysis of a fourth-order system in literal coefficients is, however, a formidable job so a third-order caster model will be developed first. Allowing the pivot to have flexibility will do this if the slip angle is constrained to vanish. This is particularly logical for casters using nearly rigid wheels instead of pneumatic tires. In such cases, an equivalent cornering coefficient would be very large indeed. In many practical cases it would be logical to assume that the wheel would have almost no slip angle until it started to slide.

Fig. 8.4 shows a top view of the caster. Because of the similarities between the equations of motion for vertical and inclined axis casters, the simpler case of a vertical axis caster will be analyzed. The new feature is the introduction of a spring with constant k at the pivot. This spring actually represents the flexibility of the structure supporting the wheel in the pivot bearing or, if the spring constant is zero, an amount of lateral play in the pivot or wheel bearings.

This analysis essentially recapitulates the analysis of the two degree-of-freedom trailer model in Chapter 5 which began with a fourth-order dynamic model using Lagrange's equations and then reduced the system to third order by assuming a very large cornering coefficient. In this case, however, Newton's laws and the imposition of a zero slip angle assumption will lead directly to a third-order equation of motion.

Observing that the lateral acceleration of the center of mass is given approximately by the expression, $\ddot{x} + a\ddot{\theta}$, Newton's laws can now be written

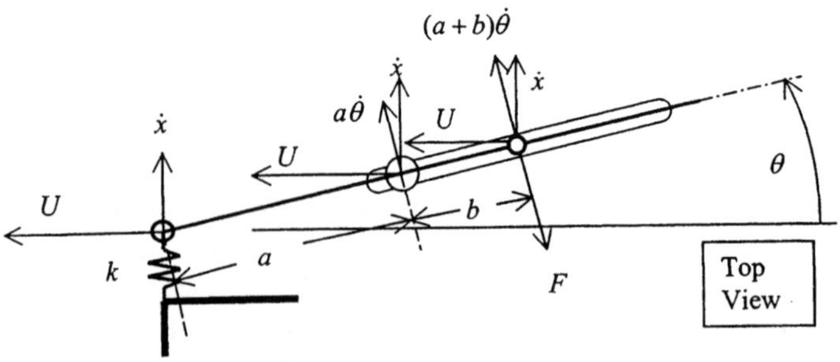

FIG. 8.4 Vertical axis caster with flexibility at the pivot point.

assuming that θ is small enough that the small angle approximations can be used.

$$m(\ddot{x} + a\ddot{\theta}) = -kx - F \tag{8.22}$$

$$m\kappa^2\ddot{\theta} = akx - Fb \tag{8.23}$$

Now instead of relating the lateral force to the slip angle, we simply assume that the slip angle or, more directly, that the lateral velocity at the wheel-road contact point is zero.

$$\dot{x} + U\theta + (a + b)\dot{\theta} = 0 \tag{8.24}$$

The lateral force can be eliminated from Eqs. (8.22) and (8.23) and, after differentiating the result, Eq. (8.24) can be used to eliminate $\dot{x}$ and $\ddot{x}$. The result is a third-order equation of motion,

$$m(\kappa^2 + b^2)\dddot{\theta} + mbU\ddot{\theta} + k(a + b)^2\dot{\theta} + k(a + b)U\theta = 0 \tag{8.25}$$

The first check for stability is to examine the coefficients that would appear in the characteristic equation to see if any could be negative. Obviously, this happens if U is negative or if b is negative. One should really check to see if Eq. (8.25) correctly yields the equation for the caster moving in reverse when U is simply given a negative value. In any case, it is in the nature of casters to turn around when moved backward so the system should be unstable for backward motion.

It is highly unlikely that one could construct a caster with a center of mass behind the wheel center so a negative b is unlikely to be a cause of instability.

On the other hand, even if all coefficients of a third-order characteristic equation are positive, Routh's criterion provides another criterion, which must be met for the system to be stable. When this criterion, Eq. (3.24), is applied to the coefficients in Eq. (8.25), a number of factors can be canceled out and the criterion for stability in this case finally reduces to a very simple inequality.

$$ab > \kappa^2 \tag{8.26}$$

The surprise is that this criterion does not involve the spring constant k. This means that in addition to flexibility in the caster structure, play in the pivot, which could be interpreted as a spring of zero spring constant, could be a cause of case instability if the criterion above were not satisfied.

A. Introduction of a Damping Moment

So far the model has shown the possibility of instability but there is no critical speed. This means that a caster is predicted to be either stable for all speeds however high or unstable even at extremely low speeds. This does not seem realistic. A potentially unstable caster is certainly stable if the speed is reduced sufficiently.

This can be explained by the lack of any friction in the model. Although the friction in the pivot is almost certainly better described by some sort of "dry" friction than by viscous friction, we will introduce a damping moment proportional to $\dot{\theta}$ in order to keep the model linear. Let the damping moment about the pivot axis be given by the equation

$$M_{\mathrm{d}} = c\dot{\theta} \tag{8.27}$$

When this moment is inserted into Eq. (8.23), and the three equations, Eqs. (8.22)–(8.24), are combined into a single equation, the only change in the final equation of motion for θ, Eq. (8.25), is in the coefficient of $\ddot{\theta}$ which becomes $(c + mbU)$ instead of just mbU. This removes one possible cause of instability at low speeds. Even if b were to be negative, at a sufficiently low speed, the coefficient would be positive and the system would not necessarily be unstable.

In addition, the Routh stability criterion changes from Eq. (8.26) to

$$\frac{c(a+b)}{mU} + ab > \kappa^2 \tag{8.28}$$

This implies that all casters are stable at sufficiently low speeds if viscous friction is included in the model as the stability criterion of Eq. (8.28) is always satisfied if U is small enough.

If the caster has the potential for instability,

$$ab < \kappa^2 \tag{8.29}$$

then there is a critical speed above which the system becomes unstable.

$$U_{\mathrm{crit}} = \frac{c(a+b)}{m(\kappa^2 - ab)} \tag{8.30}$$

This model demonstrates that casters can exhibit unstable behavior for certain combinations of parameters and that friction may stabilize the motion of a caster at least up to a critical speed. The conclusion applies to both vertical axis and inclined axis casters.

V. A VERTICAL AXIS CASTER WITH PIVOT FLEXIBILITY AND A FINITE CORNERING COEFFICIENT

A true two degree-of-freedom system results if the same model is used as above but rather than constraining the slip angle to be zero, the lateral force is related to the slip angle with a cornering coefficient, C_α. The equations are not difficult to formulate either using Newton's laws or Lagrange's equations, but the derivation of the characteristic equation is algebraically tedious. Therefore only the main results will be given.

The characteristic equation is found to be as follows:

$$a_0 s^4 + a_1 s^3 + a_2 s^2 + a_3 s^1 + a_4 s^0 = 0 \tag{8.31}$$

with

$$\begin{aligned} a_0 &= m^2\kappa^2 \quad a_1 = m(\kappa^2 + b^2)C_\alpha/U \\ a_2 &= m(\kappa^2 + a^2)k + mbC_\alpha \\ a_3 &= C_\alpha(a+b)^2 k/U \\ a_4 &= C_\alpha(a+b)k \end{aligned} \tag{8.32}$$

All the coefficients appear to be positive in practical cases. Routh's criterion for a fourth-order system delivers two criteria for stability. One is the same as Eq. (3.24) for a third-order system and the other is

$$(a_1 a_2 - a_0 a_3)a_3 > a_1^2 a_4 \tag{8.33}$$

After a good deal of algebraic manipulation, for the case at hand this criterion becomes

$$(ab - \kappa^2)((ab - \kappa^2)k(a+b)/C_\alpha + (\kappa^2 + b^2)) > 0 \tag{8.34}$$

As in the previous results for the third-order model, if $ab - \kappa^2 > 0$, the system is always stable. This is the only criterion if $k \to 0$ or $C_\alpha \to \infty$.

If $(ab - \kappa^2) < 0$, the system could still be stable if the quantity

$$(ab - \kappa^2)k(a+b)/C_\alpha + (\kappa^2 + b^2) \tag{8.35}$$

were also negative. This requires that

$$|ab - \kappa^2| > C_\alpha(\kappa^2 + b^2)/k(a+b) \tag{8.36}$$

This could happen if C_α were small and k were large.

This analysis points out that there are a number of parameter sets that can lead to instability for a caster. Just as with the third-order model, the stability criterion does not involve the speed when no friction forces or moments are included. Realistically, there is always a speed below which instability is not observed experimentally. When friction is included in the equations of motion, it is likely that the analysis will predict that there is a critical speed for a system that would be unstable in the absence of friction. In principle, however, it is possible for certain types of friction in a complex system to be incapable of stabilizing the system at any speed but this is rarely the case for low-order models.

VI. A CASTER WITH DYNAMIC SIDE FORCE GENERATION

It is known that there is a time lag between a sudden change in slip angle of a pneumatic tire and the build up of a lateral force. This has been modeled through the use of the concept of a *relaxation length*, σ (Loeb et al. 1990; Sharp, 1971). At high speeds, it turns out that the time lag becomes small and thus this time lag is usually neglected in modeling automobiles that have more important handling problems at high speeds than at low speeds. On the other hand, wobble of motorcycle front forks and shimmy of automobile wheels occur at relatively low speeds so it is possible that the time lag in side force generation is responsible for these types of instability in some cases. In this section, a model of dynamic lateral force generation will be incorporated into a caster model to study the effect on the stability of motion.

For simplicity, dynamic force generation will be studied using a vertical axis caster rather than an inclined axis caster that would be more appropriate for a motorcycle front wheel. It has already been shown that both types of casters are described by similar dynamic equations but the vertical axis caster equation is simpler.

The tire will now be assumed to have a flexible sidewall with an associated spring constant, k. In the new model, there will be a distinction between the lateral velocity of the wheel and the lateral velocity of the contact patch. In the previous analyses the slip angle definition actually had to do with the wheel motion, but in the new model there will be a new definition of a slip angle based on the lateral motion of the contact patch.

A sketch of flexible sidewall tire and a caster with such a tire is shown in Fig. 8.5. In previous studies, the slip angle was defined based on the lateral and the rolling velocities of the wheel.

$$\alpha = v_1/U = \theta + (a+b)\dot{\theta}/U \tag{8.37}$$

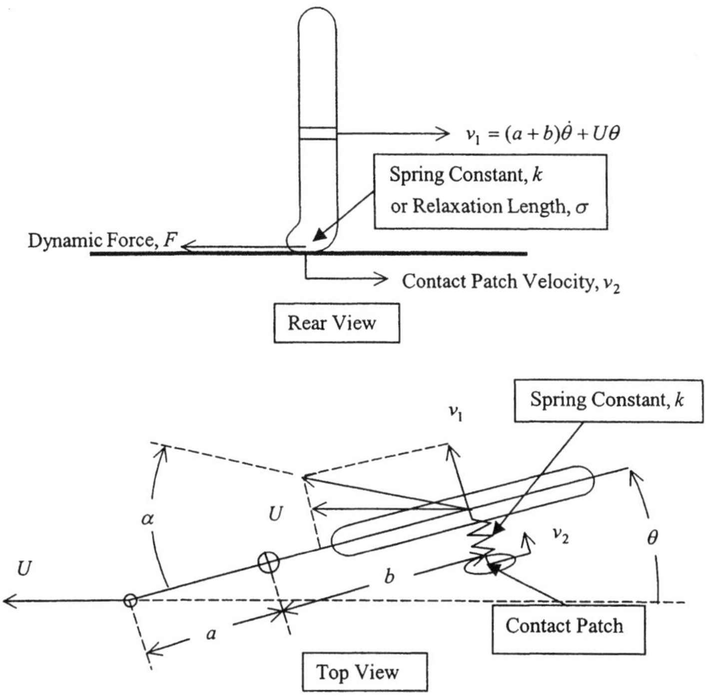

FIG. 8.5 Caster with a flexible sidewall tire.

The force on the tire was then defined as a cornering coefficient times the slip angle. Now we will consider this force as the steady-state force, F_{ss}, which would apply if the slip angle were constant for some time.

$$F_{ss} = C_\alpha \alpha = C_\alpha(\theta + (a+b)\dot{\theta}/U) \tag{8.38}$$

The relaxation length concept is that after a step change in the slip angle, the force approaches the steady-state value exponentially in the distance traveled as the tire gradually assumes a steady deflected shape. The step change in slip angle could be achieved in a tire tester by changing the angle the tire made with the direction of travel from zero to α and then letting the tire roll and recording the lateral force as it built up from zero. If the force were initially zero, the force as a function of distance would have the form

$$F(x) = F_{ss}(1 - e^{-x/\sigma}) \tag{8.39}$$

where σ is defined to be the *relaxation* length, assuming the tire reacted as a linear system. As a function of time, the corresponding relation is

$$F(t) = F_{ss}(1 - e^{-Ut/\sigma}) \tag{8.40}$$

This leads to the idea that the dynamic force lags the steady-state force as a first-order system with a time constant τ of σ/U. [The concept of a time constant was introduced in Chapter 1 in Eqs. (1.25) and (1.26).] The differential equation relating F and F_{ss} based on the time behavior in Eq. (8.40) is

$$(\sigma/U)\dot{F} + F = F_{ss} = C_\alpha(\theta + (a+b)\dot{\theta}/U) \tag{8.41}$$

In Eq. (8.41), if the dynamic force F were initially zero and if the steady-state force F_{ss} were suddenly increased by a step change in the slip angle, then the response of the dynamic force would be exactly given by Eq. (8.40). The dynamic force equation can now be incorporated into the single degree-of-freedom caster model.

The basic dynamic equation for the single degree-of-freedom caster is

$$m(\kappa^2 + a^2)\ddot{\theta} = -F(a+b) \tag{8.42}$$

This equation was used in Eq. (8.7) but now the steady relation between the force and the slip angle, Eq. (8.38), will not be used but rather the dynamic force relation, Eq. (8.41).

Using Eq. (8.42), F and $\dot{F}$ can be found in terms of $\ddot{\theta}$ and $\dddot{\theta}$. When these expressions are substituted into Eq. (8.41), a third-order dynamic equation for the caster results.

$$\begin{aligned}(\sigma/U)m(\kappa^2 + a^2)\dddot{\theta} + m(\kappa^2 + a^2)\ddot{\theta} + (C_\alpha/U)(a+b)^2\dot{\theta}\\ + C_\alpha(a+b)\theta = 0\end{aligned} \tag{8.43}$$

Before analyzing the stability properties of this equation another interpretation of dynamic force generation will be given.

A. The Flexible Sidewall Interpretation of Dynamic Force Generation

There is another way to think about how tires generate lateral forces in a dynamic way besides the introduction of a relaxation length concept. Think of the sidewall as a spring of constant k that connects the wheel to the contact patch. Then assume that the contact patch has a lateral velocity of v_2 that is only equal to the lateral velocity of the wheel, v_1, in the steady state (Fig. 8.5). Now define a contact patch slip angle α'.

$$\alpha' = \theta + v_2/U \tag{8.44}$$

and make the new assumption that the dynamic force is related to the contact patch slip angle.

$$F = C_\alpha \alpha' = C_\alpha(\theta + v_2/U) \tag{8.45}$$

In fact, Eq. (8.45) will be used to find the contact patch velocity in the form

$$v_2 = (F - C_\alpha \theta)U/C_\alpha \tag{8.46}$$

Why this should be so can be seen by making bond graph models for the relations determining F_{ss} and F and then applying causality.

As a preliminary step, consider the steady-state force, which has been used before in Eq. (8.41),

$$F_{ss} = C_\alpha \theta + C_\alpha v_1/U \tag{8.47}$$

This relationship can be represented by the bond graph fragment shown in Fig. 8.6 (Karnopp et al., 2000).

The bond graph fragment indicates that the steady force–velocity relationship has two components. One is proportional to the lateral velocity and is resistive in nature and the other is a force component related directly to the angle θ.

In Eq. (8.47), the slip angle is defined with respect to the lateral velocity of the wheel, v_1. The next step will be to define the slip angle in terms of the lateral velocity of the contact patch, v_2 in place of v_1. The bond graph representing Eq. (8.45) is identical to the bond graph in Fig. 8.6 but the velocity involved is v_2 rather than v_1.

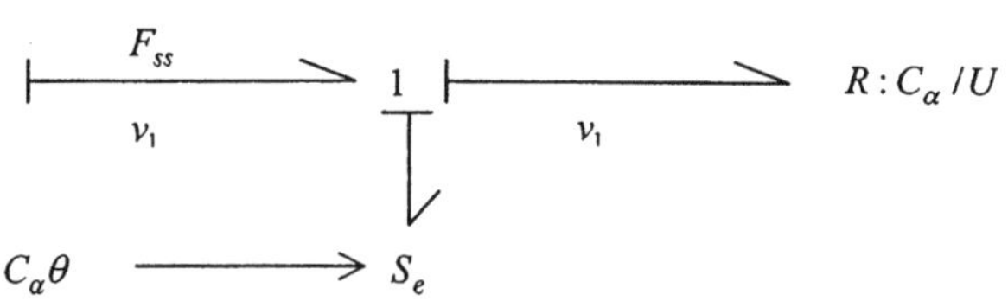

FIG. 8.6 A bond graph fragment representing the steady-state force relationship, Eq. (8.47).

We now construct a bond graph using the contact patch slip angle rather than the wheel slip angle for the force relation, Eq. (8.45), and with the sidewall spring relative velocity $\dot{x}$ being $v_1 - v_2$. This bond graph is shown in Fig. 8.7.

Note that the causal strokes on the *R*-elements in Figs. 8.6 and 8.7 are on opposite ends of the bonds. In Fig. 8.6, the *R*-element computes a force component from a velocity, while in the second bond graph, the *R*-element computes a velocity from a force. The dynamic force in Fig. 8.7 is determined by the C-element representing the sidewall spring. Fig. 8.6 represents Eq. (8.47) whereas the right side of Fig. 8.7 represents Eq. (8.46) in determining v_2 from the dynamic force *F*.

The bond graph in Fig. 8.7 has the state equation

$$\dot{x} = v_1 - v_2 = v_1 - (U/C_\alpha)(kx - C_\alpha\theta) \tag{8.48}$$

and an output equation

$$F = kx \tag{8.49}$$

After eliminating x and $\dot{x}$ from Eq. (8.48) using Eq. (8.49), the result is

$$(C_\alpha/kU)\dot{F} + F = C_\alpha(\theta + v_1/U) = F_{ss} \tag{8.50}$$

A comparison of Eq. (8.41) with Eq. (8.50) shows that the relaxation length can be related to the sidewall stiffness and to the cornering coefficient.

$$\sigma = C_\alpha/k \tag{8.51}$$

This gives an alternative explanation of the relaxation length concept and another method of predicting the relaxation length based on the cornering coefficient and the sidewall stiffness. These are two parameters that can

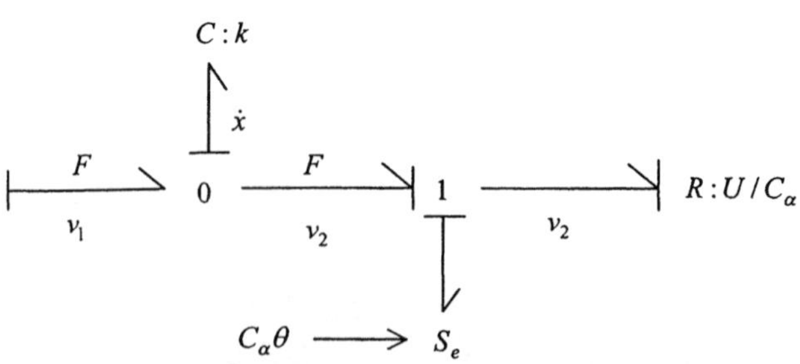

FIG. 8.7 A bond graph representation of dynamic force generation.

be measured experimentally. The direct measurement of the relaxation length is certainly possible but it requires a special piece of equipment to measure the lateral force as the tire is rolled along the roadway surface.

B. Stability Analysis with Dynamic Force Generation

Using C_α/k from Eq. (8.51) in place of σ, the equation of motion, Eq. (8.43), becomes

$$(C_\alpha/kU)m(\kappa^2+a^2)\dddot{\theta} + m(\kappa^2+a^2)\ddot{\theta} + (C_\alpha/U)(a+b)^2\dot{\theta} + C_\alpha(a+b)\theta = 0 \quad (8.52)$$

Equation (8.52) can be checked by comparing it with previous results. For example, if $U \to \infty$, the equation becomes

$$m(\kappa^2+a^2)\ddot{\theta} + C_\alpha(a+b)\theta = 0 \quad (8.53)$$

which is the same as Eq. (8.8) for the single degree-of-freedom caster under the condition that $U \to \infty$. Furthermore, when $U \to 0$, the equation becomes

$$m(\kappa^2+a^2)\dddot{\theta} + (a+b)^2k\dot{\theta} = 0 \quad (8.54)$$

This equation, after one time integration, can be interpreted to be the equation describing the angular vibrations of the stationary caster possible because of the flexible sidewall tire with spring constant k. The expression $m(\kappa^2+a^2)$ is the moment of inertia about the pivot and $(a+b)^2 k$ is a torsional spring constant related to the tire sidewall spring constant. This interpretation would be much harder to make if the equivalent equation, Eq. (8.43), using the relaxation length were used.

Using Routh's criterion for a third-order system, Eq. (3.24), a simple criterion for stability can be derived based on the coefficients in Eq. (8.52).

$$(a+b) > C_\alpha/k, \quad \text{or,} \quad (a+b) > \sigma \quad (8.55)$$

As an example, according to one estimate (Sharp, 1971) motorcycle tires have a relaxation length of about 0.25 m so it is possible that this criterion is not fulfilled in some cases. This could explain why some motorcycle front forks are prone to back-and-forth oscillations commonly called wobbling.

Of course, if pivot damping were introduced, there may either be a critical speed or no possible instability even if this criterion were violated.

It is intuitively obvious from Eqs. (8.40) and (8.41) that at high speeds, there is little difference between F and F_{ss} particularly when the changes in slip angle are not abrupt. At high speeds the time constant

$$\tau = \sigma/U = C_\alpha/kU \tag{8.56}$$

becomes short and at high enough speeds the difference between the steady state and the dynamic force will be so small that there is no point in including dynamic force generation in a vehicle dynamics model. This is the justification for neglecting the dynamics of lateral tire force generation for vehicles expected to exhibit stability problems at high speed.

9
Aerodynamics and the Stability of Aircraft

I. INTRODUCTION

This chapter will discuss several aspects of the stability and control of aircraft. Only low-speed (subsonic) aircraft will be considered, and only enough wing theory will be presented for the purposes of basic stability analysis. There are a large number of reference books and articles dealing with aerodynamics and the stability and control of aircraft. See, for example, Babister (1980), Dole (1981), Etkin (1972), Etkin and Reid (1996), Hunsaker and Rightmire (1947), Hurt (1960), Irving (1966), McCormick (1995), McReuer et al. (1973), and Vincenti (1988).

Because airplanes fly in three dimensions and are capable of a wide variety of maneuvers, there are many aspects of stability and control that could be discussed. Some are fairly obvious. For example, the vertical tail functions much as the tail of a weather vane. It tends to make the airplane line up with the relative wind and thus stabilizes the yaw rate in straight flight.

Other aspects are quite subtle as when the airplane pilot attempts to execute a so-called "coordinated turn." In this case, the airplane turns in a manner similar to a bicycle or motorcycle in a steady turn. As example of some of the subtleties associated with three-dimensional motion, consider first the initiation of a right-hand-coordinated turn. The airplane must roll to the right meaning that the left wing must generate more lift than the right through the use of the ailerons. This increase of lift on the left is accompanied by an increase in drag on the left wing compared to the right that tends to make the airplane turn to the left rather than the right as intended. The pilot must counteract this so-called "adverse yaw" effect using the rudder.

Then there is a question whether the airplane is stable in a coordinated turn or whether the pilot must stabilize it as a bicycle rider must stabilize his vehicle. Once in a tight right turn, it is clear that the right wing is moving more slowly through the air than the left (because of the circular motion) and thus tends to generate less lift. In some cases, this leads to a "spiral instability" in which the right wing drops, the turn tightens, the lift on the right wing decreases further, and so on. The result can be a downward spiral of increasing speed and decreasing radius that can end in a crash. In bad visibility, the pilot may not even realize what is happening because the spiral remains close to a coordinated turn that can feel to the pilot as if he were flying straight and level. Early aviators discovered these instability problems and had to learn to deal with them often without effective control means in the earliest aircraft.

In this chapter, only one particularly interesting aspect of aircraft stability will be considered. This concerns motion in the vertical plane when the basic motion is straight and level flight. This aspect of stability is usually called "longitudinal" stability. It has often been noted that the Wright brothers's first airplane was longitudinally unstable and that they had to practice a long time in order to learn to control it. Although there were many stability and control problems to be solved before manned flight was possible, a basic problem was that longitudinal instability was not understood at the time. Many people think that the Wright brothers's experience with bicycles, which need to be stabilized by human intervention, was crucial to developing a vehicle that at least was capable of being controlled although it was inherently unstable.

The study of longitudinal stability for airplanes can be compared to the consideration of lateral stability for automobiles that was discussed at some length in Chapter 6. In both cases, plane motion is involved, and in both cases, there are two locations at which forces are applied. For cars, the locations are at the front and rear axles. For airplanes, the locations are at the wing and at the tail. In this chapter, the focus will be on determining the size and location of the horizontal tail that will assure stability. On conventional airplanes with a large wing in front and a smaller horizontal tail, the tail is often correctly called the stabilizer.

Although the stability analyses of cars and airplanes have many features in common, the differing ways in which tires and wings generate forces lead to quite differing conclusions about how to analyze stability. In particular, in the section on wheels and wings, we will see why static stability is a more useful concept for aircraft than it is for automobiles because of the difference in the ways that pneumatic tires and wings generate forces.

II. A LITTLE AIRFOIL THEORY

For those not particularly familiar with the field of aerodynamics, a minor hurdle is the aerodynamicist's tendency to express all results in terms of nondimensional coefficients. For longitudinal stability analysis, the main concern will be with lift forces generated by wings.

Fig. 9.1 shows a typical airfoil shape with air flowing past at an angle of attack, α, as if the airfoil were in a wind tunnel. The force generated on the wing consists of a component perpendicular to the airflow direction called the lift, L, and another component parallel to the relative wind direction called the drag, D. Depending on the point at which the lift and drag forces are assumed to act and the airfoil shape, there may be a moment called M_0. Later, a particular point called the aerodynamic center will be defined at which the lift and drag forces will be assumed to act. Then the moment M_0 will represent the moment about the aerodynamic center and will be considered to be positive if it were in the nose-up direction shown in Fig. 9.1. (For an airfoil of the general shape shown, the moment would actually be negative.)

Airfoil sections are often studied in a wind tunnel where the flow is essentially two-dimensional. In this case, the lift and drag forces as well as the moment are usually given per unit length of the wing section. If a

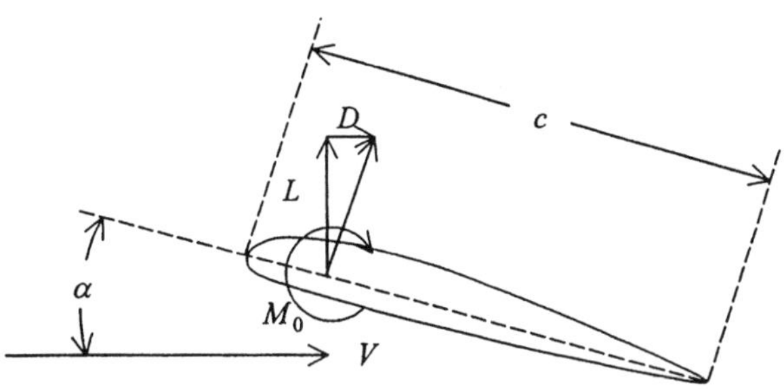

FIG. 9.1 Angle of attack, lift, drag, and moment for a wing or a wing section.

complete wing is tested in a wind tunnel, the flow is three-dimensional particularly near the wing tips and the total lift and drag forces and total moment are measured. Although the results for a unit length section of an airfoil indicate the basic properties for the airfoil shape, it is not trivial to predict the lift and drag results for a finite length wing from the sectional data. Real wings may taper in plan form and may have twisted sections. In a later section on parameter estimation, theoretical results will be used to estimate some parameters relevant for stability studies for finite length wings.

In any case, it is found that the forces on a wing or wing section are almost exactly proportional to the dynamic pressure $(1/2)\rho V^2$, where ρ is the mass density of the air and V is the speed of the air flowing toward the wing. Using the area of a wing, S, a nondimensional lift coefficient, C_L, and drag coefficient, C_D, can be defined by observing that the product of the dynamic pressure and an area has the dimension of a force. The nondimensional coefficients are

$$C_L \equiv L/(\rho V^2/2)S \tag{9.1}$$

$$C_D \equiv D/(\rho V^2/2)S \tag{9.2}$$

In the case of data for a unit length of an airfoil section, the area involved is the chord, the nose to tail length shown in Fig. 9.1, c, times the unit spanwise length. Then, when L and D represent the lift and drag forces per unit length, respectively, slightly different sectional lift and drag coefficients are defined.

$$C_l \equiv L/(\rho V^2/2)c \tag{9.1a}$$

$$C_d \equiv D/(\rho V^2/2)c \tag{9.2a}$$

The lift and drag coefficients for a complete wing can be distinguished from the corresponding sectional coefficients by the use of upper-case subscripts for the complete wing and lower-case subscripts for the wing section.

In addition, a nondimensional moment coefficient C_{M_0} can be defined for a complete wing using the dynamic pressure, the area S, and the cord c. (In this case, the dynamic pressure times the area and times the chord has the dimension of a moment.)

$$C_{M_0} \equiv M_0/(\rho V^2/2)Sc \tag{9.3}$$

Again, there is a version of the moment coefficient for a unit span when M_0 represents the moment per unit span.

$$C_{m_0} \equiv M_0/(\rho V^2/2)c^2 \tag{9.3a}$$

A dimensional analysis indicates that the coefficients in Eqs. (9.1)–(9.3) should be functions of a number of nondimensional quantities: the angle of attack or incidence, α, the Mach number (the flow velocity V divided by the speed of sound), and the Reynolds number (ρVc divided by the viscosity). As long as the flow over the wing nowhere approaches the speed of sound, these coefficients depend almost exclusively on the angle of attack, although there is a noticeable dependence on Reynolds number. In what follows, subsonic aircraft will be discussed and it will be assumed that the dependency of the lift, drag, and moment coefficients on Mach number and Reynolds number can be neglected for the stability analysis. A typical plot of lift coefficient for an airfoil section as a function of angle of attack is shown in Fig. 9.2.

It is clear that in Fig. 9.2, there is some arbitrariness in the definition of angle of attack. If the angle is measured between the airflow direction and a line in the airfoil between the nose and the tail, the lift coefficient may not be zero for a zero angle of attack. A little thought will convince one that for a symmetrical airfoil, when the airstream is flowing parallel to a line from the nose to the tail, there can be no lift merely by consideration of the symmetry of the situation.

The shape shown in Fig 9.1 is, however, not symmetrical. The top surface is curved more than the bottom surface so the centerline of the

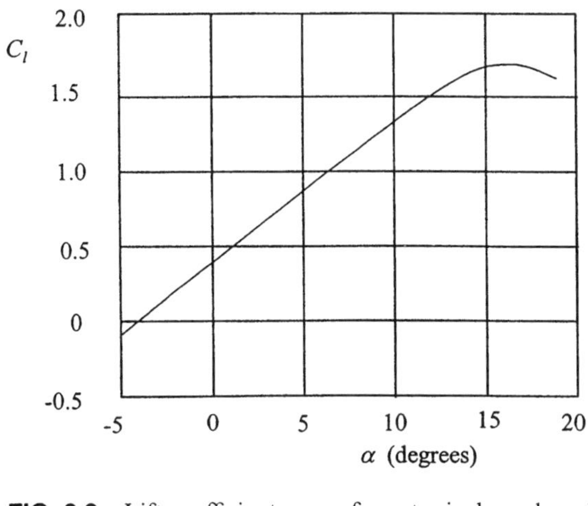

FIG. 9.2 Lift coefficient curve for a typical cambered airfoil (angle of attack, α, defined with respect to the nose-to-tail line).

airfoil shown in Fig. 9.1 has a small downward curvature known as *camber*. Airfoils with camber generate some lift even when the angle of attack defined with respect to the nose to tail line is zero. The airfoil section that produced Fig. 9.2 evidently was cambered in the sense of the section shown in Fig. 9.1.

Camber is used for the airfoils of airplanes that are not intended to fly upside-down. One effect the camber has is to shift the minimum drag point toward positive angles of attack. (For a symmetrical wing section, it is almost obvious that the minimum drag occurs for zero angle of attack and hence zero lift.) A cambered airfoil allows a designer to optimize a wing for low drag during cruise flight. At cruise conditions, the wing must operate at a positive lift coefficient in order to generate a lift to counteract the weight. At this condition, it is desirable to have minimum drag and a cambered wing is beneficial. Stunt airplanes that do frequently fly upside-down typically have symmetrical wing profiles so that upside-down flight will not be burdened with a very high drag coefficient.

Fig. 9.3 shows a typical plot of the sectional drag coefficient and moment coefficient vs. the sectional lift coefficient. Note that not only is the drag coefficient small compared to the lift coefficient, but also that it is low for usefully high positive lift coefficients. The drag coefficient rises more

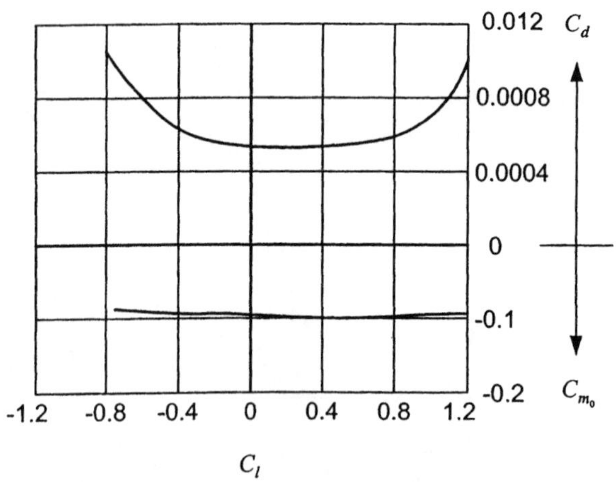

FIG. 9.3 Typical sectional drag and moment coefficients for an airfoil with camber.

steeply for negative lift coefficient values. This shows that the airfoil is less efficient in generating negative lift than positive lift.

It is also worth noting that the moment coefficient is negative and very nearly constant as the lift coefficient varies. As previously noted, for airfoil shapes such as that shown in Fig. 9.1, the moment (and the moment coefficient) turns out to be negative. The fact that the moment coefficient is constant with lift coefficient is related to the choice of the point at which the lift is supposed to act. The aerodynamic center (to be discussed below) is, in fact, the point at which the moment coefficient is theoretically constant and experimentally almost constant. For the stability analysis to follow, the drag will play no role for conventional aircraft configurations, but it is important to note the constant moment due to the wing camber.

For our later considerations, we will assume that the angle of attack is measured between the airflow direction and an imaginary line in the wing called the zero lift line. For the profile of Fig. 9.2, this would be a line inclined about -4° from the nose-to-tail line. This definition implies that when the angle of attack is zero, the lift coefficient is also zero. This definition simplifies the writing of equations for stability analysis. In equations, the angle of attack will be measured in radians, although it is conventional to plot the angle of attack in degrees as in Fig. 9.2.

In Fig. 9.2, it can be seen that there is a region for which the lift coefficient is nearly proportional to the angle of attack, but for large angles of attack, the lift coefficient stops increasing with angle of attack and even begins to decrease. This phenomenon begins to occur when the airflow streamlines no longer can follow the wing profile closely. At very large angles of attack, the wing is said to *stall* and the lift coefficient (and hence the lift) decreases significantly.

The Wright brothers recognized the importance of camber for the efficient generation of lift, but their airfoils resembled curved plates and they were only efficient for a small range of angles of attack. Modern airfoils have camber but also carefully designed rounded leading edges and profiles that avoid separation and stall until large angles of attack. Extensive data on sectional lift, drag, and moment coefficients can be found, for example, in McCormick (1995).

There is a rough analogy between the generation of lift force by a wing and the generation of lateral force by a pneumatic tire. Comparing Fig. 4.1 with Fig. 9.2, one can see this clearly. Stall for a wing is comparable to a skid for a tire. The tire slip angle and the wing angle of attack play similar roles, and one could even say that the nondimensional friction coefficient in Fig. 4.1 is analogous to the lift coefficient. As with many analogies, however, there are some significant ways in which the analogy breaks down.

In the analysis of stability for land vehicles using pneumatic tires, a cornering coefficient, C_α, was defined to be the slope of the curve of lateral force vs. slip angle in the linear region. Similarly, the slope of the lift coefficient curve vs. angle of attack in the linear region will play a prominent role in the stability analysis of airplanes. The coefficient analogous to C_α is called simply a and is thus defined as:

$$\partial C_L / \partial \alpha \equiv a \tag{9.4}$$

where the slope is to be taken near the point at which the lift coefficient is zero.

The plot in Fig. 9.2 is really a plot for a section of the wing as it might be tested in two-dimensional flow in a wind tunnel. If the same airfoil profile was made into a wing of finite length and tested in a wind tunnel, a somewhat different plot would result because near the wing tips, the flow would no longer be two-dimensional. Rather, the higher pressure under the wing would cause a flow around the tips of the wings toward the lower pressure above the wing. This produces trailing vortices sometimes seen as vapor trails streaming behind airplanes.

The important result from the point of view of stability analysis is that the slope of the lift curve is less than that of a wing section in two-dimensional flow, although the definition in Eq. (9.4) is still valid. Later, a theoretical means of estimating the lift coefficient slope for finite wings will be given.

Finally, the concepts of the *center of pressure* and the *aerodynamic center* need to be explained in some detail. First, one can define a center of pressure as that point at which the lift force (which is the sum of the pressure forces acting on the wing) appears to act. At the center of pressure, one could support the wing or wing section by counteracting the total lift force without also having to counteract a moment. For a symmetrical wing profile, thin airfoil theory predicts that the center of pressure lies behind the leading edge of the wing a distance equal to 25% of the chord, c, shown in Fig. 9.1. Therefore for a symmetrical wing section, one can consider that the lift force always acts at the center of pressure and experiment shows that the center of pressure is quite close to the quarter chord point even for practical airfoil profiles.

For cambered wing profiles, however, the center of pressure is not a convenient point at which to assume that lift is acting because as the angle of attack changes, not only does the size of the lift force change, but also the center of pressure moves back and forth along the wing section. In fact, because of the curvature of a cambered wing profile, there is a moment on the wing even when the wing is generating zero lift. In this case, the center of pressure moves toward infinity. (The moment is the zero lift force times the infinite center of pressure distance.)

One can consider the lift force to act at any point along the airfoil if one defines a moment such that the combination of lift force and moment is statically equivalent to the lift force and moment computed about any other point. For a symmetrical wing, the center of pressure is the point at which the moment is zero and this point does not move as the lift changes with angle of attack. For cambered wings, thin airfoil theory again predicts that it is convenient to pick the quarter chord point to be the point at which the lift should be assumed to act, but this point can no longer be considered the center of pressure.

For a cambered wing, the quarter chord point is called the aerodynamic center instead of the center of pressure because there will be a finite moment about the point rather than zero moment. The moment is shown as M_0 in Fig. 9.1 with the positive direction as shown. In fact, for the camber shown in Fig. 9.1, the moment would actually be negative. Again, experiment shows that the aerodynamic center for actual wing profiles is near the quarter chord point.

Experimentally, the aerodynamic center is that point along the profile at which the lift force is assumed to act with the property that the moment remains constant as the lift changes as shown in Fig. 9.3. For a symmetrical wing section, the moment is zero and the center of pressure and the aerodynamic center are identical. In the stability analysis to follow, we will assume that the moment remains exactly constant as the lift changes with angle of attack when the lift is placed at the aerodynamic center, although this is not strictly true for actual wings. For long straight wings, the aerodynamic center is found experimentally to be near the quarter chord point as it is for wing sections. For tapered and swept back wings, there is still an aerodynamic center somewhere and the references given above give methods for estimating its location.

III. DERIVATION OF THE STATIC LONGITUDINAL STABILITY CRITERION FOR AIRCRAFT

The basic idea of longitudinal static stability for airplanes is simply expressed. The first step is to consider a basic motion consisting of straight and level flight. For this basic motion to be possible, certain trim conditions must be met. The sum of all forces must be zero and the sum of all moments about the center of mass must vanish. Roughly speaking, the thrust from the power plant and the drag on the whole airplane should cancel each other out and the lift on the entire airplane should cancel out the weight. Any moment about the center of mass would cause the angular momentum about the pitch axis to deviate from the zero value that the basic motion assumes. Thus the net moment about the center of mass must vanish.

For a static longitudinal stability analysis, one then considers a perturbation in the pitch angle of the aircraft (positive in the nose-up direction). If a nose-down moment about the mass center subsequently develops, there will be a tendency for the airplane to return to the basic equilibrium motion and the aircraft is called statically stable. If a nose-up moment develops, the pitch angle will tend to increase and the airplane is said to be statically unstable.

Sometimes, the term "positive pitch stiffness" is used instead of the term "static stability." This means that if the airplane is pitched up by a disturbance, a moment will be developed proportional to the change in pitch angle and in a direction to reduce the pitch angle much as if the airplane was connected to a torsional spring that resisted pitch angle changes. A stable airplane has an equivalent torsional spring with a positive spring constant or positive stiffness. An unstable airplane has an equivalent torsional spring constant.

The pitch angle perturbation considered will increase the angle of attack of the lifting surfaces (primarily the wing and horizontal tail) and will thus increase the total lift. Although the increased lift will mean that the total lift and the weight will no longer balance and airplane will accelerate upward, static stability is concerned only with the development of a pitch moment. The pitch moment arises because the forces developed at the wing and the horizontal tail changes when the pitch angle is perturbed away from the angle that produced no net moment in the basic motion case.

The static stability criterion can be stated in terms of the direction of the change in pitch moment as the lift increases rather than as the pitch angle and thus the angle of attack of the lifting surfaces increases. In fact, the criterion is conventionally stated in terms of nondimensional coefficients rather than directly in terms of forces and moments. Rather than discussing how the moment changes when the lift changes, the stability criterion has to do with how the *moment coefficient* changes with the *lift coefficient*. This less intuitive but more universal formulation will be discussed below.

Often, the wing and body are combined as far as the main lift force is concerned, although most of the lift occurs on the wing for conventional aircraft. The horizontal tail, often referred to as the stabilizer, is considered separately. This is because the static stability depends on the size and location of the horizontal tail. The horizontal tail generally does not provide much of the lift necessary for flight. Its main purpose is to provide stability. Flaps on the trailing edge of the horizontal tail called elevators also provide a means for the pilot to control the pitch attitude of the aircraft.

A final concept involves the so-called *downwash* at the tail. As a consequence of the generation of lift to support the airplane, the wing deflects air downward behind it. When the wing increases lift after pitching up, this downward airstream deflection or downwash angle increases and this reduces the change in angle of attack at the tail due to the change in pitch angle. This means that the increase in the lift at the horizontal tail when the airplane pitches up is less than it would be if it were not behind the wing. Aircraft designers must take this downwash effect into account because it reduces the effectiveness of the tail in assuring static stability as will be seen subsequently.

An elementary static stability model of an airplane is shown schematically in Fig. 9.4. The model applies most directly to subsonic flight regimes and incorporates the concepts discussed above. Note that the wing and body are combined as far as the lift force is concerned, although the wing area is used in the definition of the lift coefficient. The horizontal tail is considered separately because the size and position of the tail will be the main parameters used to assure static stability.

In Fig. 9.4, lift coefficients are defined for the wing-body combination and for the horizontal tail as follows:

$$C_{\mathrm{Lwb}} \equiv L_{\mathrm{wb}}/(\rho V^2/2)S \tag{9.5}$$

$$C_{\mathrm{LT}} \equiv L_{\mathrm{T}}/(\rho V^2/2)S_{\mathrm{T}} \tag{9.6}$$

The moment about the center of mass can now be expressed in terms of the lift forces acting on the wing-body combination and the horizontal tail in Fig. 9.4. There is a constant moment due to wing camber and, of course, no moment about the center of mass due to the weight. The horizontal tail typically has a symmetric profile since it is not designed to supply much lift at cruise conditions and it has a smaller area than the wing so it is not

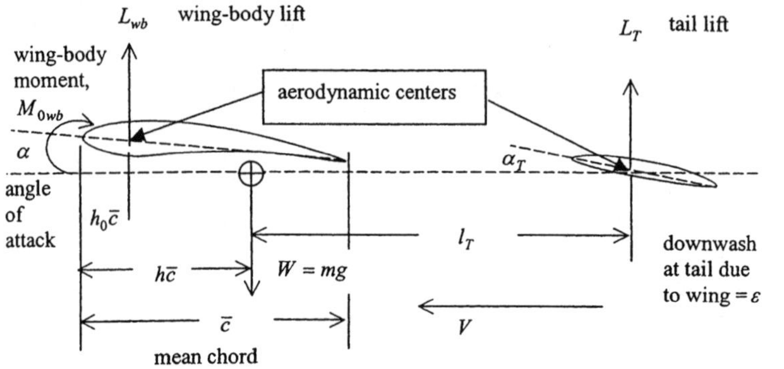

Wing: Area, S,
Lift coefficient, C_{Lwb},
Lift coefficient slope, $\dfrac{\partial C_{Lwb}}{\partial \alpha} = a$.
$M_{0wb} < 0$ due to camber

Tail: Area, S_T,
Lift coefficient, C_{LT},
Lift coefficient slope, $\dfrac{\partial C_{LT}}{\partial \alpha_T} = a_T$.

FIG. 9.4 Forces, dimensions, and definitions for an airplane longitudinal static stability study.

considered to have a camber moment. Considering the dimensions shown in Fig. 9.4, the total moment is easily expressed.

$$M_{\mathrm{G}} = M_{0\mathrm{wb}} + (h - h_0)\bar{c}\,L_{\mathrm{wb}} - l_{\mathrm{T}}L_{\mathrm{T}} \tag{9.7}$$

The lift forces in Eq. (9.7) can be replaced by expressions involving lift coefficients from Eqs. (9.5) and (9.6).

$$M_{\mathrm{G}} = M_{0\mathrm{wb}} + (h - h_0)\bar{c}\left(\frac{1}{2}\right)\rho V^2 S C_{\mathrm{Lwb}} - l_{\mathrm{T}}\left(\frac{1}{2}\right)\rho V^2 S_{\mathrm{T}} C_{\mathrm{T}} \tag{9.8}$$

At this point, the change in the net moment about the center of mass as the pitch angle of the airplane and hence the angle of attack of the wing and tail surfaces changed could be evaluated to establish a criterion for static stability. The quantities that change if the airplane pitch angle were to change would be the lift coefficients. Instead of this, it is traditional to convert Eq. (9.8) to a nondimensional form before proceeding further.

The terms in Eqs. (9.7) and (9.8) are all moments including the constant moment due to the camber of the wing and the moments given by the lift forces times their moment arms about the center of mass. Equation (9.8) is converted to nondimensional coefficient form by dividing all terms by the quantity $(\rho V^2 S\bar{c})/2$. This expression has the dimension of a moment since it consists of the dynamic pressure times an area times a length. Since the wing is the main contributor to the lift, the wing area, S, and the wing mean chord or average wing width, $\bar{c}$, are used (see Fig. 9.4).

For a straight wing, the mean chord is just the sectional chord shown in Fig. 9.1. For tapered wings, there are various ways to compute a mean chord. See any of the books referenced at the beginning of the chapter.

After dividing all terms in Eq. (9.8) by $(\rho V^2 S\bar{c})/2$, the moment coefficient version of Eq. (9.8) then becomes

$$C_{\mathrm{MG}} = C_{\mathrm{M_0}} + (h - h_0)C_{\mathrm{Lwb}} - \frac{l_{\mathrm{T}} S_{\mathrm{T}}}{\bar{c} S} C_{\mathrm{LT}} \tag{9.9}$$

where the two new moment coefficients are defined to be the original moments in Eq. (9.8) divided by $(\rho V^2 S\bar{c})/2$.

The last term in Eq. (9.9) involves the ratio of two quantities with the dimension of volume (length times area). This has been termed the "tail volume coefficient," $\bar{V}$.

$$\bar{V}(l_{\mathrm{T}} S_{\mathrm{T}})/\bar{c} S \tag{9.10}$$

The tail volume coefficient turns out to represent a major design parameter to be adjusted to insure static stability. When Eq. (9.10) is substituted into Eq. (9.9), the result is

$$C_{\mathrm{MG}} = C_{\mathrm{M_0}} + (h - h_0)C_{\mathrm{Lwb}} - \bar{V} C_{\mathrm{LT}} \tag{9.11}$$

The trim condition for the basic motion is that the lift and weight should balance and that the net moment about the center of mass should vanish. In moment coefficient form, the latter requirement is simply

$$C_{\mathrm{MG}} = 0 \tag{9.12}$$

The static stability criterion requires that the moment about the center of mass should be negative if the angle of attack or the lift should increase. In moment coefficient form, this can be stated as

$$\frac{\partial C_{\mathrm{MG}}}{\partial C_{\mathrm{L}}} < 0 \tag{9.13}$$

Now the static stability condition in the form of Eq. (9.13) will be worked out under the assumption that the tail lift is considerably smaller than the lift of the wing and body

$$L_T \ll L_{wb} \tag{9.14}$$

which is the case for most conventional airplanes.

We first take the derivative of Eq. (9.11) with respect to the angle of attack of the wing.

$$\frac{\partial C_{MG}}{\partial \alpha} = \frac{\partial C_{M_0}}{\partial \alpha} + (h - h_0)\frac{\partial C_{Lwb}}{\partial \alpha} - \overline{V}\frac{\partial C_{LT}}{\partial \alpha_T}\frac{d\alpha_T}{d\alpha} \tag{9.15}$$

The assumption that the lift is to be considered to act at the aerodynamic center means that there is a moment M_0 but that it does not change with α so

$$\frac{\partial C_{M_0}}{\partial \alpha} = 0 \tag{9.16}$$

The slopes of the lift curves in the linear regions are designated by a for the wing and a_T for the tail as indicated in Eq. (9.4) (see Figs. 9.2 and 9.4). (As noted above, these lift coefficient slopes are somewhat analogous to cornering coefficients for pneumatic tires.) With these definitions and using Eq. (9.16), Eq. (9.15) becomes

$$\frac{\partial C_{MG}}{\partial \alpha} = (h - h_0)a - \overline{V} a_T \frac{d\alpha_T}{d\alpha} \tag{9.17}$$

It has been previously mentioned that the angle of attack at the tail is affected by the downwash angle, ε, which is shown in Fig. 9.4. The downwash is due to the lift produced by the wing. In order for lift to be produced, there must be a momentum flux in a downward direction behind the wing. At the location of the tail, this downward velocity component reduces the angle of attack by the effective angle ε. When the lift increases, the downwash angle also increases so the angle of attack at the tail does not increase as much as the angle of attack at the wing. This is accounted for by assuming the following relation:

$$\frac{d\alpha_T}{d\alpha} = \left(1 - \frac{d\varepsilon}{d\alpha}\right) \tag{9.18}$$

In addition, the tail operates to some extent in the wake of the wing and body so the flow over the tail may not have the same speed as the flow over the wing. To account for this in an approximate way, an efficiency factor η_T is introduced. Equation (9.17) then becomes

$$\frac{\partial C_{MG}}{\partial \alpha} = (h - h_0)a - \overline{V}a_T\left(1 - \frac{d\varepsilon}{d\alpha}\right)\eta_T \tag{9.19}$$

Equation (9.19) leads to the most obvious form of the static longitudinal stability criterion. An airplane will be stable if when it pitches up slightly, and thus increases the angle of attack, a negative moment develops. The static stability criterion is just that the right-hand side of Eq. (9.19) should yield a negative result. Usually, however, the criterion is stated in terms of how the moment about the center of mass changes as the lift coefficient changes rather than as the angle of attack changes. Equation (9.19) can be changed to the conventional form using the reasonable approximation that for conventional airplane configurations, the lift at the tail is small compared to the lift at the wing.

If the total lift the wing contributes most of the lift,

$$L = L_{wb} + L_T \cong L_{wb} \tag{9.20}$$

then the change in lift coefficient for the complete airplane is approximately

$$dC_L \cong \frac{\partial C_{Lwb}}{\partial \alpha} d\alpha = a d\alpha \tag{9.21}$$

Upon use of Eq. (9.21), one can switch Eq. (9.19) to the final form traditionally given for the static stability criterion because

$$\frac{\partial C_{MG}}{\partial C_L} = \frac{1}{a}\frac{\partial C_{MG}}{\partial \alpha} \tag{9.22}$$

The final form for the static stability criterion involves the change in moment coefficient with a change in lift coefficient.

$$\frac{\partial C_{MG}}{\partial C_L} = (h - h_0) - \overline{V}\left(\frac{a_T}{a}\right)\left(1 - \frac{d\varepsilon}{d\alpha}\right)\eta_T < 0 \tag{9.23}$$

Although Eq. (9.23) may appear to be complicated, it merely states that a plot of the moment coefficient as a function of the lift coefficient must have a negative slope if the aircraft is to be statically stable and gives a formula for the slope in terms of airplane parameters. This situation is shown in Fig. 9.5.

The plot in Fig. 9.5 demonstrates that a statically stable airplane must have a positive moment coefficient (and moment) if it is oriented with respect to the wind in such a way that the total lift is zero. This is necessary

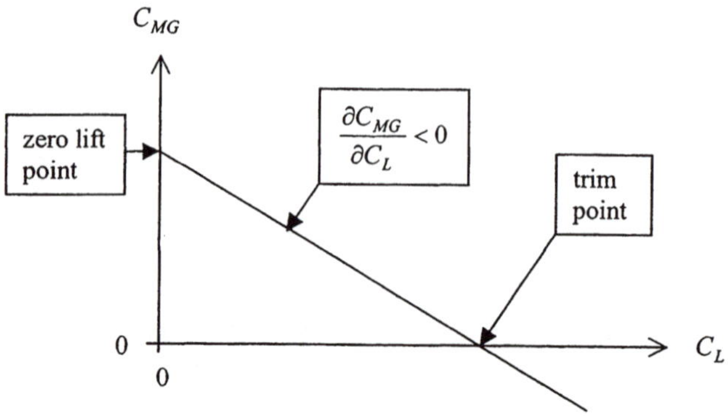

FIG. 9.5 Plot of moment coefficient vs. lift coefficient for a statically stable aircraft.

for two related reasons. First, the slope of the line relating the moment coefficient to the lift coefficient must be negative according to the criterion, Eq. (9.23), if the airplane is to be statically stable.

Second, the trim point, at which the moment coefficient is zero, must occur at a positive lift coefficient so that the lift can counteract the weight. The plot in Fig. 9.4 shows the only way that these two conditions can be met.

Considering Eq. (9.23) and looking at the dimensions and definitions in Fig. 9.3, a number of factors that influence stability can be appreciated. First, a reduction in h means that the center of mass has been moved forward. This reduces the positive term $h - h_0$ in Eq. (9.23) and makes the slope in Fig. 9.4 more negative and the airplane more stable. Shifting the center of mass toward the rear increases $h - h_0$ and makes slope less negative and the plane less stable.

A large tail far to the rear yields a large volume coefficient $\overline{V}$ and, because of the negative sign on the term involving $\overline{V}$, makes the slope in Fig. 9.4 more negative and increased stability. Roughly speaking, one can achieve static stability by having the center of mass far enough forward and having a large enough horizontal tail far enough to the rear.

Modern airplanes can carry loads that are a significant fraction of the weight of the airplane itself. This means that, when heavily loaded, an airplane's center of mass location depends importantly on how the load is distributed in the airplane. Aircraft manufacturers have devised a number of means to assure that the center of mass of a loaded airplane is not far enough rearward to cause the airplane to lose static stability. The Wright brothers could not have realized that when they mounted their motor and pusher propellers far to the rear of their 1903 Flyer that they had made the airplane unstable and harder to fly than necessary.

A. Parameter Estimation

Some of the parameters in Eq. (9.23) dealing with lengths and areas are fairly easy to estimate. Others can be estimated at least roughly from theoretical results. There are a number of results for wings of elliptically shaped wings that give at least a reasonable idea of the magnitude of some of the parameters for wings of other plan shapes.

An important shape parameter for a wing is its aspect ratio, A. For a rectangular wing of span (or length) b and chord (or width) c, the aspect ratio is simply

$$A = b/c \tag{9.24}$$

For other shapes, the aspect ratio can be defined in terms of the area S and the span b or the mean chord $\bar{c}$

$$A = b^2/S \quad \text{or} \quad A = S/\bar{c}^2 \tag{9.25}$$

For an elliptical wing shape, thin wing theory can predict the slope of the lift coefficient curve with respect to the angle of attack as a function of the aspect ratio.

$$\partial C_L/\partial\alpha = a = 2\pi A/(A+2) \tag{9.26}$$

and this theoretical result can serve a guide for other shape wings if their shape is not too different from an elliptical shape. It is of interest that for very long thin wings such as are found in sailplanes with $A \gg 2$, the slope $a \rightarrow 2\pi$ according to Eq. (9.26). The slope of the sectional lift coefficient for a wing section in pure two-dimensional flow is exactly 2π. For a high aspect ratio wing, most of the wing operates in an essentially two-dimensional flow as if it was a wing section in a wind tunnel. Only near the wing tips does the wing operate in the three-dimensional flow field.

High aspect ratio wings are particularly efficient at generating lift with low values of the so-called induced drag. The induced drag actually results from a tilting of the total lift vector toward the rear. This component of drag has nothing to do with fluid friction but does have to do with the portion of the wing in the three-dimensional flow near the wing tips. The induced drag does, however, have to be counteracted by the power plant for steady flight so that a low value of induced drag is favorable for airplane efficiency.

In addition, the downwash angle for the flow at the wing location for an elliptically shaped wing is predicted to be

$$\varepsilon_W = C_L/\pi A \tag{9.27}$$

If the tail is fairly far behind the wing, the downwash angle at the tail is theoretically about twice the downwash angle at the wing

$$\varepsilon \cong 2\varepsilon_W \tag{9.28}$$

Again, this theoretical value for the downwash angle for an elliptical wing can be used to make a rough estimate of the downwash at the tail for other wing shapes.

Combining Eqs. (9.26)–(9.28), an important term in the stability criterion, Eq. (9.23), can be estimated,

$$d\varepsilon/d\alpha \cong 4/(A+2) \tag{9.29}$$

For an aspect ratio of 6, for example, the result is $d\varepsilon/d\alpha \cong 1/2$ which represents a significant reduction in the effectiveness of the tail in stabilizing the aircraft. In this case, the stabilizing term in Eq. (9.23) involving the tail volume coefficient is reduced by half from the value it would have if the downwash effect was not considered. In other words, one could make a mistake of a factor of two in sizing the stabilizer area if the downwash was not taken into account.

Some of the remaining terms in Eq. (9.23) are fairly easy to estimate. The term a_T/a may not be much different from unity if the wing and the horizontal tail have similar aspect ratios, and the term η_T may not be much smaller than 1, perhaps 0.9 for a typical airplane.

Note that the stability criterion, Eq. (9.23), must be considered together with the trim condition. It is always possible to make the C_{MG}–C_L line slope downward and to satisfy the stability criterion by moving the center of mass sufficiently forward. However, if the moment at zero lift is not positive, it will not be possible to trim the aircraft at a positive lift coefficient so stable flight will not be possible. This should be clear from the plot in Fig. 9.5.

The type of camber shown for the wings in Figs. 9.1 and 9.4 is good for generating positive lift with minimal drag, but wings with this sense of camber have a negative moment at zero lift. This means that a flying wing (with no tail) using this type of camber cannot be stable and also trim at a positive lift coefficient. That is, if the moment coefficient at the zero lift point is not positive, the negative slope of the moment coefficient vs. lift coefficient required for stability cannot yield a trim point at a positive lift coefficient.

A flying wing with opposite camber could be stable if the center of mass were ahead of the aerodynamic center because such a wing would have a positive moment at zero lift as well as a negative slope for the curve in Fig. 9.5. It would not be a very efficient wing, however, because it would have a large drag coefficient when operating at the trim point. This points out the difficulties in designing an efficient flying wing. Actual flying wing designs typically have swept back wings such that the wing tips themselves act essentially as a stabilizer or they are electronically stabilized using automatically controlled flaps at the back of the wing.

The horizontal tail takes care of the problem of the negative moment associated with cambered wings for conventional aircraft and that is the reason it is called a stabilizer. At the zero lift orientation to the wind, the wing generates positive lift and the tail generates negative lift of equal magnitude. Together, they form a positive couple bigger than the negative moment from the cambered wing. If then the static stability criterion is satisfied, stable flight is possible.

IV. THE PHUGOID MODE

Although the discussion of aircraft stability here is focused on static stability, i.e., the initial tendency of the airplane to return to an initial equilibrium state after a disturbance, it is important to note that dynamic stability is also important. A statically stable aircraft may not be stable dynamically if it starts to return to a basic motion but then overshoots and proceeds to develop an oscillation of increasing amplitude.

Longitudinal dynamic stability has to do with pitch and horizontal and vertical translation. It is generally possible to consider three distinct dynamic modes of longitudinal motion, although for a particular airplane, the actual motions are coupled to an extent. The *first mode* or the *long-period* mode is also called the *phugoid mode* and will be analyzed in this section. This mode of motion typically has a very long period and it may be either stable or unstable for real airplanes. Although the existence of the mode is easily demonstrated experimentally, it is normally not of great importance whether it is stable or unstable because the oscillation is so slow that a human pilot or an autopilot can easily control it.

The *second mode* or *short-period mode* has a more direct connection with static stability. Generally, if an airplane has static stability, and if the controls are held fixed, there can exist a pitch oscillation with a period of perhaps 0.5–5 sec and with sufficient pitch damping to be quite stable. This mode exists at essentially constant speed. If the controls are free to move, the so-called stick-free case, then it is possible that the pitch damping could be negative and the airplane could be dynamically unstable although statically stable. This motion is also called *porpoising* and is potentially dangerous.

A *third mode* with a very short period can exist in the elevator free case. The elevator is the flap located at the rear of the horizontal tail. The pilot uses the elevator to change the pitch angle of the airplane. The third mode consists essentially of a flapping of the elevator about its hinge line. If the elevator has been designed properly, this mode is well damped and of no great importance from a safety point of view.

Although it is clear that the second mode is the most important mode dynamically, it would take too long to discuss all the design issues connected with assuring that a real statically stable aircraft is also dynamically stable. In fact, the succeeding section will show that at least for a very elementary airplane model, a statically stable airplane is predicted to be also dynamically stable. Therefore here we will only discuss the easily analyzed phugoid mode, which, incidentally, can be easily demonstrated in a test flight.

The basic motion for the phugoid mode is a straight and level flight path at a nominal altitude that will be called $h = 0$ with a constant speed V_0. The perturbed motion involves variations in pitch attitude, altitude, and airspeed but with nearly constant angle of attack. The angle of the flight path with the horizontal, θ, is assumed to be small. The sketches in Fig. 9.6 show the situation. Note that in the lower part of Fig. 9.6, the angle of the flight path θ is exaggerated. In the analysis, it will be assumed to be small.

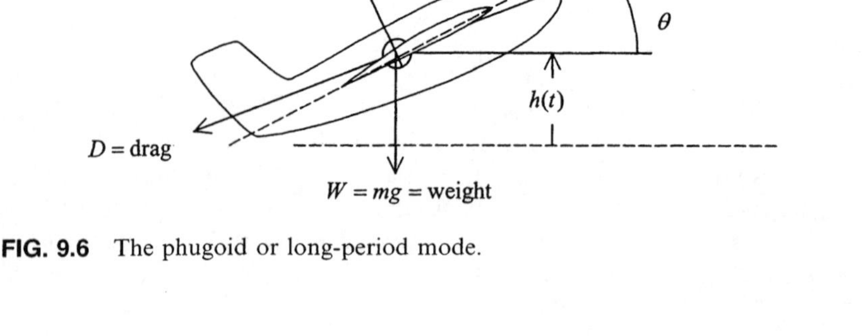

FIG. 9.6 The phugoid or long-period mode.

The phugoid mode essentially involves the conservation of the sum of potential and kinetic energy. We assume that the drag and the thrust, which act along the flight path direction, cancel each other out, $D = T$. Furthermore, because of the small flight path angle, the lift and the weight are nearly in the same direction and, for the basic motion, when the velocity $V = V_0$, they also cancel each other out,

$$L \cong W = mg \tag{9.30}$$

The lift for the entire airplane can be written as

$$L = (\rho V^2/2)C_L S_{ref} \tag{9.31}$$

where ρ is the air density, C_L is the lift coefficient, and S_{ref} is a reference area such as the wing area.

The main assumption is that the lift coefficient is constant for the perturbed motion. The lift coefficient depends primarily on the angle of attack, which is an angle between a line in the airplane (such as a line between the leading and trailing edges of the wing) and the flight path. Thus the assumption is that this angle changes little during the phugoid oscillation. The consequence of this assumption is that the lift is a constant times the square of the velocity.

The lift force is always perpendicular to the flight path (and velocity vector of the airplane) so the lift force does no work. This fact together with the other assumptions means that the total energy is constant. Remember that, for the basic motion, the speed is V_0 and the height h is zero. The conservation of energy leads to a relation between the speed for the perturbed motion in terms of the speed in the basic motion.

$$\frac{1}{2}mV^2 + mgh = \frac{1}{2}mV_0^2 \text{ or } V^2 = V_0^2 - 2gh \tag{9.32}$$

Since the lift is assumed to be proportional to the velocity squared and for the basic motion,

$$L = mg \text{ when } V = V_0 \tag{9.33}$$

the lift can be written in the following form:

$$L = mg(V/V_0)^2 = (mg/V_0^2)V^2 \tag{9.34}$$

Combining Eq. (9.34) with Eq. (9.32), the lift is

$$L = (mg/V_0^2)(V_0^2 - 2gh) = mg(1 - 2gh/V_0^2) \tag{9.35}$$

Now writing Newton's law in the vertical direction,

$$m\ddot{h} = L - mg \tag{9.36}$$

the final result upon use of Eq. (9.35) is

$$\ddot{h} + (2g^2/V_0^2)h = 0 \tag{9.37}$$

This equation represents an undamped oscillator with a natural frequency of

$$\omega_{\mathrm{n}} = \sqrt{2}(g/V_0) \tag{9.38}$$

and a period

$$T = 2\pi/\omega_{\mathrm{n}} = \sqrt{2}(\pi V_0/g) \tag{9.39}$$

The periods for practical cases are quite long. For example, for a light plane flying at 100 MPH or 44.7 m/sec, the period is 20.27 sec. At 400 MPH, the period rises to more than 80 sec.

It is harder to predict whether the phugoid mode will be stable or not but it is unlikely to be violently unstable. Because the periods are so long, pilots are often unaware of lightly unstable phugoid modes since they often make small corrections for disturbances in any case and these corrections also incidentally eliminate any growth in the phugoid mode. The low frequency of the phugoid mode also makes it easy for an autopilot to control even if the phugoid mode is unstable.

One advantage of the phugoid mode is that it is easy to demonstrate in flight. The demonstration is more interesting if the mode happens to be lightly unstable, but even if the mode is stable, it is usually possible to experience a number of oscillations before the amplitude decreases significantly. A measurement of the period will confirm that the analysis given above is essentially correct.

V. DYNAMIC STABILITY CONSIDERATIONS—COMPARISON OF WHEELS AND WINGS

For automobiles, we have written linearized dynamic equations and found that only oversteering cars can exhibit instability for lateral motion and then only when traveling at speeds greater than a critical speed. The concept of static stability is not very useful for automobiles since it has been shown that any car whatsoever is dynamically stable at low-enough speeds. In studying aircraft longitudinal stability, the concept of static stability has been found to be considerably more useful than it is for automobiles. A statically unstable airplane is basically also dynamically unstable at all speeds. Critical speeds for aircraft relate more to aeroelastic instabilities such as flutter or divergence rather than to rigid body motion. The reason why cars are analyzed for stability dynamically and airplanes are analyzed statically is connected with the different ways wheels and wings generate forces.

It was pointed out previously that there is an analogy between the manner in which pneumatically tired wheels produce lateral forces and the manner in which wings produce vertical forces. The obvious analogy between the roles of slip angle and angle of attack has been mentioned. One would think that the stability analyses for ground vehicles and for aircraft might be almost identical yet in practice they are not. This paradox can be explained by examining the way in which the analogy breaks down.

We now consider simplified models of force generation for wheels and wings to point out the essential differences between them. Fig. 9.7 recapitulates the relation between lateral force and slip angle as discussed in Chapter 4. The use of a cornering coefficient is, of course, restricted to the case in which the slip angle is fairly small.

Note that the relation between side force F_y and the lateral velocity V becomes weaker as the forward speed U increases. At low forward speeds, the wheel appears to be very stiffly connected to the ground, but as the speed U rises, the relation between F_y and V decreases. One can even think of the relation between lateral force and lateral velocity for a wheel with a pneumatic tire as a dashpot relation with the dashpot coefficient being (C_α/U). At low speeds, an applied force causes a small lateral velocity, but at high speeds, the same force causes a large velocity response.

Fig. 9.8 shows how the lift force on a wing is related to the vertical velocity, W. (Note that W is a velocity component following the conventions of Chapter 2 and not the weight as in Fig. 9.5.) When the angle of

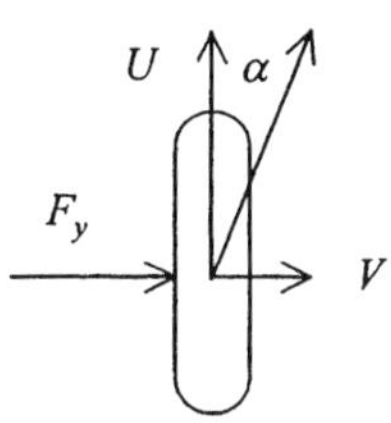

Assume $V \ll U$

$\alpha \cong V/U$, $F_y \cong C_\alpha \alpha$,

$$F_y = \left(\frac{C_\alpha}{U}\right)V$$

FIG. 9.7 Lateral force generation for a pneumatic tire.

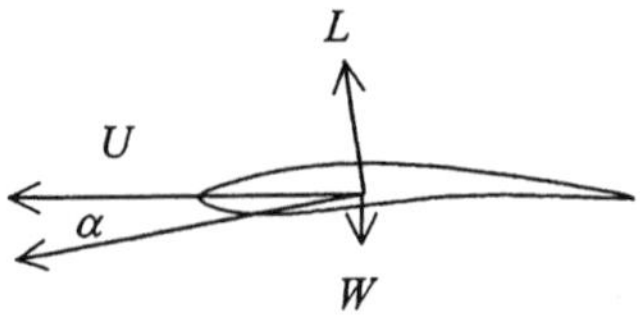

Assume $W << U$

$$\alpha \cong W/U, \quad L = \frac{1}{2}\rho U^2 S \frac{\partial C_L}{\partial \alpha}\alpha,$$

$$L = \left(\frac{1}{2}\rho \frac{\partial C_L}{\partial \alpha} SU\right)W = \left(\frac{1}{2}\rho a SU\right)W$$

FIG. 9.8 Lift force generation for a wing.

attack α is small, the lift force is substantially parallel to the vertical velocity.

In this case, the equations shown in Fig. 9.8 demonstrate that the relation between the lift (vertical force) and the vertical velocity becomes stronger as U increases. At high forward speed, the wing appears to be stiffly connected to the air. As before, a is the slope of the lift coefficient curve, assumed to be constant.

Again, one could think of the relation between the lift force and the vertical velocity as a dashpot relation but now with a coefficient ($\rho a SU/2$) that becomes larger with forward speed rather than smaller as was the case for a wheel. (Figs. 9.7 and 9.8 look a little different because L represents the force from the air on the wing while F_y represents an external force on the wheel not the force on the tire from the road. In each figure, we are considering an equilibrium situation in which there are two equal and opposite forces.)

Static stability analysis for a vehicle simply considers the moment on the vehicle as attitude of the vehicle is varied away from a trimmed condition. If the direction of the moment about the vehicle center of mass is such that the attitude would tend to return to the attitude in the trimmed condition, the vehicle is statically stable. If the moment developed is in the opposite direction, the vehicle is statically unstable. For an airplane, this analysis seems obviously valid for all speeds, but for a car, oddly enough, static instability does not mean dynamic instability except for forward speeds above a critical speed for an oversteer car.

Consider a static stability analysis for an elementary car model as shown in Fig. 9.9.

At the instant shown in Fig. 9.9, the car is supposed to be moving in the nominal path direction with the velocity U but with the car centerline inclined with respect to the direction of U at the small yaw angle, α.

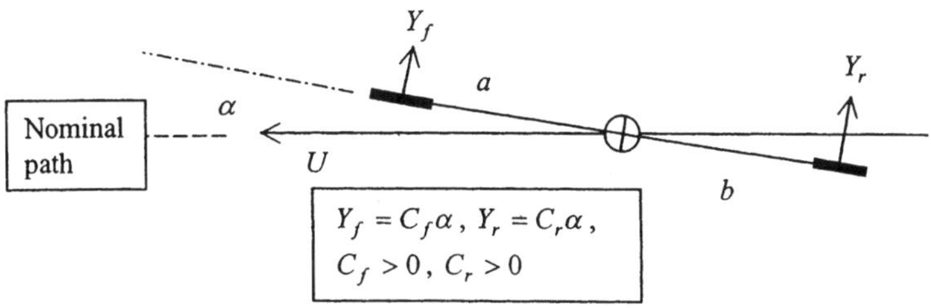

FIG. 9.9 Static stability analysis of an automobile.

This implies that the slip angle at both axles at the instant shown in Fig. 9.8 is α. The question asked in a static stability analysis is whether or not the induced forces produce a moment that will tend to reduce α. First, we compute the moment about the mass center in the direction of positive α.

$$M_c = Y_f a - Y_r b = (C_f a - C_r b)\alpha \tag{9.40}$$

A positive value for M_c means that α will tend to increase. This will be the case if $(C_f a - C_r b) > 0$ which implies the car is statically unstable. On the other hand, if $(C_f a - C_r b) < 0$, then M_c will be negative and α will tend to decrease. In this case, the car is statically stable.

From our previous dynamic analysis in Chapter 6, we know that the statically stable case corresponds to what we called understeer and understeer cars have been shown to be dynamically stable. It might seem then that static stability analysis, which is far simpler than a dynamic stability analysis, could be used for cars as well as airplanes.

The problem is that the statically unstable case corresponds to the oversteer case, but an oversteer vehicle is actually unstable only for speeds higher than the critical speed. How can this be?

Imagine dropping the car shown in Fig. 9.9 on the ground at a small angle α_0 relative to the velocity along a nominal path. Depending upon the speed and static stability, the angle α may increase or decrease as time goes on. Fig. 9.10 shows some of the paths that could be followed by various cars traveling at different speeds.

The cases indicated can be explained as follows.

1. Statically stable or understeer vehicle. The yaw rate $\dot{\alpha} < 0$ for all speeds for an understeer car so α decreases initially although the rate of decrease eventually approaches zero. The car has a lateral

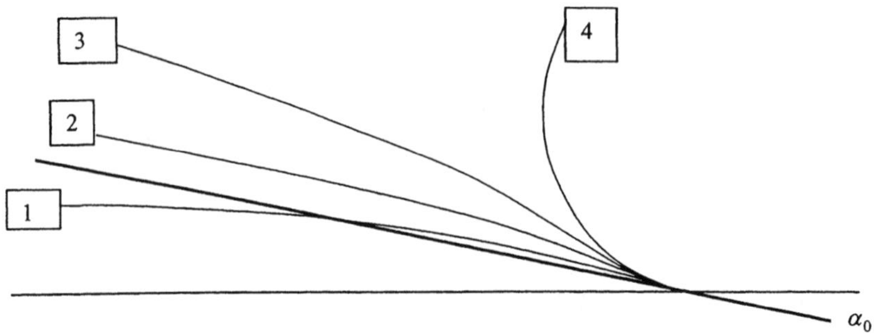

FIG. 9.10 Paths taken by a car after being dropped on the ground at an initial angle to the nominal path.

acceleration initially, which also gradually approaches zero. The car finally moves off at an angle less than α_0.

2. Neutrally stable or neutral steer vehicle. The yaw rate $\alpha = 0$ for all speeds. The neutral car also has an initial lateral acceleration that gradually diminishes. The car finally moves off parallel to the α_0 line.
3. Statically unstable or oversteer vehicle traveling at a speed less than the critical speed. Although $\dot{\alpha} > 0$ initially, and thus α increases for a time, the increase in α slows and gradually approaches zero. The car is dynamically stable and finally moves off at an angle greater than α_0.
4. Statically unstable or oversteer vehicle with a speed greater than the critical speed. The yaw rate $\dot{\alpha} > 0$ initially and increases in time because the car is dynamically unstable. Although the stability equations cease to be valid after a time, it is clear that the car will eventually spin out.

For the dynamically stable cases, $\dot{\alpha} \to 0$ eventually regardless of the sign of α initially. Only for the oversteering car traveling above the critical speed does α remain positive as the car begins to spin. For the airplane, we will show that there is no possibility of a critical speed below which a statically unstable airplane can be dynamically stable.

A. An Elementary Dynamic Stability Analysis of an Airplane

Consider now a car-type dynamic stability analysis of a very elementary airplane model. We consider forces and moments on the wing and horizontal tail but neglect forces and moments from all other sources. We also neglect downwash effects on the tail from the wing since these effects can be taken into account by reducing the effective area of the horizontal tail [see Eq. (9.23)]. The model is shown in Fig. 9.11.

The basic motion has a zero vertical velocity, $W=0$, a constant forward velocity, $U=\text{constant}$, and a zero pitch rate, $q=0$. This corresponds to straight and level flight. For the perturbed motion, the variables q and W are assumed to be small.

If L_0 and L_t are the lift forces at the wing and tail for the basic motion, the trim conditions state that the weight must be balanced by the wing and tail lift forces and the net moment about the center of mass must vanish.

$$mg = L_0 + L_{t0}, \quad M_0 + L_0 x_a = L_{t0} l_t \tag{9.41}$$

The angles of attack due to a perturbation from the basic motion at the wing and at the tail, neglecting downwash, are due to the vertical velocity of the center of mass and the pitch rate. This calculation is very similar to the calculation made in Chapter 6 for the front and rear slip angles of an elementary car model [see Fig. 6.6 and Eq. (6.20)].

$$\alpha \cong (W - x_a q)/U, \quad \alpha_t \cong (W + l_t q)/U \tag{9.42}$$

The expression for lift at the wing has a steady part that must be present when the airplane is in the trim condition and a variable due to the perturbation variables. Using Eq. (9.42), the result can be expressed as

$$L = L_0 + \frac{1}{2}\rho S U^2 \frac{\partial C_a}{\partial \alpha}\alpha = L_0 + CU(W - x_a q) \tag{9.43}$$

where a new coefficient has been defined,

$$C = \frac{1}{2}\rho S \frac{\partial C_\alpha}{\partial \alpha} \tag{9.44}$$

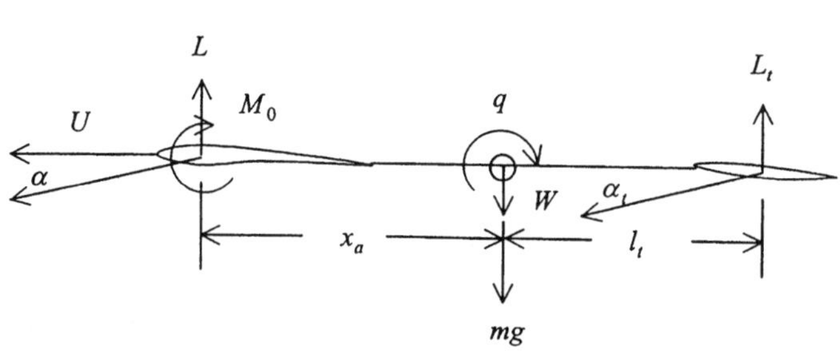

FIG. 9.11 Elementary dynamic stability analysis of an airplane.

In a similar way, the tail lift can be described by the expression

$$L_t = L_{t0} + C_t U(W + l_t q) \tag{9.45}$$

in which another coefficient C_t has been defined in a manner similar to C in Eq. (9.44).

Using Eqs. (9.43) and (9.45), and the body-centered coordinate system discussed in Chapter 2, the equations of motion are then as follows:

$$m(\dot{W} - Uq) = mg - L_0 - L_{t0} - CU(W - x_a q) - C_t U(W + l_t q) \tag{9.46}$$

$$I_{yy}\dot{q} = L_0 x_a + M_0 - L_{t0} l_t + x_a CU(W - x_a q) - l_t C_t U(W + l_t q) \tag{9.47}$$

These simplify somewhat using the trim conditions, Eq. (9.41). In final form, the two equations of motion take the form

$$m\dot{W} + U(C + C_t)W - mUq - U(Cx_a - C_t l_t)q = 0 \tag{9.48}$$

$$I_{yy}\dot{q} - U(Cx_a - C_t l_t)W + U(Cx_a^2 + C_t l_t^2)q = 0 \tag{9.49}$$

The corresponding second-order characteristic equation is

$$\begin{aligned} &mI_{yy}s^2 + U[m(Cx_a^2 + C_t l_t^2) + I_{yy}(C + C_t)]s + U^2[(C + C_t) \\ &\quad \times (Cx_a^2 + C_t l_t^2) - m(Cx_a - C_t l_t) - (Cx_a - C_t l_t)^2] = 0 \end{aligned} \tag{9.50}$$

Note that $C > 0$ and $C_t > 0$ so the coefficients of s^2 and s are positive. This means that the last term in Eq. (9.50) determines stability. After some simplification, this term can be expressed as

$$U^2\left\{CC_t(x_a + l_t)^2 - m(Cx_a - C_t l_t)\right\} \tag{9.51}$$

If the entire term in Eq. (9.51) is positive, then the characteristic equation, Eq. (9.50), will have eigenvalues with negative real parts and the airplane will be dynamically stable.

The important factor in Eq. (9.51) for the airplane is

$$Cx_a - C_t l_t = \frac{1}{2}\rho\left(S\frac{\partial C_\alpha}{\partial \alpha}x_a - S_t\frac{\partial C_{\alpha t}}{\partial \alpha_t}l_t\right) \tag{9.52}$$

If the factor in Eq. (9.52) is negative [noting the negative sign associated with this factor in Eq. (9.51)], then the simplified airplane model is statically as well as dynamically stable at all speeds. In fact, the factor could even be slightly positive meaning that the airplane would be statically unstable but the airplane could still be dynamically stable. This is because there is another inherently positive term that always exists in Eq. (9.51).

The stability does not change with speed since the square of the speed simply enters as a multiplier in Eq. (9.51) and has no effect on the sign of the complete expression.

For the automobile model studied previously in Chapter 6, the term equivalent to Eq. (9.51) appeared in Eqs. (6.12)–(6.14),

$$C_f C_r (a+b)^2 / U^2 - m(C_f a - C_r b) \tag{9.53}$$

There are strongly analogous terms in the two expressions in Eqs. (9.51) and (9.53), but the speed U enters entirely differently due to the different mechanisms of force generation for wings and wheels noted previously. The expression in Eq. (9.53) will always be positive for low-enough speeds regardless of whether the term $C_f a - C_r b$ is positive or negative, i. e., whether the car is oversteer or understeer. The expression Eq. (9.53) can produce a critical speed for the oversteer case above which the dynamic changes, but the equivalent expression for the airplane, Eq. (9.51), never changes sign with speed.

Thus the use of static stability for airplanes is conservative according to this simplified analysis since even a slightly statically unstable airplane could be dynamically stable. In any case, the dynamic stability does not change with speed as it clearly does for automobiles. For the car, a static stability analysis by itself is not very useful because it does not necessarily predict dynamic stability.

Appendix A shows bond graph representations of the car and airplane models used in this section. The rigid body bond graphs from Chapter 2 are used, and the junction structures that connect the force-producing devices with the rigid body dynamics are clearly analogous. The crucial difference between the force-generation properties of wings and wheels is clearly shown in the bond graph resistance parameters for the two cases.

VI. THE EFFECT OF ELEVATOR POSITION ON TRIM CONDITIONS

The flaps on the trailing edge of the horizontal tails of conventional airplanes are called elevators. The purpose of elevators is to change the lift force at the horizontal tail. By changing the lift force at the tail, elevators influence the pitch angle of an aircraft much as steering the wheels of a car produces changes in the yaw angle. Elevators are controlled by a human pilot or by an autopilot, and while they directly affect the pitch angle of the aircraft, they indirectly affect the trim conditions for straight and level flight.

A pilot who wishes to slow down pulls back on the stick or wheel. This raises the elevator, raises the nose of the plane, increases the lift coefficient, and allows the lift to counteract the weight at a lower speed. Of course, the thrust must also be adjusted to achieve a steady flight condition. Conversely, pushing the wheel ahead lowers the elevator, lowers the nose, lowers the lift coefficient, and allows trim at a higher speed. The speed range is limited at the lower end by stalling of the wing, which limits the magnitude of the lift coefficient at high angles of attack, and at the high end by the limit to the available thrust.

The simplified sketch of Fig. 9.12 shows the forces acting on an airplane and the definition of the elevator angle, η. The wing moment due to camber is actually in the nose-down direction for the camber as shown, although moments are considered to be positive in the nose-up direction. The lift forces are assumed to act at the aerodynamic centers of the wing and horizontal tail. The wing provides most of the lift to counteract the weight in any case, and, depending on the flight condition, the tail lift may be either positive or negative. For this reason, the horizontal tail does not usually have a camber.

In Figs. 9.13 and 9.14, there are sketches of statically stable and statically unstable airplanes in attitudes such that the total lift is zero as well as in a trim condition in which the total lift equals the weight. As noted in the discussion of Fig. 9.5, a stable aircraft will have a positive (nose up) moment when at a zero lift attitude. The negative slope of the $C_{MG}-C_L$ line then allows trim at a positive lift coefficient. An unstable aircraft will have a negative moment at a zero lift attitude, and the positive slope of the $C_{MG}-C_L$ line also allows trim at a positive lift coefficient.

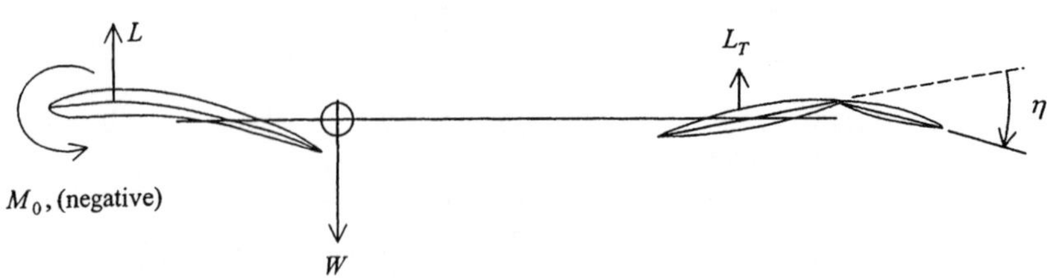

FIG. 9.12 Sketch showing elevator action.

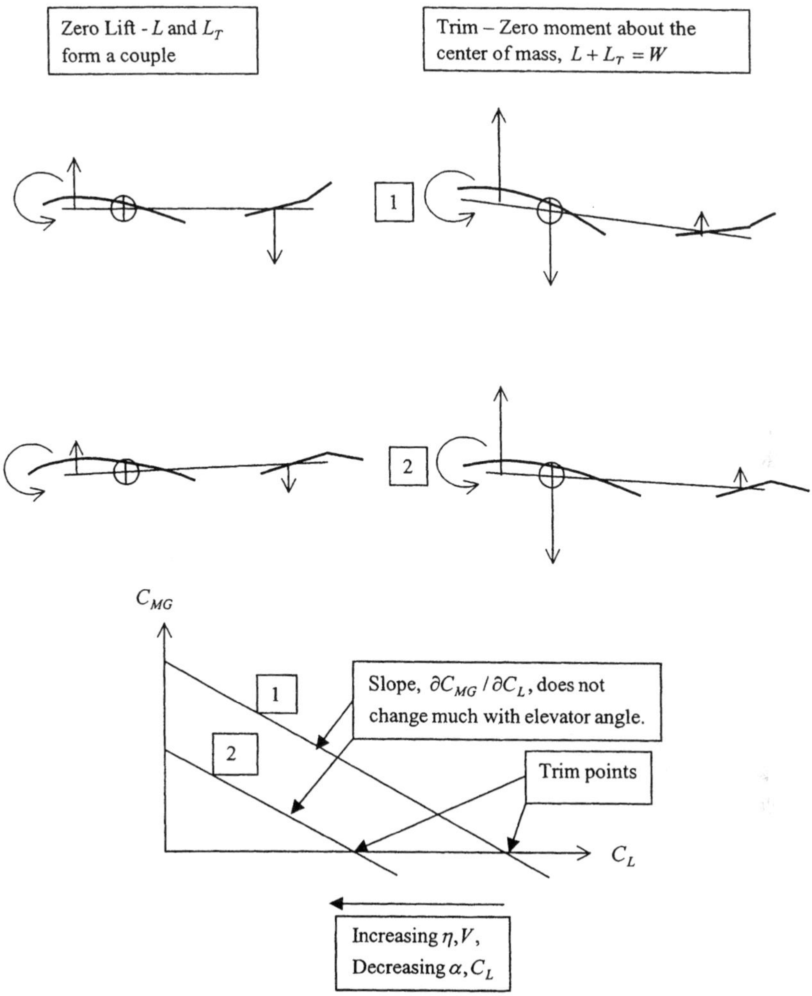

FIG. 9.13 Elevator action for a stable airplane.

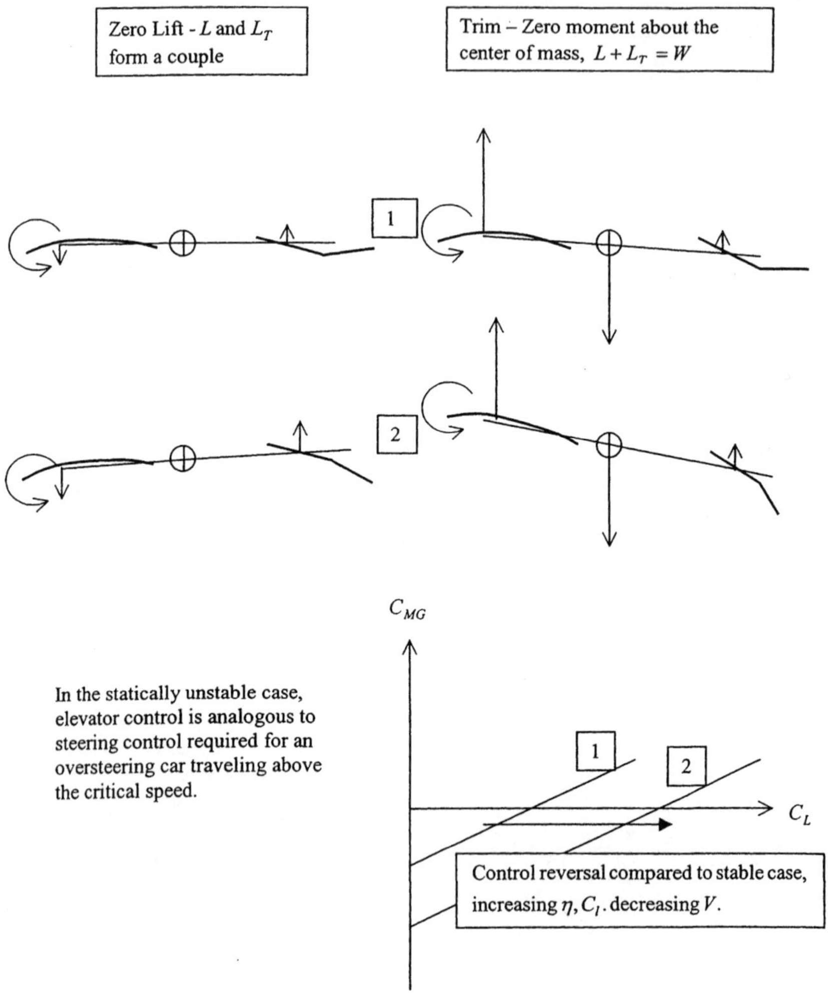

FIG. 9.14 Elevator action for an unstable airplane.

For a stable airplane, increasing the elevator angle leads to a higher speed at trim and a lower lift coefficient. The reverse is true for an unstable airplane. This is another instance of reverse action.

The pilot must not only continually manipulate the elevator to stabilize an unstable aircraft, but also, to change speed, he must find a steady elevator angle in the opposite direction than he would if the airplane was stable. This should remind one of the case of an oversteering car operating above the critical speed. Such a car is not only unstable, but also requires countersteering in a steady turn; that is, a steady right turn requires a left steer angle. This is very obvious to the driver of an open-wheeled dirt track racer.

In the case of the airplane, the reverse action may not be particularly noticeable to the pilot. The pilot cannot see the elevator angle directly, and it may not be obvious that in changing speed, the average position for the elevator does not change in the direction it would for a stable aircraft. The pilot would certainly be more aware of the need to continually manipulate the elevator control to keep the pitch attitude from diverging.

10
Rail Vehicle Dynamics

I. INTRODUCTION

The first example of a stability analysis of a vehicle in Chapter 1 concerned a rail vehicle wheelset. The idea was to show that conventional railway wheels are tapered and attached to a rigid axle to stabilize their motion as they move along a track. The results indicated that properly tapered wheels impart a self-steering effect to wheelsets that is absent for cylindrical wheels. This self-steering effect allows the tapered wheelset to steer itself toward the center of the track if it has become displaced and largely prevents contact between the flanges and the rails. Furthermore, if the wheels have a negative taper angle, the wheel set was shown to be unstable in the sense that any perturbation from a perfectly centered condition will grow until the flanges begin to contact the sides of the rails.

The analysis in Chapter 1 assumed that friction between the wheels and the rails prevented any relative motion between the wheel's contact points and the rails. Although the problem was formulated in the form of a differential equation in time, the model was not really dynamic but rather kinematic. The differential equation describing the wheelset motion can quite easily be transformed from an equation with time as the independent variable to one with distance along the track as the independent variable. The sinusoidal motion for the stable wheelset is then a function of distance, and the speed of the motion does not enter the formulation at all.

This result should come as something of a surprise because several of the vehicle stability analyses in succeeding chapters have discovered critical speeds above which a stable vehicle changes to an unstable one. One would suspect that this might be true for rail vehicles also.

In fact, the tapered wheelset with the taper of the correct sense was not shown in Chapter 1 to be strictly stable, but rather just not unstable. The predicted path of the wheelset was sinusoidal with amplitude that neither grew nor decayed in time or distance along the track. To anyone with experience in stability analysis, this sort of result is not satisfying. Any tiny effect neglected in the mathematical model might actually cause the sinusoidal motion to either grow slightly or decay slightly. Thus the analysis really discovered that the wrong sense of taper caused instability but it did not really prove that the tapered wheelset with the correct taper was really stable.

Mathematicians sometimes call the situation in which the solutions for a differential equation change character completely for infinitesimal changes in a parameter a case of "structural" or "parametric" instability. In the case of the wheelset in Chapter 1, Eq. (1.6) for the deviation from the center, y, was second order but with a coefficient of y of exactly zero. If this coefficient had a value, however small, that was positive, the conclusions about the wheelset would be that its motion was stable. A small negative value for the coefficient would have lead to the conclusion that the wheelset was unstable.

This structural instability for the wheelset model is unsatisfactory. It means that the model really cannot predict whether a tapered wheelset would actually be stable. The fact that the forward speed does not really enter the problem also suggests that the model is oversimplified and is incapable of describing problems that might be expected at high speeds. Some of these defects in the model will be remedied in this chapter.

It is known that conventional railway vehicles are limited in speed by the occurrence of lateral oscillations. The phenomenon is called *hunting*. This is an unstable behavior that only happens above a critical speed and, in certain cases, could be severe enough to cause derailment. A truly dynamic analysis is required to understand the hunting phenomenon.

Critical speeds for complete rail vehicles are determined by many vehicle parameters because railway vehicles typically have a number of subsystems such as wheelsets, trucks on which the wheelsets are mounted, and bodies mounted on spring–damper suspension units supported by the trucks. The bodies are typically long and thin and may have low-frequency bending modes that influence the dynamics in general and critical speeds in particular. The analysis of such vehicles is very complex and typically requires the use of special-purpose numerical simulation programs far too complex to be presented here. See Cooperider (1969), Garg and Dukkipate (1984), and Kortuem and Lugner (1994).

On the other hand, simplified and linearized models of rail vehicles can give insight into the hunting phenomenon and can illustrate both similarities and differences to unstable behavior of other vehicles. Here we will explain the simplest model presented in the paper by Cooperider (1969). Although only a model of a wheelset alone attached to a truck with a specified motion will be presented here, the paper also presents similar results for a truck carrying two wheelsets as well as a much more complex model of a complete railcar. The model of the two-wheelset truck appears in the problem set for this chapter, Problem 10.3.

II. MODELING A WHEELSET

Fig. 10.1 shows a single wheelset rolling on rails in a symmetrical equilibrium position. This position relates to the basic motion when the wheelset rolls down the rails at constant velocity. The dynamics of the situation when the wheelset is not exactly centered and has a small yaw angle with respect to the rails will be subsequently studied.

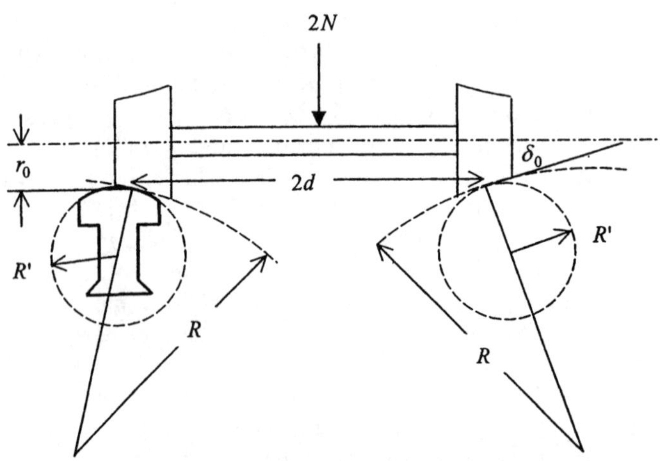

FIG. 10.1 Model of a single railway wheelset.

The wheels are tapered with a contact angle of δ_0 when the wheelset is centered on the rails as shown. The angle δ_0 is assumed to be small. In this model, the taper angle is not constant because the wheel profile is assumed to be hollow with a radius R. (Whether this sort of profile is intended or not, wheels tend to wear into this sort of hollow profile.) Similarly, the railhead profile is assumed to be approximately circular with radius R'. This means that the contact angle between the wheel and the rail will vary for the perturbed motion when the wheelset is not perfectly centered. It will be assumed that $R > R'$. The rolling radius of both wheels when they are centered is r_0, and the contact points are a distance $2d$ apart. The normal load on the wheelset is $2N$.

In Fig. 10.1, the situation is completely symmetrical so the vertical forces from the rail to the wheels are equal to N on both wheels. The horizontal forces on the wheels from the rails are assumed to be equal in magnitude and opposite in direction. The situation changes for the perturbed motion when the wheelset is no longer perfectly centered.

In Fig. 10.2, we now consider that the wheelset has moved a small distance x to the right from the equilibrium position shown in Fig. 10.1. Fig. 10.2 shows the displacement of the right-hand wheel of Fig. 10.1 in dotted lines. When the wheelset moves off center, because of the wheel taper, the right-hand wheel moves up and the left-hand wheel moves down

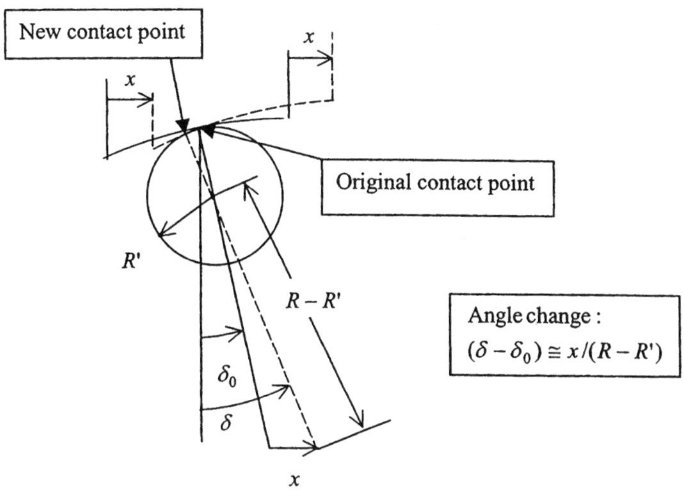

FIG. 10.2 Wheelset shown displaced by the distance x.

a certain amount but we assume that $x\delta_0 \ll 2d$ so that the wheelset does not tilt significantly. However, the rolling radii do change and the horizontal forces no longer cancel each other out completely. Fig. 10.2 shows the right-hand wheel before (solid line) and after (dotted line) the shift in position x.

In Fig 10.2, it can be seen that at both contact points (before and after the sideways shift of the amount x), the railhead circle and the wheel profile circle are tangent to each other. Therefore two lines are shown passing through the contact pints and the center of the fixed railhead circle. These lines, one solid and one dotted, are perpendicular to the two tangents at the contact points. The center of the wheel profile circle moves over the same distance the wheel does, x. This changes the contact angle as indicated in Fig. 10.2. Assuming that all the angles are small, this shift to the right by the amount x results in a change in contact angle of

$$\delta - \delta_0 \cong x/(R - R') \tag{10.1}$$

Considering the geometry of Fig. 10.2 and using Eq. (10.1), it can be seen that the contact point *on the railhead* moves to the left by an amount

$$R'(\delta - \delta_0) = R'x/(r - R') \tag{10.2}$$

However, because the wheel has moved an amount x to the right, the contact point *on the wheel* moves a total distance to the left of

$$R'x/(R - R') + x = (R/(R - R'))x \tag{10.3}$$

To a first approximation, the rolling radius changes by the equilibrium contact angle times the distance the contact point moves on the wheel. Thus the new rolling radius is approximately

$$r \cong r_0 + (R/(R - R'))x\delta_0 = r_0 + \lambda x \tag{10.4}$$

where the combined parameter

$$\lambda \equiv R\delta_0/(R - R') \tag{10.5}$$

is called the *conicity*.

Assuming that the total force between the wheel and the railhead is perpendicular to the tangent line at the contact point, it is clear that when the wheelset is centered, the vertical force is N at each wheel, and the lateral forces on each wheel (directed toward the rail centerline in each case) have magnitudes of approximately

$$F_{\mathrm{L}} \cong N\delta_0 \tag{10.6}$$

assuming that the contact angle is small.

We assume that the vertical forces at each wheel do not change appreciably when the wheelset moves over the small distance x, because x is assumed to be much smaller than the distance between the rails, $2d$. The change in the magnitude of the lateral force on one wheel is related to the change in contact angle from Eq. (10.1).

$$\Delta F_L = N(\delta - \delta_0) = Nx/(R - R') \tag{10.7}$$

The magnitude of the change is nearly the same for both wheels. However, while the force increases on the right-hand wheel, it decreases on the left-hand wheel. Because the equilibrium forces are oppositely directed, the changes at both wheels add to produce a net change in lateral force in the negative x direction. Therefore the total change in lateral force considered positive in the plus x direction for the wheelset is

$$\Delta F_{L,tot} = -[2N/(R - R')]x \tag{10.8}$$

where the term

$$2N/(R - R') \tag{10.9}$$

is called the "gravitational stiffness" because the lateral force is similar to the force that would be exerted by a spring, and the quantity in Eq. (10.9) plays the role of a spring constant in Eq.(10.8).

III. WHEEL–RAIL INTERACTION

The interaction between steel wheels and steel rails is actually quite complex but for a linear stability analysis, a simplified treatment is adequate. To fix ideas, we assume that the wheel center has a constant travel velocity of V and that there is no gross sliding at the contact point between the wheel and the rail. On the other hand, we imagine that when there are lateral and longitudinal forces between the wheel and the rail, the contact point on the wheel will exhibit rather small apparent relative velocities with respect to the rail in the lateral and longitudinal directions that we call $\dot{x}$ and $\dot{y}$, respectively.

The so-called nondimensional *creepages*, ξ_x and ξ_y, similar to the slip angles and longitudinal slip used in the description of pneumatic tire interactions with a roadway in Chapter 4 are defined. The nondimensional creepages are velocity ratios

$$\xi_x \equiv \dot{x}/V, \quad \xi_y \equiv \dot{y}/V \tag{10.10}$$

where $\dot{x}$ and $\dot{y}$ are the apparent lateral and longitudinal velocities at the wheel contact point. These apparent velocities will be related to the model variables describing the wheelset motion when deriving the equations of motion.

Finally, the lateral and longitudinal forces on the wheel from the rail are assumed to be proportional to the creepages in a way similar to how lateral and longitudinal forces were related to slip angles and longitudinal slip for pneumatic tires in a linear range.

$$F_x = f_x \xi_x, F_y = f_y \xi_y \tag{10.11}$$

The paper by Cooperider (1969) presents a discussion of how one could compute the creep coefficients f_x and f_y but eventually it will be shown that only their ratio will be important for the stability analysis. Fortunately, the ratio is not very different from unity so a detailed computation of the creep coefficients is not absolutely necessary for a stability analysis.

The creepages are related to the deflections that arise near the contact point when lateral and longitudinal are being transmitted. Just as is the case for pneumatic tires, the existence of a creepage does not imply a sliding velocity between the wheel and the rail. As one might imagine, the creepages themselves are small and the creep coefficients are large for steel wheels running on steel rails when compared to the equivalent parameters for rubber tires.

IV. CREEPAGE EQUATIONS

Fig. 10.3 shows a top view of the wheelset connected by lateral and longitudinal springs to a truck that is assumed to move along the rails at a constant velocity V. The variables for the two degrees of freedom are the lateral position of the wheelset away from the centered position, q_1, and the yaw angle, q_2. For the basic motion, both of these variables vanish and for the perturbed motion they are both assumed to be small.

Note that q_1 plays the role of the displacement x in Fig. 10.2. (The view of Fig. 10.2 can be related to the view of Fig. 10.3 if one imagines looking at an elevation view of the wheelset in Fig. 10.3 as it approaches. This view interchanges the meaning of right and left but otherwise causes no problems.)

Using the x–y coordinate system indicated in Fig. 10.3, one can now find the velocity components of the wheel contact points with the rails in terms of the variables $q_1(t)$ and $q_2(t)$ that describe the position of the

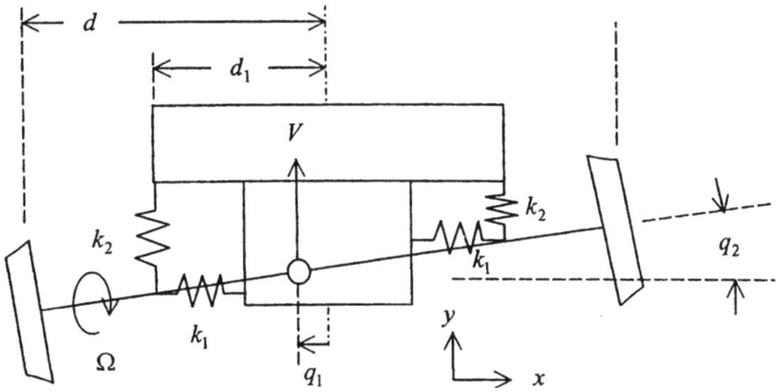

FIG. 10.3 View from the top of a wheelset attached to a truck by springs.

wheelset with respect to the truck. The center of mass of the wheelset has the velocity V in the y direction and $-\dot{q}_1$ in the x direction.

The velocities of the wheel contact points are composed of the center of mass velocity plus the effect of the yaw rate $\dot{q}_2$ and the angular velocity Ω with which the wheelset is rotating around its axle. The velocity components for the center of mass and the contact point for the right-hand wheel in Fig. 10.3 are shown in Fig. 10.4.

The terms "lateral" and "longitudinal" actually will refer to the wheelset, which is rotated through the small angle q_2 with respect to the x–y coordinate system aligned with the rails. However, because the equations of motion will be written in the x–y system, the results will be expressed in the x–y system using the approximations $\sin q_2 \cong q_2$ and $\cos q_2 \cong 1$.

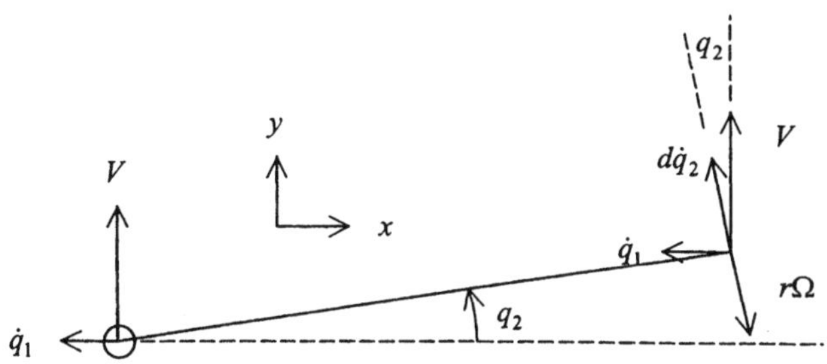

FIG. 10.4 Velocity components for the contact point of the right-hand wheel in Fig. 10.3.

The contact points of both wheels have the same lateral velocity with respect to the slightly yawed wheels. The lateral velocity consists of a component of the forward velocity V due to the yaw angle q_2 of the form $V \sin q_2$ and a component due to the rate of change of the sideways displacement, $-\dot{q}_1 \cos q_2$. Using the small-angle approximation, the final result expressed in the x–y frame is

$$\dot{x} = q_2 V - \dot{q}_1 \tag{10.12}$$

Note that this lateral velocity really is with respect to the wheel. It can be expressed in the x direction with respect to the rails because the yaw angle q_2 is small enough that the approximation $\cos \dot{q}_2 \cong 1$ can be used again.

The lateral creepage, following Eq. (10.10), for both wheels is then

$$\xi_x = (q_2 V - \dot{q}_1)/V = q_2 - \dot{q}_1/V \tag{10.13}$$

The longitudinal creepage is somewhat more complicated to compute because it involves the wheelset rotation Ω about the axle (Fig. 10.3). This degree of freedom is actually uncoupled from the other two. This uncoupling of Ω from the perturbation degrees of freedom was also noted in the simpler analysis of Chapter 1.

It will be assumed that the rate of rotation about the axle always has the proper average value

$$\Omega = V/r_0 \tag{10.14}$$

It can be shown then that the incremental torques about the axle due to small values of q_1 and q_2 are of opposite sign at the two wheels and cancel each other out. For the right-hand wheel in Fig. 10.3, the longitudinal velocity of the contact point shown in Fig. 10.4 is approximately

$$\dot{y}_{RH} = V - r\Omega + d\dot{q}_2 \tag{10.15}$$

This then yields the longitudinal creepage upon use of Eqs. (10.4), (10.5), and (10.14).

$$\begin{aligned} \xi_{yRH} &= \dot{y}_{RH}/V = 1 - r/r_0 + d\dot{q}_2/V = (r_0 - r)/r_0 + d\dot{q}_2/V \\ &= \lambda q_1/r_0 + d\dot{q}_2/V \end{aligned} \tag{10.16}$$

Using a similar argument, the longitudinal creepage for the left-hand wheel in Fig. 10.3 can be found.

$$\xi_{yLH} = -\lambda q_1/r_0 - d\dot{q}_2/V = -\xi_{yRH} \tag{10.17}$$

These creepages can be used with the creep coefficients to compute wheel lateral and longitudinal contact forces according to Eq. (10.11).

V. THE EQUATIONS OF MOTION

The equations of motion are written in a straightforward manner by relating the change in linear momentum to the forces in the x direction and the change in the angular momentum to the moments about the center of mass. With m_w being the mass of the wheelset and I_w the centroidal moment of inertia, the linear momentum equation is

$$m_W\ddot{q}_1 = -(2k_1 + 2N/(R - R'))q_1 - 2f_x(\dot{q}_1/V - q_2) \tag{10.18}$$

where one can recognize the spring forces having to do with the spring constant k_1, the gravitational spring-like forces from Eq. (10.8), and finally the lateral creepage forces from Eqs. (10.11) and (10.13).

The angular momentum equation is

$$I_W\ddot{q}_2 = -d2f_y(d\dot{q}_2/V + \lambda q_1/r_0) - 2k_2d_1q_2d_1 \tag{10.19}$$

where the first term on the left is the moment due to the longitudinal creepages on the two wheels, Eqs. (10.11) and (10.16) and the last term is the moment due to the longitudinal springs with spring constants k_2. The two equations can be rearranged in the form of Eq. (3.7) as follows:

$$\begin{aligned} m_W\ddot{q}_1 + \frac{2f_x}{V}\dot{q}_1 + (2k_1 + \frac{2N}{R - R'})q_1 - 2f_xq_2 = 0 \\ I_W\ddot{q}_2 + \frac{2f_yd^2}{V}\dot{q}_2 + 2k_2d_1^2q_2 + \frac{2f_y\lambda d}{r_0}q_1 = 0 \end{aligned} \tag{10.20}$$

VI. THE CHARACTERISTIC EQUATION

After putting the equations in a matrix form using the Laplace transform or assuming exponential forms for the variables as outlined in Chapter 3, the fourth order characteristic equation can be derived in the usual way, Eq. (3.22). The results after a certain amount of algebra are as follows:

$$a_0s^4 + a_1s^3 + a_2s^2 + a_3s^1 + a_4s^0 = 0 \tag{10.21}$$

with

$$a_0 = m_WI_W \tag{10.22}$$

$$a_1 = \frac{m_W 2f_y d^2}{V} + \frac{I_W 2f_x}{V} \tag{10.23}$$

$$a_2 = m_w 2k_2 d_1^2 + \frac{4f_x f_y d^2}{V^2} + 2I_w\left(k_1 + \frac{N}{R - R'}\right) \tag{10.24}$$

$$a_3 = \frac{4f_x k_2 d_1^2}{V} + \frac{4f_y d^2}{V}\left(k_1 + \frac{N}{R - R'}\right) \tag{10.25}$$

$$a_4 = 4k_2 d_1^2\left(k_1 + \frac{N}{R - R'}\right) + \frac{4f_x f_y \lambda d}{r_0} \tag{10.26}$$

VII. STABILITY ANALYSIS AND CRITICAL SPEED

Equations (10.21)–(10.26) represent a fairly complex, fourth-order characteristic equation so accomplishing a stability analysis in literal coefficients is not a trivial undertaking. The first step in determining if the system is stable is to examine the coefficients to see if all of them are positive. This appears to be the case although one can see that a_1 and a_3 become small as the speed increases. Previous experience with trailers and automobiles in Chapters 5 and 6 indicates that when coefficients decrease with speed, there is a good chance that a critical speed exists above which the system is unstable.

When the Routh criterion is applied to Eq. (10.21), assuming that the coefficients are positive, the following expression must be satisfied if the system is to be stable:

$$(a_1 a_2 - a_0 a_3)a_3 > a_4 a_1^2 \tag{10.27}$$

This result was to be derived in Problem 3.3 and the result has been previously given in Eq. (8.33).

Again, if the expressions for $a_0, \ldots, a_4$ are substituted in the criterion [Eq. (10.27)] the result seems hopelessly complex. However, a close inspection of the characteristic equation coefficients shows that in two cases, terms involving the product $f_x f_y$ appear in combination with terms that in all probability are considerably smaller in magnitude.

As previously mentioned, the creep coefficients for steel wheels on steel rails are certainly large. Just as is the case for cornering coefficients for pneumatic tires, these coefficients do not have anything to do with gross slipping but rather with distortion of the wheels and the rails near the area of contact. For steel wheels on steel rails, the distortion and hence the creepages are very small, and the creep coefficients are large so that finite forces result from the product of the creep coefficients and the creepages.

Thus it is logical to approximate the coefficients a_2 and a_4 in Eqs. (10.24) and (10.26) by the following expressions,

$$a_2 \cong 4f_x f_y d^2/V^2, \quad a_4 \cong 4f_x f_y \lambda d/r_0 \tag{10.28}$$

but with the coefficients a_0, a_1, and a_3 remaining unchanged.

Now considering the fourth-order Routh stability criterion [Eq. (10.27)], one can observe that a_1a_2 involves creep coefficient products $f_xf_y^2$ and $f_yf_x^2$ while a_0a_3 involves only f_x and f_y. Assuming that both f_x and f_y are large compared to the other parameter groups, this leads to the idea that a_0,a_3 can be neglected in comparison to a_1a_2. This then simplifies the Routh criterion of Eq.(10.27) to just

$$a_2a_3 > a_4a_1 \tag{10.29}$$

The critical speed, V_c, occurs when $a_2a_3 = a_4a_1$. When the approximate coefficients from Eq. (10.28) are substituted into this equation, and one solves for the velocity, the result is

$$V_c \cong \left(\frac{r_0 d}{\lambda}\right) \frac{2k_2 \dfrac{f_x}{f_y} d_1^2 + d^2\left(2k_1 + 2N/(R - R')\right)}{d_2 m_w + I_W \dfrac{f_x}{f_y}} \tag{10.30}$$

A fortunate feature in this result is that the stability of the wheelset is found to depend only on the ratio of the creep coefficients rather than on their absolute values. Because the coefficients are not easily calculated with great accuracy, this means that only the ratio needs to be estimated and one can easily imagine that the ratio will not be far from unity.

One can note that for cylindrical wheels with $\delta_0 = 0 = \lambda$, an infinite critical speed is predicted. However, this is not a satisfactory solution to hunting instability because as we have shown in the introductory example in Chapter 1, nontapered wheels have no self-centering properties and such wheelsets have been shown to drag their flanges almost continuously, resulting in heavy wear.

As tapered wheels wear, the conicity, λ, increases, which tends to decrease the critical speed; however, the gravitational stiffness term increases, which tends to increase the critical speed. The spring stiffnesses play roles in this change of critical speed and the paper by Cooperider (1969) points out that, except for very soft suspensions, worn wheels generally lead to lower critical speeds.

In this case, at the critical speed, two of the four eigenvalues of Eq. (10.21) find themselves on the imaginary axis of the s-plane. The frequency of the oscillations at the critical speed can be found by observing that at the critical speed, the two critical eigenvalues will have values of $\pm j\omega_c$ and that these values must satisfy the characteristic equation. Using the substitution $s = j\omega_c$ in the characteristic equation, the result is an expression having both real and imaginary parts and both parts must vanish. The vanishing of the complex part leads to the equation

$$a_1(-j\omega_c^3) + a_3 j\omega_c = 0 \tag{10.31}$$

After substitution of the expressions in Eqs. (10.23) and (10.25) for the coefficients, the result is

$$\omega_c \cong \frac{2k_2 \dfrac{f_x}{f_y} d_1^2 + d^2\left(2k_1 + 2N/(R - R')\right)}{d_2 m_w + I_W \dfrac{f_x}{f_y}} \tag{10.32}$$

The vanishing of the real part leads to

$$a_0\omega_c^4 + a_2(-\omega_c^2) + a_4 = 0 \tag{10.33}$$

When the a_0 term is neglected because it contains no creep coefficients while a_2 and a_4 contain $f_x f_y$, the result is a relation between the critical speed and the critical frequency

$$V_c^2 = \frac{r_0 d}{\lambda}\omega_c^2 \tag{10.34}$$

which is consistent with the previous results.

A model of a simple truck carrying two wheelsets is also given in Cooperider (1969), and its equations of motion resemble those of the single wheelset analyzed here. The general behavior of the critical speed with parameter changes is also similar (see Problem 10.3). More complex truck models and models of complete railcars yield characteristic equations that are complex enough that they must be analyzed numerically rather than in literal coefficients.

It should be noted that the analysis given above did not consider any damping forces related to the relative velocity between the wheelset and the truck. A complete rail vehicle not only has many more degrees of freedom, but also has suspension elements that provide a damping function for the railcar body motions. The introduction of damping elements into the model might well result in a change in the computed critical speed. A problem with the introduction of damping is that some types of damping elements used in the railroad industry use dry friction devices that are distinctly difficult to represent in a linearized model.

The invention of the tapered wheel made railroads practical, but as speeds increased, the analysis of hunting instability became important. Extremely high speed trains can only be confidently designed using sophisticated computer simulation techniques. The elementary analysis given in this chapter gives a good idea of the complexities involved in the stability analysis for rail vehicles but also shows that there are some similarities between the stability problems of railcars and other vehicles.

11
Electronic Stability Enhancement

I. INTRODUCTION

This chapter will deal with the possibilities of changing the dynamic properties of vehicles through electronic means. In particular, the idea that the stability of a vehicle can be enhanced by active control means will be discussed. The previous chapters have all concentrated mainly on analyzing the inherent stability properties of a number of types of vehicles. By the terms "inherent stability" we mean the ability of the vehicle to return to a basic motion after a disturbance without the aid of a human operator or an automatic control system.

As we have seen in Chapters 5, 6, 8, and 10, for many ground vehicles stability problems do not arise until a critical speed is exceeded. This fact meant that for most ground vehicles, until power plants were developed that allowed high speeds to be obtained, little attention was paid to vehicle dynamics in general and to stability in particular.

An exception discussed in Chapter 7 involves two-wheeled vehicles that are basically unstable and need to be stabilized by the rider, particularly at low speeds. Bicycles are examples of useful vehicles that absolutely require active stabilization in order to function at all.

As it happens, human beings are especially good at balancing having had a long history of walking upright. After a short learning period, most children become very competent bicycle riders. Even when bicycles are deliberately modified to make them even more violently unstable than they normally are, it is found that under the active control of a rider, they can be quite easily stabilized (Jones, 1970). As has been noted in Chapter 7, it is significantly more difficult for a rider to stabilize a bicycle by steering only the rear wheel (Schwarz, 1979), but even this is possible with practice.

It was pointed out in Chapter 9 that the situation is quite different with airplanes. Airplanes are not automatically stable and easily controllable at low speeds as automobiles are, so stability and control issues were important from the very beginning. Early experiments with unmanned gliders were obvious total failures if the gliders were not designed to be inherently stable, so early experimenters tried very hard to devise stable unmanned aircraft. Of course, at the time, the achievement of a stable configuration was typically more the result of intuition and trial and error rather than mathematical analysis.

Many pioneers in aviation concentrated on making their aircraft stable and paid relatively little attention to the means for the pilot to control the motion. Of course stability without controllability is not necessarily a good thing in a vehicle. A stable aircraft headed for an obstacle or even the ground will crash if there is not an effective means for changing the flight path. It has been suggested (Vincenti, 1988) that the success of the Wright brothers was related to their profession as bicycle mechanics. They realized that a human pilot would have to control an airplane actively and, just as with a bicycle, it might be possible to learn how to do that even if the airplane was inherently somewhat unstable. After testing replicas of the 1903 Wright Flyer it appears that it was, in fact, quite unstable but the Wrights were able eventually to learn to control its motion effectively, Culick and Jex, 1985.

Although the Wright brothers would probably have preferred a more stable airplane than their Flyer if they had known how to build one, there may be situations in which an unstable vehicle has some desirable properties that a stable version would not have. If then an active stabilization scheme could be developed, the result could be a superior vehicle design. In other cases, a stability enhancement system might come into play in emergency situations in which the human operator's ability to control a vehicle might be overtaxed.

II. STABILITY AND CONTROL

In most vehicles, there is a conflict between stability on one hand and controllability or maneuverability on the other hand. In Chapter 6, it has been shown that an understeering car is stable, but its yaw rate gain decreases for speeds higher than its characteristic speed and the lateral acceleration gain approaches a limit at high speeds. This means that a car with a low characteristic speed may be very stable at high speeds, but, on the other hand, it may not respond well to steering inputs at high speed. Although stability in a vehicle is generally regarded as a positive attribute, excessive stability usually comes at the price of reduced maneuverability and controllability. Thus it was that the Wright brothers were more successful than their predecessors by concentrating on active control more than on inherent passive stability. This can be considered to be an early case in which an unstable vehicle under active control (by a human being) was made stable. Recent progress in electronics has given rise to the idea that many stability problems may be solved through active automatic control techniques.

The compromise between stability and controllability has been recognized for many years and the degree of stability of most vehicles has been tailored to their intended uses. One would not expect a fighter plane to be as stable as a commercial passenger plane or a sports car to be as stable as an 18-wheel tractor-trailer. In the first part of the last century, the recognition of the importance of stability and controllability mainly affected the physical design of vehicles. Later, as the principles of automatic control became understood and as sensors, actuators, and various forms of computers or controllers were developed, it became clear that stability and automatic control could become linked. This opened the possibility that if a reason existed to design an inherently unstable vehicle, it might be possible to use active means to stabilize it. There are examples of fighter planes that are very maneuverable but in the absence of automatic control would be unstable and very difficult or even impossible for a pilot to fly.

Many innovations involving active stability and control found their first applications in aircraft. The autopilot and the first antilock brakes were fairly simple automatic control devices concerned with stability and control. The autopilot was able to keep an airplane on course without the attention of the pilot. This is particularly useful for an airplane without a high degree of stability because such airplanes normally require the pilot's attention continuously during gusty weather. On the other hand, the attempt to design a high degree of stability into an airplane often compromises other aspects of its performance.

Similarly, it was often difficult for a pilot to brake sharply upon landing without locking the wheels. This is a consequence of the manner in which pneumatic tires generate longitudinal forces as we have seen in Chapter 4. It is easy to appreciate from Fig. 4.5 that because the traction force on a tire decreases when the braking slip exceeds the slip that generates the maximum braking force, this leads to an unstable situation in which the slip increases further and the traction force further decreases. Whenever too much torque is applied to the brakes and the longitudinal slip reaches a high value, the force from the pavement decreases and the wheel decelerates to a locked state quickly.

This unstable behavior is particularly dangerous for aircraft that land at high speeds because locked wheels can easily lead to tire failures. The first aircraft antilock braking systems simply prevented wheel locking from happening. Later, when antilock brakes were developed for automobiles, the control schemes began to include sophisticated algorithms to maximize the braking force and to improve directional stability in addition to preventing locking.

Now that the principles of automatic control have become well established, and sensors, actuators, and digital electronics have been developed to be powerful and cost effective, the idea of active stability control has spread to many types of vehicles. For aircraft, there has been a gradual shift from mechanical controls used at first to power assisted controls and later to "fly-by-wire" controls. In the latter case, the only connection between the pilot's input motions in the cockpit and the aircraft control surfaces is by means of computer-generated electrical signals. This progression has partly been necessary as airplanes became so large that a human could not be expected to move the control surfaces without the help of a powered actuator. As a consequence it became possible for the control surfaces to be influenced not only by the pilot's inputs but also by a controller reacting to sensors measuring the airplane's motion. This means that the dynamic behavior of an airplane can be influenced not just by its basic design but also by the design of the controller providing commands to the control surface actuators.

There are now a number of military aircraft designed on purpose to be inherently unstable but very maneuverable that are stabilized by active means. Similarly, the space shuttle, which at one point in its mission must act like an airplane, would be very difficult to fly without active control during the landing phase of its mission. The necessity of being both a spacecraft and an airplane makes severe compromises on the basic design inevitable. Using active means, the shuttle can be made to emulate the behavior of a commercial aircraft so that the astronauts can pilot it with confidence.

The fly-by-wire concept not only opened up new possibilities for modifying the stability and control properties of aircraft, but also brought some new challenges. When mechanical linkages between the pilot and the control surfaces were replaced with computers and actuators the questions of reliability and fail-safe designs became important. Often, the only convincing way to assure safety in the event of an electronic failure was through the use of redundancy. This approach, while effective, was costly and complex. In addition, pilots were used to having kinesthetic force feedback through mechanical linkages to their hands and feet that helped them judge the severity of a maneuver or the strength of external disturbances. Artificial force feedback systems have been developed to replicate the feel of mechanical control linkages, but they again increase the complexity and cost of the system.

These challenges were important for aircraft, which are expensive vehicles typically produced in limited numbers, but they have been even more significant when fly-by-wire concepts were extended to drive-by-wire for relatively inexpensive, mass-produced vehicles such as automobiles. The advances in electronic devices and automatic control techniques that have resulted allowed the introduction of a number of active control techniques to be applied to even such fairly inexpensive vehicles. This trend appears to be poised to be even more important in the future (Anon., 2003; Birch, 2003).

To illustrate the basic idea of active control of vehicle dynamics in general and stability in particular some examples from the automotive area will be studied here. Basic models developed in previous chapters will be used although it must be appreciated that actual production systems are based on more extensive mathematical models as well as on experimental results.

III. FROM ANTILOCK BRAKING SYSTEM TO VEHICLE DYNAMIC CONTROL

Although there are countless examples of the use of active means to influence vehicle dynamics we will concentrate first on a familiar example from the automobile sector. It is particularly impressive because it involves important safety aspects of a mass-market product, which means that it must be cost effective, and fail-safe. In addition, it deals directly with stability issues in a sophisticated way. What will be described is an active automobile stability control system that is a development of ABS, the Antilock Braking System. There are many trade names and special acronyms such as ESP, Electronic Stability Program, used to describe these electronic stability enhancement systems, but a fairly generic term is VDC, Vehicle Dynamic Control (van Zanten et al., 1998; Hoffman et al., 1998).

The original antilock brakes for airplanes were intended at first merely to prevent brake locking when landing and thus to lessen the chances of catastrophic tire failure. The idea was to monitor wheel angular velocity and to detect when it began to decrease abruptly indicating when the wheel was about to lock. When imminent locking was detected, the brake hydraulic pressure was released until the wheel angular speed accelerated again under the influence of the friction force on the tire from the runway. At that time when the wheel speed returned to near its original value, the brake pressure was reapplied. This resulted in a cycling of the brake pressure and of the wheel angular acceleration until the pilot no longer required maximum braking. The average braking force during the cycling behavior may not have been as large as it would have been if the pilot had been able to exert just enough pressure on the brake to achieve the optimum longitudinal slip, but the important point was to prevent the wheel from locking and the tire from skidding along the runway surface.

In the course of time, the antilock braking control strategy increased in sophistication. By the time it began to be applied to cars, the antilock system was capable of not only preventing wheel lockup, but also of searching for the optimal longitudinal slip for maximal deceleration. This extension of the function of ABS brought along its own stability problems.

In the case in which a car was traveling on a so-called split-μ surface, i.e., a surface with different coefficients of friction on the right and left sides, optimal braking on one side would generate a larger retarding force than on the other side. This would cause a yaw moment on the car and the car would tend to swerve. In this case the system that prevented instability in the wheel angular speeds during braking caused vehicle instability in yaw. This problem was ameliorated but not really solved by adopting strategies such as the "select low" principle at the back axle. Select low meant that the ABS action was applied equally to both rear wheels when either one began to lock. Thus some of the braking force was sacrificed at a rear wheel that was on a high friction surface in order to reduce the yaw moment associated with unequal braking forces at the two rear wheels.

The original ABS systems were only capable of releasing and reapplying brake pressure being supplied in response to the driver's foot pushing on the brake pedal. (Of course, many cars had power brakes that provided higher hydraulic pressure than could be generated by the driver's force on the brake pedal. It is still true for such systems, that if the driver is not pushing on the brake pedal, there is no pressure to modulate at any of the wheels.) Even for simple ABS systems, however, a number of components must be present. There have to be wheel-speed sensors, high-speed electrohydraulic valves, and a digital control system. In addition, there had to be a small pump to replace the brake fluid discharged from the brake lines during an ABS episode.

It was soon realized that most of the ABS equipment could be used to solve another problem. Cars with an open differential on the drive axle deliver the same torque to both wheels. If one wheel happens to be on a surface with poor traction, it cannot generate a large traction force without spinning. In such a situation, better traction at the other drive wheel cannot be utilized because the spinning wheel limits the torque to both wheels.

One solution to this problem is a limited slip differential that can increase the torque to the nonspinning wheel. This requires the addition of a more complex and expensive component in place of the simple open differential. Another solution is to apply the service brake at the spinning wheel. This has two effects. As we have seen in Chapter 4, a spinning wheel with nearly 100% longitudinal slip generally produces a smaller traction force than a wheel with a smaller slip and also is capable of generating a smaller lateral force for a given slip angle. A reduction in the angular speed of a spinning wheel generally helps increase the possible lateral and longitudinal tire forces.

The other effect is to apply the total torque applied to the spinning wheel through the differential to the wheel with better traction. The total torque at the spinning wheel consists of whatever torque is caused by the traction force plus the applied brake torque. Thus even when one wheel is on ice with an extremely low coefficient of friction, the other wheel can receive a large torque from the differential equal to at least the brake torque applied at the spinning wheel. Thus the nonspinning wheel can generate as large a traction force as possibly allowed by the pavement upon which it is resting.

As ABS systems monitor wheel speed, it is possible to detect wheel spin as well as imminent locking. If the system has a source of hydraulic pressure other than that provided by the driver's foot, then the controller can be programmed to control wheel spin by applying the brake at the spinning wheel. Traction control systems, abbreviated as TCS, typically use the brakes and ABS components in combination with throttle control. In low traction situations, the brakes can react more quickly than the throttle control to reduce wheel spin, but on a longer time scale, it is logical to reduce engine power. In any case, a TCS system can make the car react in a more stable and predictable way in low traction situations.

As ABS and TCS systems were developed into commercially viable products, engineers began working on a true stability enhancement system, the VDC. With the addition of a supply of hydraulic pressure independent of the braking action of the driver for use with TCS, it became evident that it was possible to apply any one of the four brakes on a car at any time. This opened up the possibility of monitoring the motion of the car and detecting any unusual situations. Then the brakes could be used to correct any deviations from the motion the driver might have expected under normal conditions.

IV. MODEL REFERENCE CONTROL

The VDC systems have some similarity to model reference control systems used in aviation. In these systems, the pilot's inputs to the airplane are measured and applied to a computer model of a reference airplane. The response of the computer model is computed in real time and compared to the sensed response of the actual airplane. Then deviations between the model response and the actual response are used as error signals and used by a controller to actuate control surfaces in such a way as to reduce these deviations. In rough terms, this is the way a space shuttle is made to fly like a commercial transport aircraft. The general idea of model reference control applied to the steering dynamics of an automobile is shown schematically in Fig. 11.1. We will return to this example of model reference control in the section on active steering but now the basic idea will be applied to VDC systems.

For an automotive VDC, some additional sensors beyond those used for ABS and TCS are usually required. Typically, a yaw rate sensor, a lateral acceleration sensor, and a steering wheel position sensor are included. The ABS systems usually already contain a longitudinal acceleration sensor. The wheel-speed sensors together with the longitudinal accelerometer can be used to deduce the vehicle speed even when the wheels begin to lock.

Although there are many refinements to VDC systems to account for special circumstances, it is fairly easy to see how to construct a strategy for using the brakes to improve the dynamics of the vehicle. The simple bicycle model used in Chapter 6 could be used as a reference model to predict the response to steering wheel inputs, for example. Then, knowing the speed and measuring the yaw rate and the lateral acceleration it would be easy to

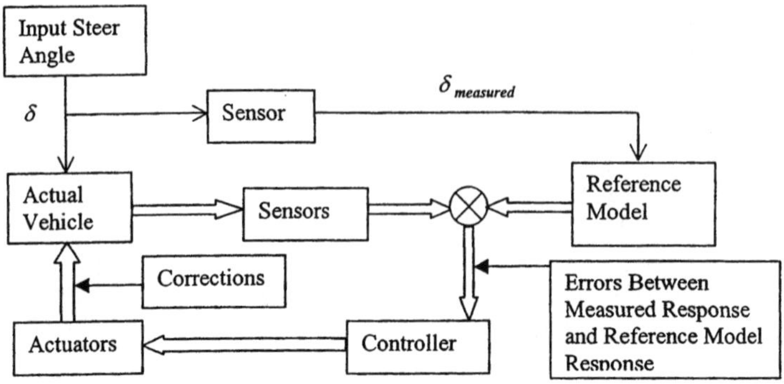

FIG. 11.1 Model reference control scheme for influencing the steering dynamics of an automobile.

see if the car was responding in a manner similar to the reference model response.

A basic scheme for using the brakes to correct an excessive oversteer or understeer condition is shown in Fig. 11.2.

Suppose, for example, that the measured yaw rate in a turn were more than the reference model yaw rate predicted by the computer with the same speed and steering angle. This would indicate an excessive oversteer condition, or in other words that the rear slip angle was larger than normal.

This could be caused by any number of conditions. It could be an underinflated rear tire, a slippery spot in the road being under the rear tires, or an unusually heavy load concentrated at the rear of the car. The control system could then apply braking to the outside front wheel. As the rear slip angle is unusually large, the indication is that the front wheels have more traction than the rear and can safely generate a longitudinal braking force in addition to the necessary lateral force. The important point is that the braking force on the outside wheel will generate a yaw moment tending to reduce the excess yaw rate.

On the other hand, if the measured yaw rate is less than the reference model yaw rate, the inside rear wheel brake would be applied. This would be an understeer condition in which the front slip angle was unusually large. The rear wheels then appear to have excessive traction compared to the front, and braking the inside rear wheel will produce a yaw moment tending to increase the yaw rate and make the car turn more rapidly.

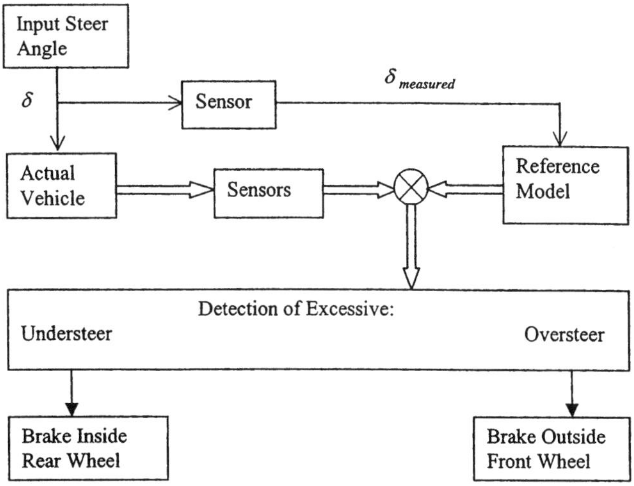

FIG. 11.2 A basic version of vehicle stability control using ABS and TCS hardware and model reference control.

As might be expected, this explanation of VDC is highly simplified and the several commercial versions differ in many details from one another. The point is that the stability properties of vehicles under active control may differ significantly from their inherent properties without control. Particularly in emergency situations, drivers are often not able to control their vehicles effectively. Heavy braking, for example, may turn a normally stable vehicle into an unstable one as the car pitches forward and the normal force at the front increases and the normal force at the rear decreases. As we have seen in Chapter 6, this transient change in normal forces will increase the lateral force generating capabilities at the front and reduce them at the rear. The tendency is to change a normally understeer and stable car into one that is at least temporarily oversteer and possibly unstable. The VDC system can restore stability to the car in a variety of situations and can make the car react in a manner much closer to the way the driver would expect under normal conditions.

One could certainly classify VDC systems as safety systems as well as stability enhancement systems. This is because the only actuators used are the brakes and any action of VDC systems results in the vehicle slowing down. In most emergency situations, it is beneficial to slow down the car as well as to increase its stability. Even in this form of active stability control, there are differences in emphasis between maneuverability and safety or controllability. Sports car drivers often prefer a less active VDC system than drivers of family sedans because it leaves them with more flexibility to drive aggressively (Markus, 1996). Some cars offer the option of turning the VDC system off to let the driver control the car without electronic intervention.

V. ACTIVE STEERING SYSTEMS

In principle, there are several other ways one could change the stability properties of an automobile besides selectively applying the brakes at individual wheels. For example, it is possible to steer the rear wheels of a car using an actuator and a controller responding to sensed deviations of motion from a reference vehicle response. A number of prototype vehicles have been produced with such schemes. Recently, the trend toward various types of drive-by-wire has resulted in the production of active steering systems but most of these systems have acted on the front wheels.

In fact, pure drive-by-wire systems are less common in automobiles than fly-by-wire systems in aircraft. This has several reasons. In the first place, automobiles are typically less massive than many airplanes and lightweight cars can actually do without power assistance for their controls. Secondly, cost and complexity considerations are extremely important for all but the most expensive automobiles. Finally, there are a number of laws relating to automobiles that vary from market to market. These may make some active systems illegal at least in certain parts of the world.

The law in some countries requires that there must be a direct mechanical connection between the driver's steering wheel and the road wheels. This would seem to make an active steering drive-by-wire system impossible. This is not quite the case, however. If a differential element is inserted in the steering column, then the steer angle of the road wheels can be the sum of two angular inputs. One can be the steer angle commanded be the driver through the steering wheel and the other can be the angle from a rotary actuator, such as an electric motor and gear train, commanded by a computer.

Such a system retains the required mechanical connection but allows the controller to apply a correction to the driver-commanded steer angle at the road wheels much as a true steer-by-wire system would. If the actuator is designed so that it will not back drive because of torques in the steering column (e.g., by the use of a self-locking worm gear drive), then in the case of an actuator failure, the actuator will remain at a fixed position and the system will revert to a conventional mechanical steering system. This is a fairly old idea that only recently has been developed into a production system (de Benito et al., 1989; Fonda, 1973; Anon., 2003; Birch, 2003).

Such steering systems are particular examples of an even more general idea of electronically controllable chassis and suspension systems (Birch, 2003). These include active or semiactive suspension systems and systems for lane keeping or collision-recognition systems. The general topic of chassis system integration is too complex to be handled in any detail here, but the basic ideas of active steering either by wire or by other fail-safe means will be introduced and related to the previous study on automobile dynamics in Chapter 6.

In an active steering system, there need be no fixed relationship between the steering wheel angle and the angle of the road wheels. Not only can the effective steering ratio be varied with speed, for example, but also the road wheel angles can be controlled by a combination of driver and computer inputs. This means that the response of the vehicle to steering wheel inputs can be varied and, just as the space shuttle can be made to respond much as a commercial airliner, an automobile can be made to respond to steering inputs much as a reference vehicle would. In such a case, the stability properties of a car could be made to be essentially constant even when there were physical changes in the vehicle.

Major changes in load, tire pressure, or surface traction pose a significant challenge to a driver because the steering response of the car can change drastically from the response under normal conditions. With active steering, a model reference type of controller can make the car respond in a more normal fashion even when circumstances have changed the inherent stability properties of the vehicle.

The problem of steering automobiles was solved early in automotive history by providing a steering wheel directly linked to the front road wheels. Some very early automobiles used a tiller similar to those used today on small sailboats instead of a wheel, but even in these examples it apparently seemed obvious that cars should be steered from the front rather than from the rear as many boats are. Today, only some specialized vehicles such as street sweepers, forklift trucks, and lawn mowers are steered from the rear. As has been seen in Chapter 6, the dynamics of rear-wheel steering is quite different from the dynamics of front-wheel steering. This means that human operators need some practice before they can control a rear-steering vehicle safely and it means that an active steering system operation on the rear wheels is more complicated to design than one operating on the front wheels.

Historically, the introduction of power steering reduced the effort required at the steering wheel, but it did not disturb the fixed relationship between the hand wheel angle and the angle of the road wheels. Somewhat surprisingly, the steering ratios of cars offered with both manual steering and power steering are usually not very different although power steering would permit much "faster" ratios. This is partly due to the desire to make the car steerable should the power steering system fail.

From time to time, the conventional situation is called into question. See, for example, Cumberford (1991). It is clear from examples of motorcycles, bicycles, and fly-by-wire airplanes that human operators can steer vehicles using less motion than is commonly necessary for steering automobiles. In fact, as noted in Cumberford (1991), it has been shown in developing controls for fighter aircraft that pilots can control their vehicles using devices that essentially do not move at all but merely react to pressure. Experiments with automobiles have indicated that under some conditions very fast steering ratios can lead to more accurate steering than conventional steering ratios.

This leads to the possibility that active steering systems may not only improve the stability of a vehicle but may also be adapted to provide better controllability for particular maneuvers. A simple example of this idea that is now commercially in several forms is speed-dependent steering. At slow speed, the steering is made more responsive so less motion of the steering wheel is required for parking but at high speeds the steering ratio is made "slower" so that the driver does not feel that the car reacts too violently.

Steering the rear wheels in the opposite direction to the front wheels at low speeds and not steering them at high speeds can accomplish a similar effect. The main reason for doing this is usually to reduce the minimum turning circle but in effect the steering ratio is also changed. A more direct way to change the ratio is to steer the front wheels either "by-wire" using an actuator alone or by using a differential in the steering column to allow the road wheels to be turned by a combination of driver input and actuator input. This way of accomplishing a variable steering ratio or changing the dynamic response to driver inputs will be briefly discussed below.

The logic of these schemes has to do with the lateral acceleration and yaw rate gains that tend to increase with speed as discussed in Chapter 6 depending, of course, on the degree of understeer or oversteer the car exhibits. The idea is that as the acceleration and yaw rate gains increase with speed, it seems logical to reduce the ratio of steering wheel motion to road wheel motion. Of course, if the steering ratio were so high that the system was nearly pressure controlled rather than motion controlled, one might cast the argument in terms of a gain between pressure exerted on the wheel (or joy stick) and the angular rate of change of the steer angle of the road wheels rather than in terms of a conventional steering ratio.

A. Stability Augmentation Using Front, Rear, or All-Wheel Steering

It is clear from the results presented in Chapter 6 that an automobile can be steered from the front, rear, or both axles and thus to improve the handling dynamics and stability of a car one could consider a wide variety of active systems. However, the transfer functions derived in Chapter 6 [Eqs. (6.32)–(6.37)] relating the front and rear steer angles to the lateral velocity, yaw rate, and lateral acceleration did indicate significant dynamic differences between front and rear wheel steering for the elementary vehicle model. Notable among the differences was the fact that the lateral acceleration transfer function for rear-wheel steering had a right half-plane zero, indicating a nonminimum phase response to a sudden change in rear steer angle. This dynamic effect remained even when the transfer functions were simplified by assuming that the car was neutral steer and that

$$I_z = mab \qquad (6.46)\ \text{repeated}$$

The transfer functions in this special case were Eqs. (6.48–6.51).

In Karnopp and Wuh (1989), a general study of active steering using the elementary vehicle model of Eq. (6.29) and state variable feedback was presented. Fig. 11.3 shows the control scheme. The driver gives a steering angle input using the steering wheel called δ_{fref} and a controller with a gain matrix $[K]$ operating on the state variables V and r determines the actual steer angles δ_f and δ_r. The $[A]$ and $[B]$ matrices represent Eq. (6.29). The

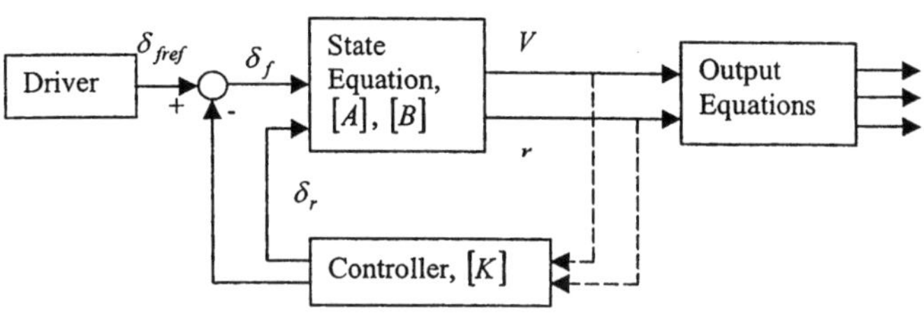

FIG. 11.3 State variable feedback active steering.

output equations may be used to find any variables in the system that depend on the state variables. A general study of this type can point out basic possibilities and limitations of various types of active steering systems.

In the absence of rear-wheel steering, the rear axle force, which depends on the rear slip angle, depends on the lateral velocity at the rear axle. For a sudden change in front-wheel steer angle commanded by the driver, the front slip angle and hence the front axle force will respond quickly. (The model neglects any lag in the dynamic tire force as discussed in Chapter 8.) However, the axle force at the rear may not respond quickly as the lateral velocity at the rear must build up before there is a slip angle and hence a force. The dynamics of the lateral velocity build up in the absence of direct rear-wheel steering is more complicated than one might imagine.

For the simple model of an automobile being used, it is easy to see that if a force is suddenly applied at the front axle, the immediate effect on the lateral velocity at the rear axle is related to the classical problem of determining the center of percussion of a body in plane motion. For the special case of Eq. (6.46) it turns out that the center of percussion is exactly at the rear axle. Therefore the immediate effect of a sudden change in steer angle at the front is no immediate change in the rear axle lateral velocity at all and thus no immediate slip angle or rear axle force.

Using previous results from Chapter 6, one can also see this fact. As shown in Fig. 6.7, the lateral velocity at the rear axle is $V-br$. Considering that the total velocity at the rear axle is represented in a body-centered frame, the lateral acceleration at the rear axle is given by the expression $(\dot{V}-b\dot{r}+rU)$. Using the transfer functions given in Eqs. (6.32) and (6.33) and the denominator Δ from Eq. (6.31), a transfer function relating the rear axle lateral acceleration to the front steer angle can be derived. (The transfer functions for $\dot{V}$ and $\dot{r}$ are just the transfer functions for V and r multiplied by s.) The result is

$$\frac{(\dot{V}-b\dot{r}+rU)}{\delta_f}=\frac{(I_z-mab)C_f s^2+(a+b)C_f C_r}{\Delta} \tag{11.1}$$

The immediate response of the acceleration at the rear axle to a step change in the front steer angle is found by letting $s\to\infty$ in Eq. (11.1) (Ogata, 1970). The result, using Eq. (6.31), is

$$\left.\frac{(\dot{V}-b\dot{r}+rU)}{\delta_f}\right|_{s\to\infty}=\frac{(I_z-mab)C_f}{mI_z} \tag{11.2}$$

One can now see that if Eq. (6.46) is satisfied, there is no immediate lateral acceleration at the rear axle. In time, a slip angle will develop but just after the step in front steer angle, there will be no slip angle and no force at the rear axle in this special case.

When

$$I_z > mab \tag{11.3}$$

the lateral acceleration is positive at the rear, the lateral velocity starts to increase in the positive direction, and the slip angle at the rear becomes positive. Then, using the force-slip angle relation, Eq. (6.21), the force is initially negative or in opposition to the direction for the force at the front axle. When the vehicle has a large moment of inertia, the forces at the front and rear initially combine to rotate the vehicle more than to accelerate it. Eventually, the slip angle will become negative and the front and rear forces will both become positive in a steady turn.

On the other hand, if

$$I_z < mab \tag{11.4}$$

the initial lateral acceleration at the rear axle is negative, which leads to a negative slip angle and a positive force at the initial instant. When the moment of inertia is small, the initial response is mainly for the forces at the front and rear to combine to produce lateral acceleration of the vehicle. Again, in a steady right turn the front and rear forces will be positive.

For the case of Eq. (11.3), i.e., for vehicles with a large moment of inertia, there will always be a reverse action for the force at the rear axle if the rear wheels are not steered. Although active front-wheel steering can alter the dynamics of the vehicle significantly, active front-wheel steering cannot eliminate this reverse action.

The situation is different if the rear wheels can also be steered as the rear steer angle can be used then to control the slip angle and hence the force directly. Karnopp and Wuh (1989) show typical results demonstrating that all wheel steering cannot only increase the speed of response and the stability of a car but that it can eliminate the reverse response of the rear axle force if it exists.

It is not likely that the rear side force reverse action is particularly important in practice except for vehicles with and unusually large moment of inertia. This effect, however, is one of the few theoretical advantages of active rear wheel steering.

Another aspect of vehicle dynamics that can be influenced by active steering is disturbance rejection. This is the ability of the vehicle to maintain a desired trajectory when acted upon by external forces such as wind gusts. Obviously, a vehicle that is barely stable will be easily disturbed from its desired path and may take a fairly long time to recover a steady motion. This is an easily observed phenomenon for trailers traveling on a highway during a period of gusty winds. The trailer may be strictly stable but is in a continuous state of oscillation due to the wind forces. A similar phenomenon occurs for cars that are not particularly stable when they are subjected to wind gusts. An active steering system might be able to increase the degree of stability and considerately damp the response to the disturbing wind gust forces.

This is a situation in which the introduction of an extra measure of stability may still result in unsatisfactory behavior. An interesting example is found in Tran (1991), in which an attempt was made to reduce the perceived lateral acceleration of an automobile to crosswind gusts through rear-wheel active steering. In this case the concept was to use a feed forward control technique reacting to measured pressure fluctuations on the vehicle.

The results should have been predictable based on the discussion of the transfer function between lateral acceleration of the center of mass and the rear axle steer angle [Eq. (6.37)]. As discussed in Chapter 6, this transfer function is of the nonminimum phase type, which means that its response has a reverse action behavior.

In order to generate a force toward the right to counteract a gust force to the left, for example, the rear steer angle has to steer to the left initially in order to rotate the car to establish the correct front slip angle. Ultimately, both the front and rear axle forces are directed toward the right to cancel the gust force. In the short term, however, although the aim of the control was to reduce the perceived acceleration to gusts, the basic nature of rear-wheel steering required a temporary increase in acceleration after a gust in order to later reduce the effect of the gust.

The system was indeed able to reduce the path deviation due to crosswind gusts, but many drivers reported a "poor subjective assessment" of the rear-wheel active steering system compared to the same vehicle without active steering. In this instance it would appear that active control of front-wheel steering would have been a better choice because the relevant transfer function, Eq. (6.36), has no reverse action associated with it.

B. Feedback Model Following Active Steering Control

The basic idea of model reference or model following control was introduced in Fig. 11.1. The idea is to force an actual vehicle to respond nearly as an ideal reference vehicle would with the same inputs. In the present example, we consider the simple automobile model developed in Chapter 6 and consider active front-wheel steering only. The input will be the steering angle commanded by the driver but the actual steering angle at the road wheels will be dynamically determined by the control system. If the controller is successful, the vehicle will respond much as the ideal vehicle would if given the same steer command.

To the extent that this control system functions as intended, it can solve many problems. For example, if the ideal model represents the real vehicle under normal conditions, the controller will attempt to maintain normal steering responses even when the real vehicle has become unstable. This could happen if a pickup truck were overloaded or if a rear tire were to lose air. Another possibility is that a vehicle may have poor steering characteristics because some other necessary aspects of its design preclude optimizing the dynamic responses. In this case the controller may render the vehicle much easier to drive than would otherwise be the case.

Furthermore, disturbance rejection is automatically a part of a model reference control. When the real vehicle is acted upon by disturbances, it will not follow the model response and the active system will act to bring the two responses closer together. Thus it can be expected that a model reference active steering system has the potential to react less to disturbances than the uncontrolled vehicle.

Finally, it is conceivable that the model parameters can be varied to suit either the task at hand or the driver's preference. During emergency maneuvers, the system could work with a VDC system as described above to do a better job of making the vehicle respond to the driver's commands. It is also conceivable that the driver could choose the response characteristics of the reference model either to make the controlled vehicle respond rapidly for negotiating a curvy road or to make it more stable for cruising a freeway (Anon., 2003).

It should be obvious that the active system will have physical limitations. If the model response deviates too much from the actual vehicle response it may be that the front wheels are simply not capable of generating enough force to bring the two responses together. Furthermore, for fail-safe operation the authority of the controller may be purposefully limited. In the case of the steering control using a differential in the steering column, the actuator will normally be limited in the amount of steering correction that may be added to the driver-commanded steer angle under most conditions. This is to limit the correction steer angle possible should the computer call for a very large correction angle due to a computer error.

All of these considerations obviously mean that stability cannot be guaranteed in all situations. For example, on a very low friction surface no possible active control can make a vehicle follow the responses of a model that presumes a normal friction surface. There will also always be some level of disturbance that will cause significant deviations between the model and the real vehicle. The designer of an active system must always perform computer or actual experiments to make certain that the controller does not make the situation worse under extreme conditions when its effectiveness is much reduced.

C. Sliding Mode Control

In this section, a particular type of active steering system will be discussed. Although the example to be presented will use linear models for the reference model as well as for the actual vehicle, and thus classical linear control techniques such as that shown in Fig. 11.3 could be used, it is clear that these models are only linearized approximations to the nonlinear dynamics actually involved. Thus it is reasonable to consider control techniques developed to be robust to variations in the linearized coefficients in the models as well as to disturbances. A robust control should continue to function even when the system is operating in the nonlinear region of some of its components. In the case of automotive vehicle lateral dynamics, the nonlinearities of the tire force generation are not severe until the car is maneuvering well beyond the normal range of acceleration in most cases. Therefore there is a good chance that a robust control technique will yield useful improvements in most practical cases.

In this case we will use the so-called "sliding mode control" technique (Muraca and Perone, 1991; Slotine, 1984; Slotine and Sasfry, 1983; Slotine and Li, 1991). The sliding mode theory will be presented only for the case of a single input linear system. For simplicity, the active steering application will use the front steer angle as the input and linearized models as developed in Chapter 6 for the actual vehicle and the electronic reference model.

As an introduction to the sliding mode concept, consider a dynamic system represented by the state equation

$$\dot{x} = Ax + bu + d \tag{11.4a}$$

where x is an n-dimensional vector of states, u is a scalar input, and d is a bounded vector of disturbances. [The equations of motion, Eqs. (6.22) and

(6.23), will be used in the active steering example.] Let $x_d(t)$ be a desired trajectory that the state $x(t)$ is intended to follow.

A function S is defined by the relation

$$S = c^T(x - x_d) \tag{11.5}$$

where c^T is an n-dimensional row vector.

Clearly, when $x = x_d$, $S = 0$. Then $S(x,x_d) = 0$ defines a surface in the state space that is called the sliding surface. This surface varies in time as $x_d = x_d(t)$. A controller is now constructed such that it will tend to drive S toward zero or, equivalently, that it will tend to drive x toward x_d. In the steering example, the desired trajectory x_d will be supplied by an electronic reference model that receives essentially the same steer angle input that the driver gives to the real vehicle. If S can be driven toward zero, then the system state variables will approach the variables of the desired trajectory. For the active steering example, if S can be driven toward zero, the actual vehicle will react to the driver's steering command much as the reference model vehicle would if it had the identical steering inputs.

Consider the following behavior for $\dot{S}$:

$$\dot{S} = -\rho S - \alpha \operatorname{sgn} S = -\rho S - \alpha \frac{|S|}{S} \tag{11.6}$$

where ρ and α are positive constants. Then the derivative of S^2 is easily found.

$$\frac{d}{dt}\frac{S^2}{2} = S\dot{S} = -\rho S^2 - \alpha|S| \tag{11.7}$$

upon use of Eq. (11.6). Equation (11.17) states that the derivative of the square of S is always negative, no matter whether S is positive or negative. This is a sufficient condition to guarantee that S converges to zero (Muraca, 1991).

A control strategy is derived by first differentiating Eq. (11.5) and substituting the result into Eq. (11.6).

$$\dot{S} = c^T(\dot{x} - \dot{x}_d) = -\rho S - \alpha \operatorname{sgn} S \tag{11.8}$$

Now Eq. (11.4) is used for $\dot{x}$ with the temporary assumption that the disturbance d is zero and Eq. (11.5) is used for S. When these relations are substituted into Eq. (11.8), the result is

$$c^T Ax + c^T bu - c^T \dot{x}_d = -\rho c^T(x - x_d) - \alpha \operatorname{sgn} c^T(x - x_d) \tag{11.9}$$

When Eq. (11.9) is rearranged, a control law emerges.

$$u = -\frac{1}{c^{\mathrm{T}}b}\left[c^{\mathrm{T}}Ax - c^{\mathrm{T}}\dot{x}_{\mathrm{d}} + \rho c^{\mathrm{T}}(x - x_{\mathrm{d}}) + \alpha \ \mathrm{sgn}\, c^{\mathrm{T}}(x - x_{\mathrm{d}})\right] \tag{11.10}$$

In Eq. (11.10), it must, of course, be assumed that

$$c^{\mathrm{T}}b \neq 0 \tag{11.11}$$

The robustness of the control law, Eq. (11.10), can be illustrated by using it with the system equation, Eq. (11.4), but now including the disturbance $d(t)$. When this is done, instead of Eq. (11.8), a slightly different equation for S is found.

$$\dot{S} = -\rho S - \alpha \ \mathrm{sgn}\, S + c^{\mathrm{T}}d(t) \tag{11.12}$$

Now if the disturbance is bounded,

$$\left|c^{\mathrm{T}}d(t)\right| \leq \beta \tag{11.13}$$

Eq. (11.7) becomes

$$S\dot{S} \leq -\rho S^2 - (\alpha - \beta)|S| \tag{11.14}$$

This implies that as long as

$$\alpha > \beta \tag{11.15}$$

the convergence to the sliding surface $S=0$ is still guaranteed.

This means that for finite-sized disturbances, one can always make $x \to x_{\mathrm{d}}$ by making α large enough. A similar proof shows that if α is large enough, convergence to the sliding surface $S=0$ will be assured even when the parameters of the system to be controlled vary in a bounded way (Muraca, 1991).

For the application to active steering, the conclusion is that even if the real vehicle responses vary because the vehicle responds as a nonlinear system depending on the situation in which it finds itself, the robust sliding mode controller should be able to force the system to respond much as the reference model does to the same steering inputs.

Although the control law, Eq. (11.10), is robust with respect to disturbances and parameter variations when α is sufficiently large, a chattering phenomenon may occur as the control law switches discontinuously whenever the sliding surface is crossed. (The signum function, sgn $S = S/|S|$, jumps between $+1$ and -1 whenever S changes sign.) In many cases this jerky behavior of the control variable is undesirable. In the case of steering control, it would not be desirable for the correction steer angle to chatter back and forth around some nominal value.

The chattering problem is often solved while retaining the essential robustness of the controller by using a function that has a continuous transition between +1 and −1 for a small region near $S=0$ in place of the signum function. An example of such a function would be $S/(|S|+\varepsilon)$ where ε is a small positive constant.

Finally, there is the task of choosing the components of c^{T} to assure reasonable behavior of the system when it is forced to remain on the surface $S=0$ in the state space. One possibility is to consider only the linear part of the control law, Eq. (11.10), by setting $\alpha=0$. With the assumptions that $x_{\mathrm{d}}=0$, $d=0$, one can then choose c^{T} such that the system has good stability properties. With these assumptions, when Eq. (11.10) is substituted into Eq. (11.4), the result is

$$\dot{x} = \left[A - \frac{bc^{\mathrm{T}}}{c^{\mathrm{T}}b}A - \rho\frac{bc^{\mathrm{T}}}{c^{\mathrm{T}}b}\right]x \tag{11.16}$$

One eigenvalue turns out always to be $s=-\rho$ as Eq. (11.6) becomes

$$\dot{S} = -\rho S \tag{11.17}$$

if only the linear part of the control law is used. The remaining eigenvalues depend on the choice of the components of c^{T} (Slotine and Sasfry, 1983).

Another possibility is to choose c^{T} so that only easily measured components of x are involved in S and then to check to make sure that the system can have good stability properties with this choice of components. Although all the components of the state vector may be involved in the control law, Eq. (11.10), because of the term $c^{\mathrm{T}}Ax$, the other terms $\rho c^{\mathrm{T}}(x-x_{\mathrm{d}})$ and α sgn $c^{\mathrm{T}}(x-x_{\mathrm{d}})$ would not then involve the state variables that are hard to measure.

In the example to follow, we will see that there are sometimes certain state variables that are not necessarily important in the control law. When this is the case, a simplified controller results.

State variables not directly measurable may, of course, be estimated using observers, but it is probably better to make as little use of observed variables as possible for the sake of simplicity and reliability. The use of an observer to estimate a state variable that is difficult to measure directly will be illustrated below.

D. Active Steering Applied to the "Bicycle" Model of an Automobile

The model of the lateral dynamics of an automobile with front-wheel steering developed only in Chapter 6 is represented by Eqs. (6.22) and (6.23). The state variables are the lateral velocity of the center of mass, V, and the yaw rate, r. The control input is the front steering angle, here to be denoted δ. For the present purpose the steer angle will be assumed to be a combination of a driver-supplied steer angle, δ_f, and a correction steer angle supplied by the controller, δ_c.

$$\delta = \delta_f + \delta_c \tag{11.18}$$

When Eqs. (6.22) and (6.23) are put in the form of Eq. (11.4) the result is

$$\begin{bmatrix} \dot{V} \\ \dot{r} \end{bmatrix} = \begin{bmatrix} A_{11} & A_{12} \\ A_{21} & A_{22} \end{bmatrix} \begin{bmatrix} V \\ r \end{bmatrix} + \begin{bmatrix} C_f/m \\ aC_f/I_z \end{bmatrix} (\delta_f + \delta_c) + \begin{bmatrix} d_1 \\ d_2 \end{bmatrix} \tag{11.19}$$

where

$$\begin{aligned} A_{11} &= -\frac{(C_f + C_r)}{mU}, \quad A_{12} = \left[-\frac{(aC_f - bC_r)}{mU} - U\right], \\ A_{21} &= -\frac{(aC_f - bC_r)}{I_z U}, \quad A_{22} = -\frac{(a^2 C_f + b^2 C_r)}{I_z U} \end{aligned} \tag{11.20}$$

The electronic reference model will be described by similar equations but with generally different parameters.

$$\begin{bmatrix} \dot{V}_d \\ \dot{r}_d \end{bmatrix} = \begin{bmatrix} A_{11d} & A_{12d} \\ A_{21d} & A_{22d} \end{bmatrix} \begin{bmatrix} V_d \\ r_d \end{bmatrix} + \begin{bmatrix} C_{fd}/m \\ aC_{fd}/I_z \end{bmatrix} \delta_f \tag{11.21}$$

in which it is assumed that the driver's steering angle input δ_f is measured and used as an input to the reference model.

The parameters of the model can be freely chosen so that the response of the model represents a desired response. For example, it might be logical to choose the parameters of the reference model to represent the actual vehicle when it is operating under design conditions. Thus no corrections from the controller would be necessary unless the vehicle began to deviate from the type of response that the designers had in mind. This philosophy allows the model reference controller to function as a fault detector. It could give the driver a warning that the car was no longer responding in a normal fashion while at the same time correcting the steering response to keep the response closer to the design response than it would be without active control.

Another philosophy has to do with defining a desired response characteristic different from the response characteristic inherent to the uncontrolled vehicle. Often, there are practical reasons why a certain vehicle type will not have inherently desirable steering responses. In such a case, the model reference controller can improve the response. As it has previously been discussed that a neutral steer vehicle is often thought to represent a sort of optimum vehicle in terms of steering, it might be logical to specialize the results for the neutral steer case and to consider these parameters for use in the reference model.

For neutral steer,

$$aC_\mathrm{f} = bC_\mathrm{r} \tag{11.22}$$

so C_r can be eliminated in favor of C_f as was done in deriving the neutral steer transfer functions in Eqs. (6.41)–(6.45). A neutral steer reference model has the following simplified set of parameters to be used in Eq. (11.21):

$$\begin{aligned} A_{11\mathrm{d}} &= -\frac{(a+b)}{bmU}C_\mathrm{f}, \quad A_{12\mathrm{d}} = -U, \\ A_{21\mathrm{d}} &= 0, \quad A_{22\mathrm{d}} = \frac{a(a+b)}{I_zU}C_\mathrm{f} \end{aligned} \tag{11.23}$$

In general, the sliding surface function S will involve all the state variables. In the case at hand, the yaw rate is easy to measure and many production vehicles have a yaw rate sensor. In contrast, the lateral velocity is not simple to measure directly. The lateral acceleration is often measured in production vehicles but this signal is not as straightforward to use in a control system for a number of reasons (see, for example, Nametz et al., 1988). For this reason, and for the sake of simplicity in presentation, we will develop a pure yaw rate controller.

E. Active Steering Yaw Rate Controller

The vehicle state variables and the reference model state variables for the bicycle model of Eqs. (6.19) and (6.20) are

$$x = \begin{bmatrix} V \\ r \end{bmatrix}, \quad x_\mathrm{d} = \begin{bmatrix} V_\mathrm{d} \\ r_\mathrm{d} \end{bmatrix} \tag{11.24}$$

To design a pure yaw rate controller, the sliding surface of Eq. (11.15) is made a function of the yaw rate alone.

$$S = c^\mathrm{T}(x - x_\mathrm{d}) = (r - r_\mathrm{d}) \tag{11.25}$$

which is achieved by choosing

$$c^{\mathrm{T}} = [0\ 1] \tag{11.26}$$

Then some of the terms in the control law, Eq. (11.10), are

$$c^{\mathrm{T}}b = \frac{aC_{\mathrm{f}}}{I_{\mathrm{z}}} \text{ and } c^{\mathrm{T}}Ax = A_{21}V + A_{22}r \tag{11.27}$$

The final form of the yaw rate control law using Eqs. (11.25)–(11.27) is

$$u = \delta_{\mathrm{f}} + \delta_{\mathrm{c}} = -\frac{I_{\mathrm{z}}}{aC_{\mathrm{f}}}\left[(A_{21}V + A_{22}r) - \dot{r}_{\mathrm{d}} + \rho(r - r_{\mathrm{d}}) + \alpha \operatorname{sgn}(r - r_{\mathrm{d}})\right] \tag{11.28}$$

We can note that because of the choice of c^{T} in Eq. (11.26) the lateral velocity appears only once in the control law while the rest of the feedback terms involve the measured yaw rate r.

Fig. 11.4 shows a block diagram of the yaw rate control together with the vehicle and the reference model. Note that although a model of the real vehicle is shown, the signals for lateral velocity and yaw rate that are shown flowing to the controller are actually measured signals coming from sensors on the actual vehicle. The mathematical model of the real vehicle only is presented to visualize the system and does not accurately represent the real vehicle in all cases, particularly when the vehicle is operating in a nonlinear region.

As the stability of the system depends on the choice of the c^{T} coefficients in the definition of the sliding surface it is worthwhile to use Eq. (11.16) to see whether the pure yaw rate control will result in a stable system when $S=0$ considering only the linear part of the control law. From Eq. (11.19) the column vector b is

$$b = \begin{bmatrix} C_{\mathrm{f}}/m \\ aC_{\mathrm{f}}/I_{\mathrm{z}} \end{bmatrix} \tag{11.29}$$

and, using Eq. (11.26),

$$bc^{\mathrm{T}} = \begin{bmatrix} 0 & C_{\mathrm{f}}/m \\ 0 & aC_{\mathrm{f}}/I_{\mathrm{z}} \end{bmatrix} \tag{11.30}$$

Using Eqs. (11.27) and (11.30) the terms in the closed-loop system matrix in Eq. (11.16) can be evaluated.

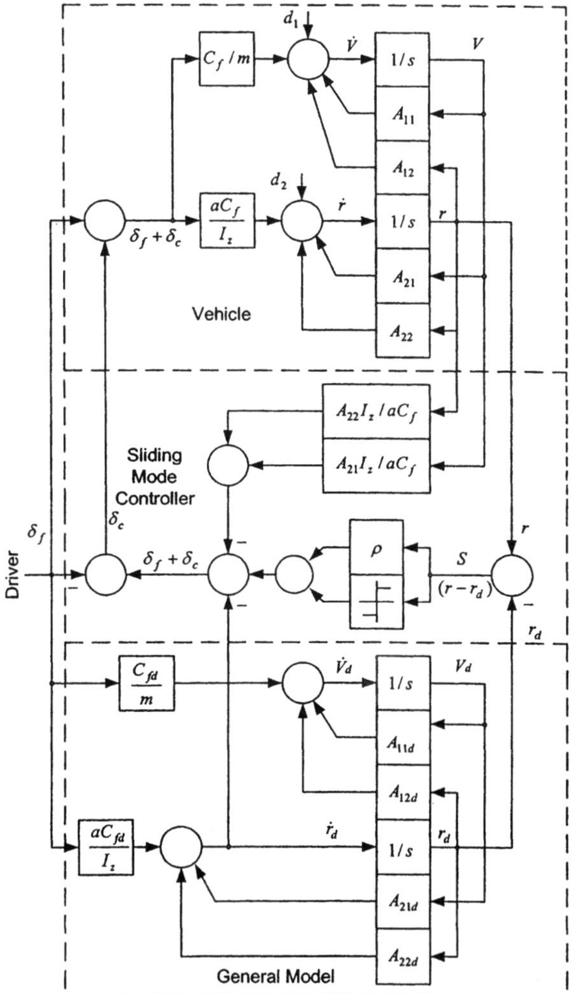

FIG. 11.4 Yaw rate model following active steering control using sliding mode technique (general case).

$$\begin{bmatrix} A_{11} & A_{12} \\ A_{21} & A_{22} \end{bmatrix} - \begin{bmatrix} 0 & I_z/am \\ 0 & 1 \end{bmatrix} \begin{bmatrix} A_{11} & A_{12} \\ A_{21} & A_{22} \end{bmatrix} - \rho \begin{bmatrix} 0 & I_z/am \\ 0 & 1 \end{bmatrix}$$
$$= \begin{bmatrix} A_{11} - A_{21}I_z/am & -A_{12} + A_{22}I_z/am - \rho I_z/am \\ 0 & -\rho \end{bmatrix} \qquad (11.31)$$

The eigenvalues corresponding to the matrix in Eq. (11.31) are found from the characteristic equation,

$$\det \begin{bmatrix} s - A_{11} + A_{21}I_z/am & -A_{12} + A_{22}I_z/am + \rho I_z am \\ 0 & s + \rho \end{bmatrix} = 0 \quad (11.32)$$

When written out, Eq. (11.32) becomes

$$(s + \rho)(s - A_{11} + A_{12}I_z/am) = 0 \qquad (11.32a)$$

This yields the eigenvalues

$$s_1 = -\rho, \; s_2 = A_{11} - A_{21}I_z/am \qquad (11.33)$$

As was predicted in Eq. (11.17), the first eigenvalue depends only on the control parameter ρ and has to do with the speed with which the system converges to the sliding surface $S=0$. The second eigenvalue relates to the stability of the system when it is in the surface.

Using the vehicle parameters of Eq. (11.20), the eigenvalue s_2 can be evaluated.

$$s_2 = -\frac{(C_f + C_r)}{mU} - \left[\frac{(aC_f - bC_r)I_z}{I_z Uam}\right] = -\frac{(a+b)C_r}{amU} \qquad (11.34)$$

All the parameters in Eq. (11.34) are positive so s_2 is negative. This negative eigenvalue shows that the controlled system is stable.

As can be seen from Fig. 11.4, because of the particular form chosen for c^T (a pure yaw rate controller) V_d is not needed for the controller, but V_d itself must be computed in order to find r_d which is needed for the controller. Fig. 11.5 shows a further simplification when a neutral steer reference model is assumed. In this case only a simplified first-order yaw model will suffice because the parameter A_{21d} in Eq. (11.23) vanishes.

This robust sliding mode controller shown in Fig. 11.5 can be related to previously studied model reference controllers of the type shown in Fig. 11.1. One idea is to use only the yaw rate error, $r-r_d$, as a control signal and to compute the *steady* yaw rate corresponding to a given δ_f as the reference yaw rate, r_d (Nametz et al., 1988).

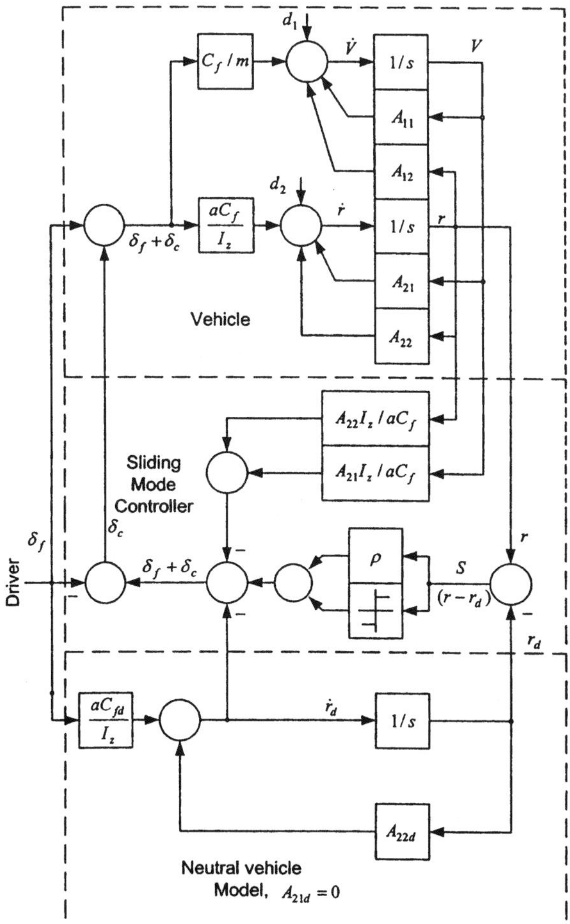

FIG. 11.5 Yaw rate model following active steering control using sliding mode technique (neutral steer reference model).

Using Eqs. (11.21) and (11.23) for the neutral steer reference model, the desired steady yaw rate is found to be

$$r_{\text{dss}} = \frac{aC_f\delta_f}{I_zA_{22d}} = \frac{U\delta_f}{(a+b)} \tag{11.35}$$

As Figs. 11.4 and 11.5 show, a model reference controller using only yaw rate error and a steady-state yaw rate for the desired yaw rate is essentially a simplification of the robust sliding mode control discussed above.

The idea of the model reference controller was to use Eq. (11.35) for the reference model. This is a static model rather than a dynamic one and it does not consider a desired lateral velocity. Similarly, the neutral steer dynamic reference model in Fig. 11.5 does not produce a desired lateral velocity. Furthermore, this model reference controller was a pure yaw rate controller and did not consider the real vehicle lateral velocity at all.

The sliding mode controllers in Figs. 11.4 and 11.5 both do, however, show a dependence on the real vehicle measured lateral velocity at one point. In fact, the real vehicle lateral velocity is actually used only to compute the yaw acceleration, $\dot{r}$, which is used as a signal in the sliding mode controller. If a yaw accelerometer could be used to determine $\dot{r}$ then there would be no need to use the lateral velocity V.

The concept for the simple yaw rate controller using only the steady-state yaw rate computed from the driver input, δ_f, is that a yaw rate controller that effectively controls the vehicle so that $r \cong r_d$ will also implicitly control V and hence the lateral acceleration as well. That such active steering control systems have been realized successfully and can be regarded as simplifications of the sliding mode controllers of Figs. 11.4 and 11.5 makes the argument that robust controllers should be able to have even better performance (see Nametz et al., 1988).

A question remains about the importance of estimating V in general. This signal is used only once in the sliding mode yaw rate controller where it is multiplied by the term $A_{21}I_z/aC_f$ and used to compute $\dot{r}$. This term $A_{21}I_z/aC_f$ is proportional to the understeer coefficient, which would be zero if the vehicle were actually neutral steer.

This suggests that, as the controller is supposed to be robust with respect to parameter variations, it may be permissible to assume that the term is zero as it would be for a neutral steer vehicle. This would mean that no estimate of the lateral velocity would be necessary for the yaw rate controller. The fact that other models following controllers have proved to be successful without using lateral velocity information strengthens this argument.

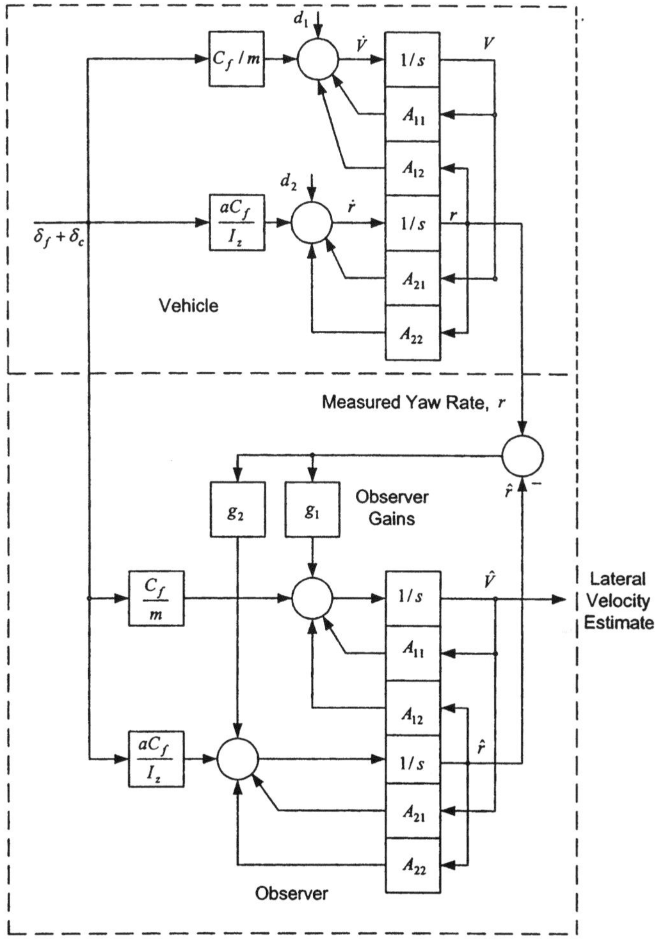

FIG. 11.6 Luenberger observer for estimating lateral velocity from measured yaw rate.

On the other hand, a strongly understeering or oversteering vehicle may benefit by using an estimate of V based on a mathematical model of the vehicle and measurements of r and $\delta_f + \delta_c$ using a Luenberger observer or Kalman filter techniques (Luenberger, 1971; Ogata, 1970). The estimate of V could be used in either a yaw rate controller or a more general controller using both yaw rate and lateral velocity error signals.

See Fig. 11.6 for an example of the general scheme of a Luenberger observer. Using only the actual steer angle and the measured yaw rate, the observer computes an estimate of the lateral velocity. This estimate can then be used in place of a measured lateral velocity in a sliding mode controller. The two gains g_1 and g_2 adjust the speed at which the estimate of the lateral velocity $\hat{V}$ converges to the actual lateral velocity, V. The well-known separation theorem indicates that, at least under some assumptions, the observer and the controller dynamics remain independent even when the observer variables are used in the feedback controller rather than the actual variables.

We have shown that it is feasible to construct a robust model following steering correction controller using sliding mode techniques and yaw rate feedback. It is also possible to use a static reference model for a simplified model reference control system.

Active steering control systems can be realized in several ways. Yaw rate sensors are in common use and it is possible to use yaw accelerometers or observers for the lateral velocity. The resulting systems can be stable and can make the vehicle respond much as the reference model would respond. The handling properties of the real vehicle can be changed by varying the parameters of the reference model, and the controlled vehicle will automatically tend to reject external moment disturbances from wind gusts or braking on split friction coefficient surfaces.

Of course, this presentation is highly simplified and real control systems will have a number of special features necessary to assure good behavior under the many special circumstances the vehicle and the driver will encounter. These examples are intended only to illustrate the philosophy of active control of handling dynamics.

VI. LIMITATIONS OF ACTIVE STABILITY ENHANCEMENT

It would be a mistake to think that the inherent dynamics of a vehicle are of little importance if active control is to be applied. One reason has to do with failure modes. An unstable vehicle may be stabilized actively but should there be a failure in the control system, it would be preferable if the operator could still control the vehicle. It is probably not a good idea to design a very dangerous vehicle and rely on an active control system to correct its bad behavior.

Also, there is the question of control authority. Any attempt to change the dynamic behavior of a vehicle with active means must consider the amount of control effort that is available. In theory, a system can be stabilized with virtually no control effort. In our analysis of vehicle stability, we considered only infinitesimal deviation from a basic motion. For an infinitesimal deviation from a basic motion, it takes only an infinitesimal force to bring the vehicle back to the basic motion. The control problem is to assure that the control force is properly related to the vehicle dynamics such that it will actually stabilize the system but the amount of control effort is always proportional to the deviation. In reality finite-sized disturbances will affect the vehicle and thus the controller must have enough authority to overcome destabilizing influences.

In some cases, large disturbances can be anticipated and this means that large control forces will be required to stabilize the system. If these forces are limited, perhaps in order to assure fail-safe behavior, it may be that on occasion, the controller will be unable to stabilize the vehicle. In this case a drastically unstable vehicle that up to a point seemed stable would pose a severe problem for a human operator.

Appendix: Bond Graphs for Vehicle Dynamics

Bond graphs are a useful way to describe dynamic systems particularly if they involve a variety of forms of energy, Karnopp et al., 2000. Figures 2.3 and 2.4 demonstrated that rigid body dynamics, when described using a body-fixed coordinate system, can be elegantly described using bond graphs, Karnopp 1976. Later in the text, Figs. 8.6 and 8.7 used bond graphs and the analysis of causality to help explain the relationships between relaxation length models of dynamic tire force generation and a model including sidewall flexibility.

Bond graphs are particularly useful when actuators using electrical, hydraulic or pneumatic power are to be integrated into an active vehicle system as described in Chapter 11. A number of studies vehicle dynamics including active steering systems, active or semi-active suspension systems, automatic transmissions and other power transduction systems for vehicles are well suited for bond graph analysis, Pacejka and Toi, 1983, Pacjka, 1985, Zeid and Chang, 1989, Hrovat and Ascari, 1994, Sanyal and Karmaker,1995, Margolis and Shim, 2001.

Of course, when the equations for a system model are already known, it may seem pointless to develop a graphical model that could be processed to yield another version of the system equations. The main point of bond graphs, after all, is to provide graphical models of components that can be linked together to construct a system model. After the components are linked, a causal analysis can be done to see whether of not there are any problems associated with the connections between the components. Finally, bondgraphs provide manual or automated techniques for deriving state equations suitable for analysis or computer simulation.

On the other hand, bond graphs are often helpful in experimenting with different versions of a vehicle dynamic model. In particular, the bond graph is often capable of predicting the effect on the relevant differential equations of the addition or elimination of physical effects that may or may not be important to understanding the vehicle dynamics.

It is often the case that it is not a simple job to combine computer models of vehicles and actuators developed by specialists in electrical, hydraulic or pneumatic devices to see how active systems can be controlled to influence vehicle stability. Causal studies of the interaction between various components in bond graph model form can be very useful in this regard. Furthermore, bond graph connections are power connections rather than mere signal connections. Bond graph elements handle power and energy in appropriate ways often avoiding the inadvertent construction of system models that violate the laws of thermodynamics.

In this appendix, some bond graph models of the systems studied in the text will be illustrated. The focus will be on representations of the laws of mechanics for rigid bodies as well as on elements that represent the force-generating components of vehicles. Examples will be given of vehicles represented in both inertial coordinate frames and in Body-fixed frames.

The bond graphs in these cases do little more than give a graphical and interactive power representation of the basic equations derived using the principles of mechanics. For those who have some appreciation of bond graphs, the examples will indicate that the equations derived do obey the basic power conservation principles built in to the bond graph language. The bond graphs provide yet another illustration that the method used to formulate equations influences the form that the final equations of motion take. Furthermore, the bond graph models are readily modified to include effects such as dynamic tire force generation. They also could be extended to allow actuators to be coupled to the vehicle models to allow studies of active stability enhancement systems.

I. A BOND GRAPH FOR THE TWO DEGREE-OF-FREEDOM TRAILER

As a first example, the two degree-of-freedom model of a trailer analyzed in Chapter 5 will be represented in a bond graph. The trailer was shown in a sketch in Fig. 5.5 and some of the kinematic variables that were necessary to find the slip angle were shown in Fig. 5.6.

As will become clear, the bond graph representation does not necessarily produce the same equations of motion that other methods produce. The Lagrange equation method used in chapter 5 resulted in two coupled second order differential equations in the position variables x and θ and their derivatives, Eqs. (5.30). Linearized equations of motion written using Lagrange's equations turn out to be in the form of Eq. (3.21). The direct use of Newton's laws typically results in equations similar to but often not identical to the equations resulting from Lagrange's equations. In contrast, the standard bond graph techniques produce coupled first order equations in the form of Eqs. (3.4–3.6). Of course, unless an error has been made in the equation formulation, all possible equation sets are equivalent and the conclusions about the stability of the vehicle do not depend on the formulation technique used.

The variables used to describe the motion are also somewhat different for the bond graph technique when compared to other methods in applied mechanics. The state variables used in bond graphs are the general energy variables called momenta and displacements. In the trailer example, there will be two momentum variables, the lateral (translational) momentum of the trailer, P_c, and the angular momentum, H_c. The two displacement state variables are the spring deflection, x, and the angular displacement, θ. The two displacement variables used in the bond graph are the same variables used in Eq. (5.30) but instead of the derivatives of the displacement variables, the bond graph uses the corresponding momentum variables.

The trailer examples in Chapter 5 used inertial coordinate frames because the basic motion was movement in a straight line so that it was easy to describe the accelerations in a frame in which Newton's laws are easily expressed without the correction terms associated with a rotating frame. In such a frame, the rigid body dynamics for plane motion are easily represented in bond graph form, Karnopp et al., 2000. The translational motion of the center of mass is decoupled from the rotation and both effects are represented by simple inertia elements. The more complicated case when a body-fixed coordinate system will be illustrated in the following two examples.

Figure A.1 shows a simplified version of the trailer shown in Figs. 5.5 and 5.6. The two wheels have been shown as a single equivalent wheel much in the manner used for the bicycle car model of Chapter 6 since it was demonstrated that the track l had little effect on the slip angle expression for a stability analysis.

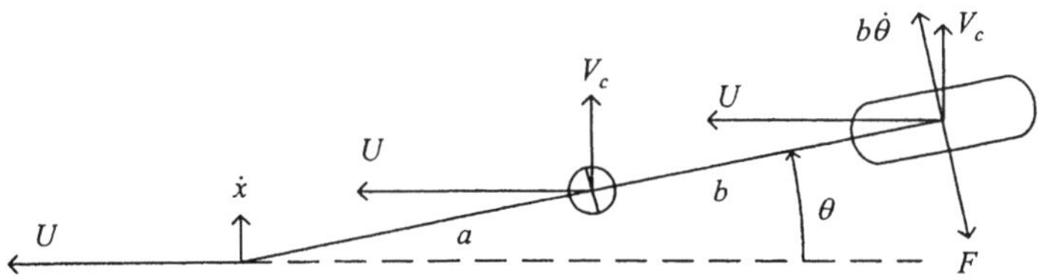

FIG. A.1 A sketch of the two degree-of-freedom trailer model corresponding to Figs. 5.5 and 5.6.

The significant difference between the velocity components shown in Fig. A.1 and those shown in Fig 5.6 is due to the focus on motion of the center of mass in Fig. A.1. The velocities at the equivalent tire are to be determined in terms of the velocity of the center of mass rather than the velocity of the nose of the trailer as was done in Chapter 5.

The lateral velocity of the wheel will be expressed in terms of the sideways velocity of the center of mass, V_c, instead of the sideways velocity of the nose of the trailer, $\dot{x}$, as was done in Eq. (5.19).

$$\alpha = U\sin\theta + V_c \cos\theta + b\dot{\theta} \cong U\theta + V_c + b\dot{\theta} \tag{A.1}$$

Since the rolling velocity remains approximately the same as in Eq. (5.18), simply U, the approximate expression for the slip angle is

$$\alpha = \theta + (V_c + b\dot{\theta})/U \tag{A.2}$$

The lateral force on the tire is then

$$F = C_\alpha \alpha = C_\alpha \theta + (C_\alpha/U)(V_c + b\dot{\theta}) \tag{A.3}$$

in which the cornering coefficient is positive when F is defined to be positive in the direction shown in Fig. A.1.

The sideways velocity of the nose of the trailer, $\dot{x}$, can now also be expressed in terms of the velocity of the center of mass.

$$\dot{x} = V_c - a\dot{\theta} \tag{A.4}$$

The linear velocity of the center of mass, V_c and the angular velocity, $\dot{\theta}$, will actually be related to the momentum variables P_c and H_c in the bond graph.

$$V_c = P_c/m \tag{A.5}$$

$$\dot{\theta} = H_c/I_c \tag{A.6}$$

A Bond Graph for the trailer appears in Fig. A.2.

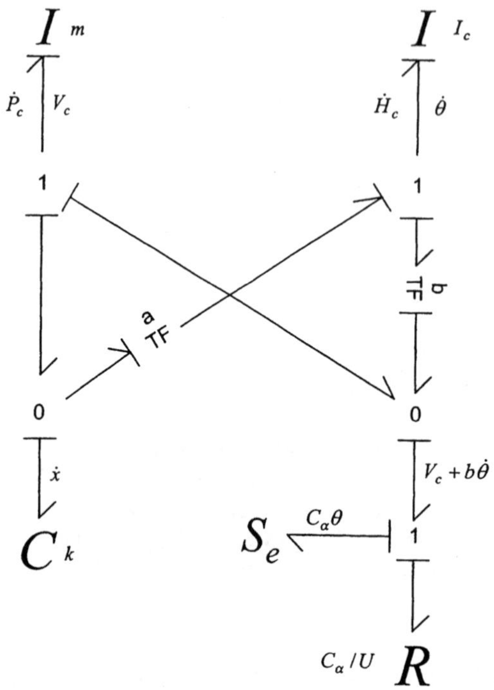

FIG. A.2 A bond graph representation for the two degree-of-freedom trailer model.

The causal marks of the two I-elements indicate that the velocity and angular velocity are related as in Eqs. (A.5 and A.6) to the momentum variables. The C-element computes the spring force in terms of the spring deflection, x. The bond graph fragment at the lower right corner of Fig. A.2 is basically the same as the fragment in Fig. 8.6. It represents the force law of Eq. (A.3).

As was pointed out in Chapter 8, this force expression really represents a steady state lateral force. This piece of the bond graph could be replaced with a bond graph representation similar to that in Fig. 8.7. This then would introduce a dynamic force with a relaxation length or a sidewall stiffness as an extra parameter. Since trailers are generally stable at low speeds and the distinction between the steady state force and the dynamic force grows smaller with increasing speed, this extra complication is probably not worthwhile. The ability to modify a model by substituting component models is, in any case, a useful feature of the bond graph method. A study of causality after a component model substitution will reveal how the order of the system has changed and whether and algebraic difficulties are associated with the change in the model.

The transformer structure is derived by considering Eq. (A.4) and the term $V_c + b\dot{\theta}$ in Eq. (A.2). A useful feature of bond graphs is that if kinematic relationships are represented using power-conserving elements such as junctions and transformers, then the force and torque relationships will automatically also be correctly represented. Thus the transformer structure in Fig. A.2 computes the net for force on the center of mass and the net torque on the moment of inertia.

The equations of motion for the bond graph of Fig. A.2 are easily written by hand or they could be written automatically by a computer bond graph processor such as CAMP-G, Granda, 2002.

A bond graph processor takes a bond graph drawn on a computer screen, analyzes it for computational inconsistencies and then produces an input file for another program to analysis or simulate the equations of motion.

If the equations are written by hand from the bond graph of Fig. A.2, the results are as shown below.

$$\begin{aligned}
\dot{x} &= P_c/m - aH_c/I_c \\
\dot{\theta} &= H_c/I_c \\
\dot{P}_c &= -kx - C_\alpha\theta - (C_\alpha/U)(P_c/m + bH_c/I_c) \\
\dot{H}_c &= akx - b(C_\alpha\theta) - b(C_\alpha/U)(P_c/m + bH_c/I_c)
\end{aligned} \tag{A.7}$$

Equations (A.7) could be put in the matrix form of Eq.(3.6) but they have been left in the form that they arise naturally by following the causal marks on the bond graph. These first order equations are particularly useful for computer simulation since many simulation programs are organized to solve sets of first order equations. It makes relatively little difference as far as simulation is concerned if the bond graph has elements with nonlinear constitutive laws or whether all the elements have been linearized as in Eq. (A.7) for the purposes of stability analysis. The four first order state equations of Eq. (A.7) correspond quite closely to the two second order displacement equation in Eq. (5.30).

II. A BOND GRAPH FOR A SIMPLE CAR MODEL

Figure 2.3 showed bond graphs for general three-dimensional motion of a rigid body. In this case there were three translational forces and velocities and three rotational moments and angular velocities. Figure 2.4 showed a major simplification that resulted when the heave velocity, W, and the pitch rate, q, were assumed to vanish. Much of Chapter 6 was concerned with the even simpler case of plane motion with heave velocity, W, pitch angular velocity, q, and roll rate, p, all assumed to be zero. When this is the case, another part of the bond graph in Fig. 2.4 involving p can be eliminated, further simplifying the general bond graph of Fig. 2.3.

The car model of Figs. 6.5 and 6.6 involves only the velocities U, V and the yaw rate r. The equations of motion for V and r are given in Eqs. (6.18 and 6.19). Although it was assumed that the forward velocity U was constant, in fact, if the force in the x-direction is assumed to vanish, there is an equation for U as noted in the shopping cart example on Chapter 1, Eq. (1.31). For stability analysis this equation could be neglected because the product of two perturbation variables, rV, is a second order term that can be neglected in a linearized analysis.

Figure A.3 is a bond graph representation of the "bicycle" car model described in a body-fixed coordinate system. The bond graph incorporates the dynamic equations for the three variables, Eqs. (1.31, 6.18 and 6.19). The right hand side is clearly a simplified version Figs. 2.3 in general and 2.4 in particular.

The left hand side of Fig. A.3 incorporates a calculation to the slip angles, Eqs. (6.20), and the lateral force laws, Eqs. (6.21). The two linear resistive R-elements produce forces as the product of lateral velocities at

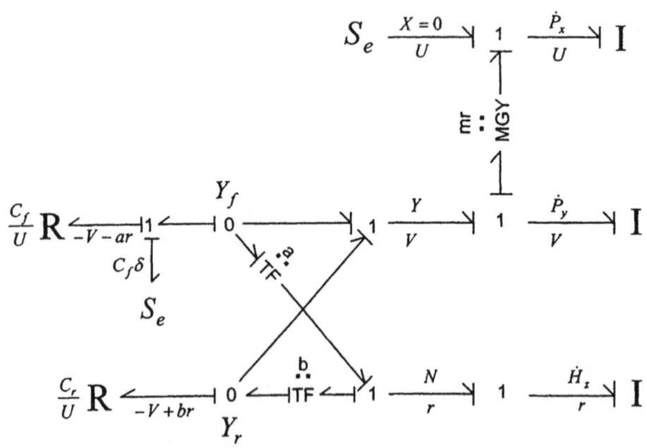

FIG. A.3 Bond graph model for the bicycle car model of Fig. 6.6.

the two axles times cornering coefficients divided by the forward velocity. This bond graph R-element representation was seen previously in Fig. 8.6. The idea that a pneumatic tire is connected to the ground by what appears to be a dashpot-like element was mentioned in Chapters 4 and 9. This concept assumes concrete form in the bond graph representation using R-elements.

The transformer structure in the middle of the bond graph relates the main variables V and r to velocities at the wheels, (more precisely at he equivalent wheels representing the axles). Simultaneously the same structure relates the total lateral force and moment to the wheel lateral forces. It is a feature of bond graphs that the power conservation properties of junction structures assures that kinematic velocity and kinetic force transformations are automatically compatible. Errors often occur when the two types of relation are performed independently as was done (correctly) in Chapter 6.

As was noted in Chapter 2 and in the first example, the basic variables of bondgraphs are momenta and their rates rather than velocities and accelerations and this can be seen in the Fig. A.1 in which $\dot{P}_y$ and $\dot{H}_z$ appear in place of $\dot{V}$ and $\dot{r}$ in Eqs. (6.22 and 6.23). The equation sets are, of course, exactly equivalent and the same results obtain no matter which set is used.

III. A BOND GRAPH FOR A SIMPLE AIRPLANE MODEL

In Chapter 9, a simplified model of the longitudinal dynamics of an airplane was developed to illustrate the differences between the lateral stability of an automobile and the longitudinal stability of aircraft. A sketch of the aircraft model appears in Fig. 9.11. It is not surprising that the bond graph representation of these two plane motion models should show some similarity. Both models should incorporate simplifications of the rigid body bond graphs in Figs. 2.3 and 2.4 although the variables retained in the plane motion models are different. Both models assumed that the forward velocity was constant, but the car model used yaw rate and lateral velocity as variables while the airplane model used pitch rate and vertical velocity as variables.

Figure A.4 shows a bond graph representation of the simplified airplane model developed in Chapter 9. On the right side of Fig. A.4 we see what remains of the general bond graph of Fig 2.3 when the assumption that the lateral velocity, V, the roll rate, p, and the yaw rate, r, are all assumed to vanish and only the vertical or heave velocity, W, and the pitch rate, q, are to be used as variables for a stability analysis.

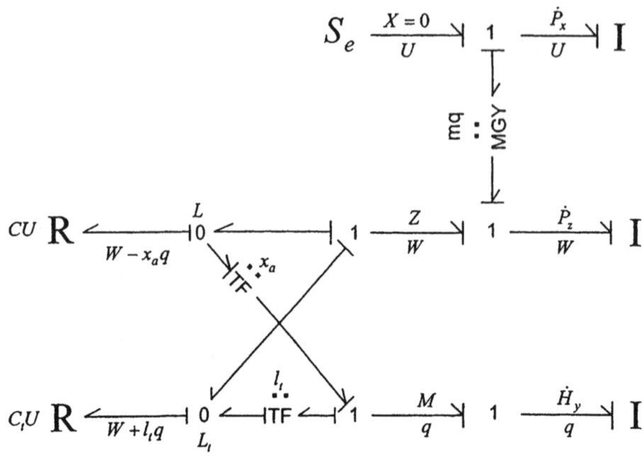

FIG. A.4 Bond graph model of the simplified airplane of Fig. 9.11.

As in Fig. A.3, the velocity U in the forward is not exactly constant even when the thrust and drag are assumed to cancel so that the net longitudinal force, X, is assumed to vanish. The rate of change of momentum has a term mqW, which for the purposes of stability analysis can be neglected as long as the perturbation variables q and W remain small.

The derivation of the equations for the airplane model considered the weight and the steady components of lift at the wing and the tail and resulted in Eqs. (9.46 and 9.47). These forces disappear when the assumption is made that he airplane is flying in a trimmed condition. The bond graph of Fig. A.4 is equivalent to Eqs. (9.48 and 9.49) in which the steady components of the forces no longer appear. The steady parts of the forces could readily be incorporated in the bond graph but it would then have a more cluttered appearance.

The transformer structure in Fig. A.24 again has two effects, as did the analogous transformer structure in Fig. A.2 and Fig. A.3. It computes the vertical velocities at the wing and at the tail and it distributes the incremental lift forces L and L_t at the wing and at the tail to the total vertical force Z and the pitch moment M. The vertical velocities determine the changes in angle of attack at the wing and at the tail. These in turn determine the incremental lift forces at the wing and tail through the coefficients defined in Eqs. (9.43 and 9.44). A relationship between a velocity and a force is represented by a resistive R-element on bond graphs so there is an automatic similarity between the R-elements in Fig. A.2 and Fig. A.3 representing lateral tire force generation and the R-elements in Fig. A.4 representing lift force generation.

Among the interesting differences between the car model of Fig. A.3 and the airplane model of Fig. A.4 is in the linear resistance parameters in the two cases. The car parameters were cornering coefficients *divided* by the forward speed, while the airplane parameters were coefficients from Eqs. (9.43 and 9.44) *multiplied* by forward speed.

This difference in the role of speed U in the two cases manifested itself in the stability analysis for the two cases. The terms in Eqs. (9.51 and 9.53) that had to be positive if the corresponding vehicles were to be stable behaved quite differently from one another as a function of speed. For the airplane, static stability assured dynamic stability at all speeds for the simple model. The car model proved to be more complicated. Static stability did assure dynamic stability at all speeds but static instability meant dynamic instability only for speeds above a critical speed.

The difference in the behavior of the bond graph resistance parameters as functions of speed for the two cases shows immediately differences between wheels and wings. The resistance parameter decreases for the pneumatic tire representation as speed increases. The equivalent bond graph resistance parameter for a wing increases as the speed increases. This reinforces the remark made earlier that as speed increases, the pneumatic tire seems to be ever less stiffly connected to the ground while a wing seem ever more stiffly connected to the air.

Problems

CHAPTER 1

1.1 Why does the variable θ not enter into Eqs. (1.1) and (1.2)?

1.2 Show that Eq. (1.3) is correct by considering a stick moving horizontally with velocities of V_U and V_L at the two ends as shown in the sketch.

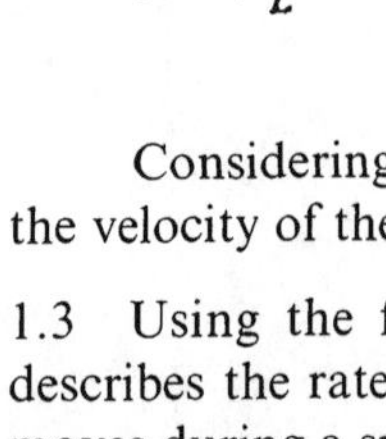

Considering the motion during a small interval of time Δt, determine the velocity of the center point of the stick.

1.3 Using the figure in Problem 1.2, show that Eq. (1.4) correctly describes the rate of rotation of the stick. Again, consider how the stick moves during a small interval of time Δt.

1.4 Consider the tapered wheelset discussed in Chapter 1. Show that the wavelength for perturbed motion is $\lambda = r_0 \Omega T = 2\pi r_0 \sqrt{l/2\Psi r_0}$, where T is the period of the oscillation, $T = 2\pi/\omega_n$, and the other parameters were defined in the chapter. Note that λ is independent of the angular velocity Ω, and hence is independent of the forward speed $\dot{x}$. See if you can reformulate the dynamic equation Eq. (1.6) as an equation in space, rather than in time. That is, find an equation for d^2y/dx^2, rather than d^2y/dt^2. This equation would yield the wavelength directly without any reference to speed.

1.5 Consider the first-order differential equation $dy/dt + a_1 y = 0$. Discuss the stability of systems described by this equation in terms of the value of the coefficient a_1. Sketch solutions for both the stable and the unstable cases, and show how a_1 is related to the time constant. Indicate how one could find the time constant from a plot of $y(t)$.

1.6 Prove that, for an exponential decay such as that sketched in Fig. 1.4, a straight line tangent to the curve starting at *any* point will intersect the zero line after a time interval equal to the time constant.

1.7 Referring to Fig.1.3, find the velocity of the center of the rear axle in the direction of the center line of the cart for the perturbed motion. Use the small angle approximations for the trigonometric functions of ϕ.

1.8 Suppose a point mass is traveling around a circular path at a constant speed U. The x–y coordinate system is attached to the point mass and rotates such that the y-direction remains along a radius of the circle. The coordinate system is similar to that used in Fig. 1.5. Note that for the x–y–z axis system to be right-handed, the z-axis must point down. The angular velocity around the z-axis is called r and is positive in the direction shown.

The radius of the circular path is R and the yaw rate for the coordinate system is actually negative as indicated in the figure below.

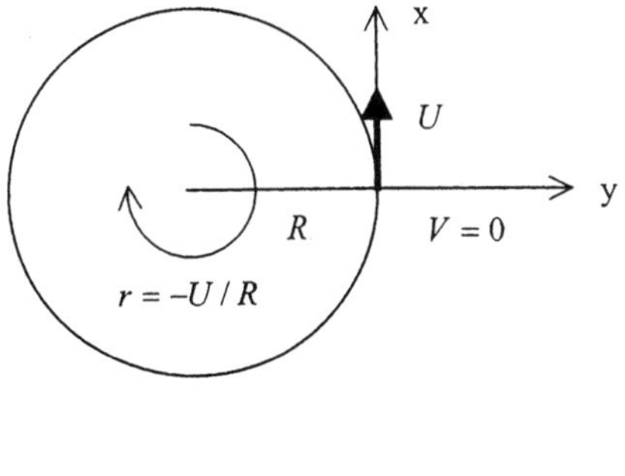

The lateral velocity in the y-direction remains constant at zero. Using Eq. (1.30), compute the acceleration in the y-direction. Does this result agree with the formula for centripetal acceleration given in books on dynamics?

1.9

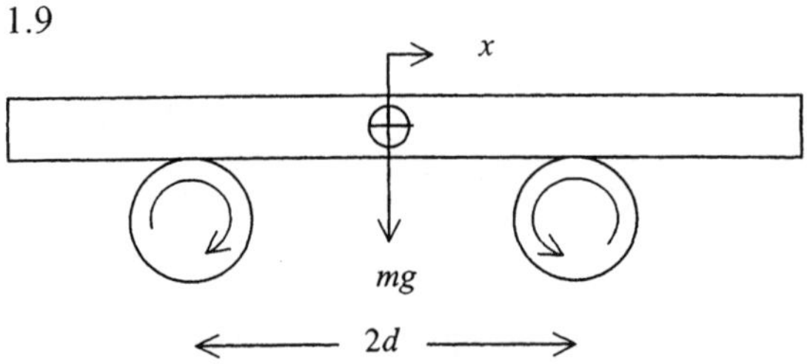

A bar rests on counterrotating rolls as shown. Dry friction causes opposing forces on the bar, which is in equilibrium at $x=0$. Assume coulomb friction, i.e., that the friction forces are proportional to the normal forces on the rolls. Show that the net force on the bar depends on x and is spring-like in character. That is, that the net force on the bar from the rolls is proportional to x and for the rotation senses shown, tends to return the bar to the center position.

The rolls have a radius R and are rotating in the directions shown with an angular speed of Ω. The separation distance between the rolls is $2d$. For the first part of the problem, assume that the rolls are turning fast enough that $R\Omega > \dot{x}$.

What is the natural frequency of the system and at what sinusoidal amplitude of motion would the relative velocity between the bar and the rolls, $\dot{x} \pm R\Omega$, first become zero?

The resulting oscillation of the bar can be stable or unstable depending on details of the friction force law. Sketch laws that would produce stable and unstable oscillations. (Hint: Think how the relative velocity between the bar and the rolls, $\dot{x} \pm R\Omega$, might influence the friction laws in a stable or unstable way.)

Is the system stable if the rolls rotate in the opposite direction shown in the figure?

1.10 The figure below is identical to Fig. 1.5, except the separation of the wheels on the back axle of the cart has been designated by the symbol t for "track."

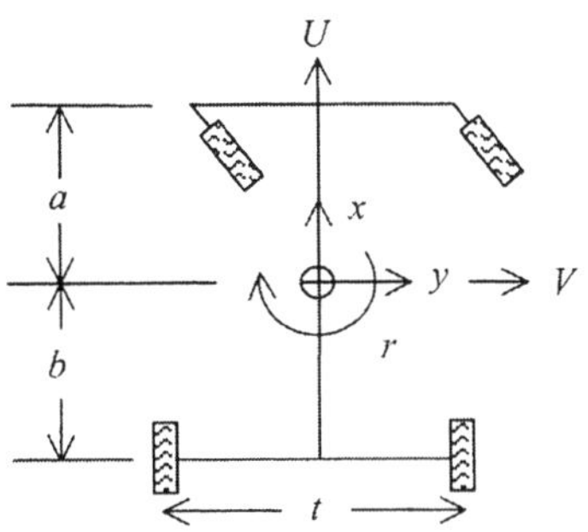

The sideways velocity for the center point of the axle was given in Eq. (1.34) as $V-br$. The total velocity of the center was found by applying Eq. (1.15) starting from the velocity of the center of mass with components U and V and then adding the term $\vec{\omega} \times \vec{r}_{AB}$, where the magnitude of $\vec{\omega}$ is r and the length of the vector $\vec{r}_{AB}$ is just b. Equation (1.34) considered only the sideways component of the total velocity and constrained it to be zero.

To find the total velocity of a wheel, one can again apply Eq. (1.15), but now $\vec{r}_{AB}$ is a vector with one component in the minus x-direction of length b and another in the plus or minus y-direction of length $t/2$.

Find the total velocity at either wheel, and confirm that the sideways velocity component at the wheel location is still $V-br$.

Because this result is independent of the track, this confirms that Eq. (1.34) correctly constrains the wheels to move only in the direction they are pointed. It also shows that all points on the axle have the same velocity in the y-direction.

1.11 The claim is made that Eqs. (1.32, 1.33, and 1.34) can be combined to yield Eq. (1.19). Show that this is true. If, instead of eliminating Y and V

to find an equation for r, one were to find an equation for Y by eliminating V and r, what do you think the result would be?

CHAPTER 2

2.1. The sketch shows a case of plane motion with velocity components U and V referred to x–y coordinates moving with a body as discussed in the chapter. The angular rate of rotation is the yaw rate r.

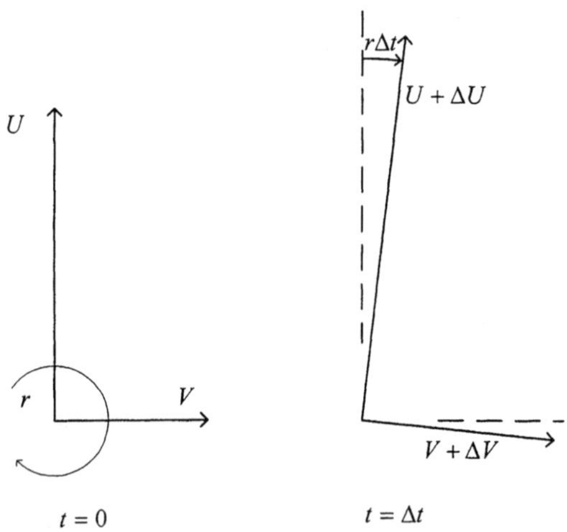

By superimposing the two cases, find the rate of change of velocity in the x-direction and y-direction, and thus verify the acceleration terms in Eqs. (2.22) and (2.23).

2.2 In elementary dynamics, it is common to see an analogy between $F = ma$ for linear motion, and $\tau = I\alpha$ for rotary motion, where τ is a torque, I is a moment of inertia, and α is angular acceleration. Equation (2.19) is very complicated but it ought to reduce to the simple form $\tau = I\alpha$ under restricted conditions. Suppose a body is moving in plane motion in the x–y plane. Show that the simple form then does apply and identify τ, I, and α from the terms used in Eq. (2.19).

2.3 Use the following moving coordinate system to describe the motion of a land vehicle:

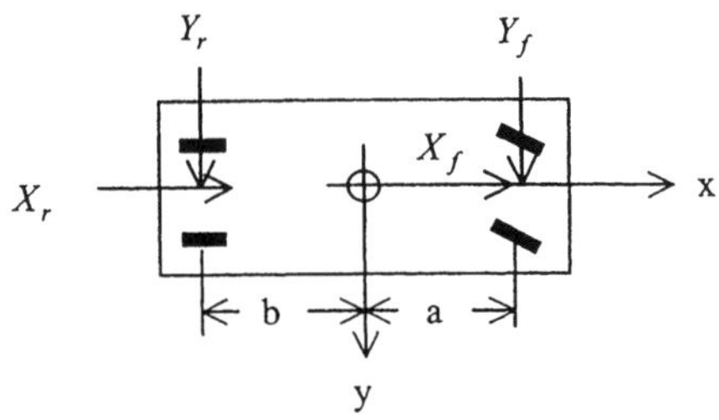

Coordinates Attached to Vehicle	x y z
Unit Vectors	$\vec{i}\ \vec{j}\ \vec{k}$
Center of Mass Velocities	U V W
Angular Velocities	p q r
Forces	X Y Z
Moments	L M N

Assume that the tire forces act at the front and rear midway between the wheels as shown. The Mass parameters are: m, I_{xx}, I_{yy}, I_{zz} and the products of inertia are all zero, i.e., it is assumed that the $x-y-z$ axes are principal axes.

Specialize the general equations given in the chapter by assuming no heave motion ($W=0$), no roll motion ($p=0$), and no pitch motion ($q=0$) to derive the following equations of motion:

$$m(\dot{U} - rV) = X_{\mathrm{f}} + X_{\mathrm{r}}, \qquad m(\dot{V} + rU) = Y_{\mathrm{f}} + Y_{\mathrm{r}},$$
$$I_{zz}\dot{r} = aY_{\mathrm{f}} - bY_{\mathrm{r}}$$

2.4 Show that if the body in question happened to be a sphere, Eq. (2.12) could be written in vector form $\vec{\tau} = I\dot{\vec{\omega}}$, much as Eq. (2.11) was written.

When Eq. (2.16) is applied to $\vec{\omega}$, however, the result is much simpler than when it was applied to the linear velocity to derive Eq. (2.17). Why?

2.5 Consider an airplane such as that illustrated in the chapter but moving only in the $x-z$ plane. What linear and angular velocity components would be needed to describe the motion? Assuming further that the $x-y-z$ coordinate system is also a principal coordinate system, write out the equations of motion, Eqs. (2.17 and 2.19), for this special case.

2.6 It is shown in the text that a rigid body in a moment-free environment can spin steadily about any principal axis (and that this motion will be stable if the axis has the largest or the smallest of the three principal moments of inertia). Suppose the body is in an environment that is not only moment-free, but also force-free. This could occur in outer space, for example.

Assuming that $p = \Omega =$ constant, $q = r = 0$, use Eq. (2.17) to derive the equations below for the velocity components V and W, assuming that all forces are zero.

$$\dot{V} - \Omega W = 0$$

$$\dot{W} + \Omega V = 0$$

Derive a single equation for V from these two equations and show that a solution for this velocity component could be expressed as

$$V = V_0 \sin(\Omega t + \phi)$$

where V_0 and ϕ are constants.

Given this expression for V, what is the corresponding expression for W?

There is a simple interpretation of these results. What is it?

CHAPTER 3

3.1 Consider the second-order equation $\frac{d^2y}{dt^2} + a_1 \frac{dy}{dy} + a_2 y = 0$. For what values of a_1 and a_2 is the system stable? How are the natural frequency and the damping ratio related to a_1 and a_2?

3.2 Suppose you wanted to support a bowling ball on a stick. It would be simple enough to compute the stress in the stick by dividing the ball weight by the cross-sectional area of the stick. One could then make sure that material of the stick could handle the stress.

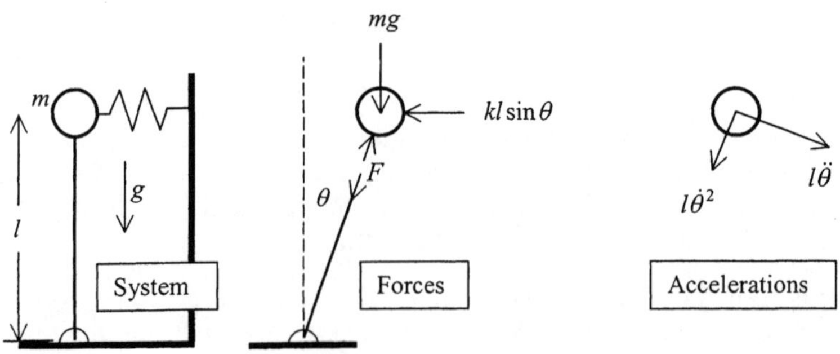

This is about all there is to designing the stick if it is to be short enough. A long stick, however, may exhibit a buckling instability in which the ball will fall over and likely break the stick although, if the ball remained upright, the stick stresses would be low. Providing a lateral support for the ball that could stabilize the system could solve the problem. The odd fact is that it is the support *stiffness* that counts, but the support would not need to provide any force when the ball and stick are in the upright position.

A natural question is "Why does there need to be a support, if the support provides no force in the upright position?"

The real problem we have described involves the buckling instability of a beam, but a simpler problem of the same character involves a stick of negligible mass pivoted at the base with a support of spring constant k to provide stability as shown in the sketch.

What you will find after analyzing this simple system is that the stability of the system depends upon the value of k, but that when the system is in its upright position there is actually no force in the spring. This is typical of many stability problems. When the parameters are correctly adjusted, the system is stable and no forces are required to hold the system in its basic motion state. If the parameters are miss-adjusted, the system is unstable and large forces arise spontaneously.

1. By resolving the forces along the stick and perpendicular to it, write two dynamic equations valid when the angle θ is not necessarily small using Newton's law.
2. Noting what the equilibrium position is when $\theta = 0$, linearize the equations for the case when θ is very small. It may not be obvious to you at present, but the nonlinear term $l\dot{\theta}^2$ *is negligible for small motions near the equilibrium.*
3. You should now be able to show that, for this stability analysis, $F \cong mg$, and $ml\ddot{\theta} + (kl - mg)\theta = 0$.
4. Now you can argue that the system is only stable if $k > mg/l$. Why?
5. Now you can resolve the paradox. The spring force vanishes in equilibrium (so it seems that the spring is not necessary), but for a small value of θ, the spring has to be stiff enough to push the ball back to equilibrium.
6. Because there is no friction in the model of the system, don't you think that the condition on the spring constant only makes sure

that the system is not unstable but does not assure that the system will ever return to $\theta = 0$ and stay there?

3.3 For practice, work out Routh's stability criterion for third- and fourth-order characteristic equations in terms of the coefficients of the powers of s in Eq. (3.9). The results are given in Eqs. (3.24) and (8.33).

3.4 A feedback control system has the characteristic equation $(s+1)^3 + K = 0$, where K is a positive gain. Using Routh's criterion, determine for what range of values of K the control system will be stable.

3.5 Suppose a system is described by equations in the form of Eq. (3.4) with the matrix [**A**] given as in Eq (3.6) as $\begin{bmatrix} -1 & 0 & 0 \\ 0 & -1 & a \\ 0 & 0 & -2 \end{bmatrix}$, where a is a number that can take on any value. Find the characteristic equation for the system and determine the values of a for which the system will be stable.

3.6 A fourth-order system is described in the form of Eq. (3.21). The matrices involved are

$$[\boldsymbol{M}] = \begin{bmatrix} m_{11} & 0 \\ 0 & m_{22} \end{bmatrix}, \quad [\boldsymbol{B}] = \begin{bmatrix} b_{11} & 0 \\ 0 & b_{22} \end{bmatrix}, \quad [\mathrm{K}] = \begin{bmatrix} k_{11} & k_{12} \\ k_{21} & k_{22} \end{bmatrix}$$

Find an expression for the characteristic equation but don't write it out in detail. Assuming that all coefficients are positive, give one condition on the k-coefficients necessary for the system to be stable.

3.7 The following differential equations describe a dynamic system:

$$a_{11}\ddot{x}_1 + a_{12}\ddot{x}_2 + b_{11}\dot{x}_1 + c_{11}x_1 + c_{12}x_2 = 0$$

$$b_{21}\dot{x}_1 + b_{22}\dot{x}_2 + c_{22}x_2 = 0$$

1. Find the characteristics equation for this system equation for this system of equations.
2. Using Routh's criterion, express the conditions on the a, b, and c coefficients that would assure that the system would be stable. Do this in terms of the coefficients in the characteristic equation.

3.8 In the block diagram below, a simple feedback control system is shown. The variables x and y are in the Laplace transfer domain with s the

Laplace variable. The proportional gain is G and $1/s$ represents an integration so the "plant" is a cascade of n integrators.

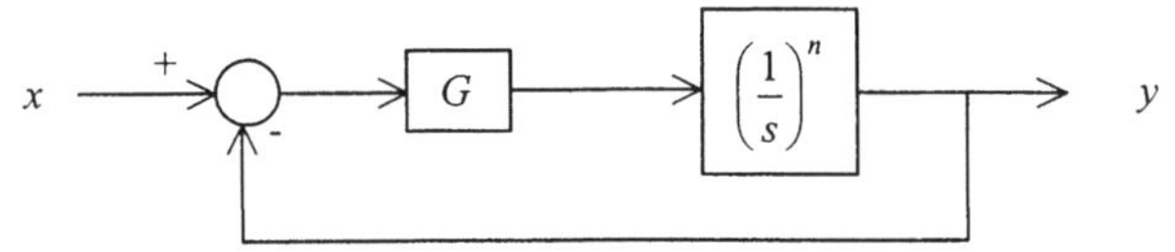

Find the transfer function between y and x either by block diagram algebra or simply by expressing y as $G(1/s)^n$ times the difference between x and y itself.

The denominator of any transfer function is a polynomial in s which, when set equal to zero becomes the characteristic equation for the system.

Prove the following facts:

For $n = 1$, the system is stable for $G > 0$ with a time constant $\tau = 1/G$.
For $n = 2$ and $G > 0$, the system is neither stable nor unstable.
For $n \geq 3$, the system is unstable for any value of G.

CHAPTER 4

4.1 In stability analyses, the relation between the lateral force in a tire F_y and the slip angle α is given by the formula $F_y = C_\alpha \alpha$, where C_α is the cornering coefficient. The dimensions of the cornering coefficient are N/rad, and it can be interpreted as the slope of the curve relating the lateral force to the slip angle near zero slip angle.

In Fig. 4.2, the lateral force is determined in terms of a friction coefficient $\mu_y(\alpha)$ that varies with the slip angle, $F_y = \mu_y F_z$, where F_z is the normal force between the tire and the road surface. Near zero slip angle, the friction coefficient μ_y is also seen to be proportional to the slip angle.

Defining the slope of the friction coefficient curve as C', one can write $\mu_y = C'\alpha$. Using this coefficient, the lateral force law for low slip angles becomes $F_y = C'F_z\alpha$, which implies that $C_\alpha = C'F_z$. The dimensions of the cornering coefficient are N/rad.

Estimate the cornering coefficient for the tire characteristic shown in Fig. 4.1 for dry concrete. Assume that $F_z = 5kN$. In estimating the slope of the curve in Fig. 4.1, to find C', be sure to convert the slip angle from degrees to radians.

4.2 Use the plots in Fig. 4.6 to estimate a cornering coefficient. Remember that the cornering coefficient relates the lateral force to the slip angle when there is no longitudinal force and is defined for small slip angles. The dimensions of the cornering coefficient are N/rad.

4.3 It is stated in the text that if the maximum lateral cornering coefficient for a tire is 0.8, and if the normal force on the tire is due only to the weight supported by the tire, then the maximum lateral acceleration possible is also 0.8 times the acceleration of gravity. Set up a free body diagram for this situation and derive a proof for the assertion.

4.4 Use sketches such as those in Fig. 4.7 to explain the 70° slip-angle curve in Fig. 4.6. Discuss why the curve on the braking side has an end point with so little braking force, while the end point on the traction side has so much traction force.

4.5 Using Fig. 4.6, confirm or deny the statement that lateral forces and traction or braking forces have very little interaction if they are both small. That is, when negotiating a gentle curve requiring a small lateral force, applying a small braking or driving torque to a wheel will have almost no effect on the lateral force or slip angle. How would one define "small"?

4.6 Suppose one axle of a vehicle supports a total weight of 2000 pounds, that both wheels on the axle support 1000 pounds and that both are operating at slip angle of 4°. If both wheel used the tires shown in Fig. 4.3, the 4° curve indicates that each wheel would generate a lateral force of 600 pounds for a total axle lateral force of 1200 pounds. This means that the axle cornering coefficient (the total lateral force per unit slip angle for the axle) would have a value of 300 pounds per degree.

Now suppose, as a result of unsymmetrical loading or the effects of steady cornering, one wheel supports 600 pound and the other 1400 pounds. (The axle still supports a total of 2000 pounds.) Show that the total lateral force on the axle under these circumstances is less than 1200 pounds. What would the axle cornering coefficient be in this case?

The 4° slip-angle curve in Fig. 4.3 has a pronounced curvature. Show graphically that if this curve were a straight line—that the lateral force were strictly proportional to the normal force—that the total lateral force on an axle would not change if the increase in normal force on one wheel were balanced by an equal decrease in normal force on the other. In this case, the axle cornering coefficient would also remain unchanged.

This means that the effective cornering coefficient for an axle depends not only on the total weight carried by the axle, but also on the distribution of normal forces at the two wheels if the relation between the lateral force for a constant slip angle has significant curvature as shown in Fig. 4.3. The axle cornering coefficient is an important parameter in the stability analyses to be discussed in Chapters 5 and 6.

CHAPTER 5

5.1 The cornering coefficient for a single tire, C_α, is defined as the slope of a force vs. slip angle curve in Fig. 5.2. The same symbol is used in Eq. (5.6), but it is meant to apply to the trailer model of Fig. 5.1 that has two wheels with essentially identical slip [Eq. (5.6)], relate this combined cornering coefficient to C_α that applies to each of the two tires on the trailer considered separately.

5.2

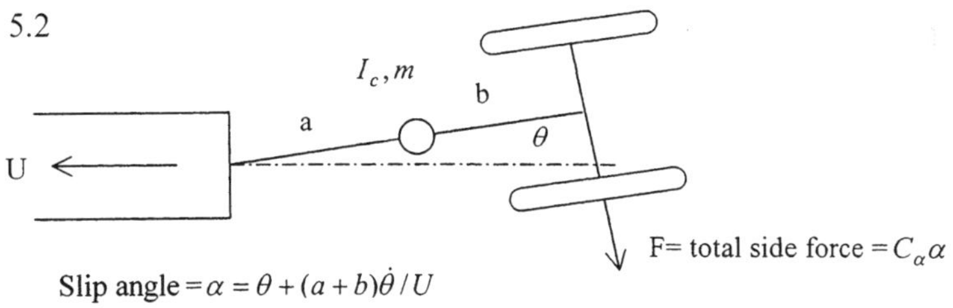

The equation of motion for this single degree-of-freedom trailer model can be derived by using Lagrange's equation [Eq. (5.15)].

$$(I_c + ma^2)\ddot{\theta} + C_\alpha(a+b)^2\dot{\theta}/U + C_\alpha(a+b)\theta = 0$$

Derive this equation by using $\Sigma\overline{F} = m\vec{a}$ for the center of mass and $\Sigma M_c = I_c\ddot{\theta}$ for the moment–angular acceleration relation about the center of mass. Some algebraic manipulation will be required to reduce the two equations to a single dynamic equation for θ.

5.3 The equation in Problem 5.2 is of the form $m\ddot{x} + b\dot{x} + kx = 0$, where the coefficient b depends on velocity U. The eigenvalues of this system arise by solving the characteristic equation $ms^2 + bs + k = 0$ for complex values of s.

Sketch the root locus locations for this equation in the complex plane as b varies from 0 to ∞. The root locus is simply the path in the complex s-plane that the eigenvalues take as the coefficient b takes on various values. (Note that you can accomplish this without knowing any root locus techniques because you can simply use the quadratic formula to find the eigenvalues in terms of m, b, and k). Note that in Problem 5.2, the b-coefficient varies from ∞ to 0 as U varies from 0 to ∞.

Using this root locus plot, discuss the stability of the trailer as a function of speed. Qualitatively sketch the response to perturbations at various speeds.

Can you find the value of b or the equivalent value of U in the equation of Problem 5.2 for which the two eigenvalues have equal negative real roots? This value of b is called critical damping constant in the case of a vibratory system, and it marks the transition between the value of the damping constant for systems that have damped oscillations and the value for systems that do not oscillate at all.

5.4 It is almost obvious that the trailer model in Problem 5.2 should be unstable if the towing vehicle is backing up. However, the equation for a backing trailer is not correctly found simply by letting substituting $-U$ for U in the equation given in Problem 5.2 because the slip angle expression is then no longer correct.

Derive the correct equation of motion for a reversing towing vehicle by using the figure in Problem 5.2, but with the arrow representing the velocity U reversed. You should see from the new equation of motion that a backing trailer is unstable at all speeds, but it is not the coefficient of $\dot{\theta}$ that becomes negative for a reversed velocity.

5.5 The figure below shows an inventor's scheme for making it easier to back trailers up. The idea is to steer the trailer wheels using an actuator so that the steer angle is proportional to the angle between the centerlines of the towing vehicle and the trailer. The proportionality factor or gain is called g.

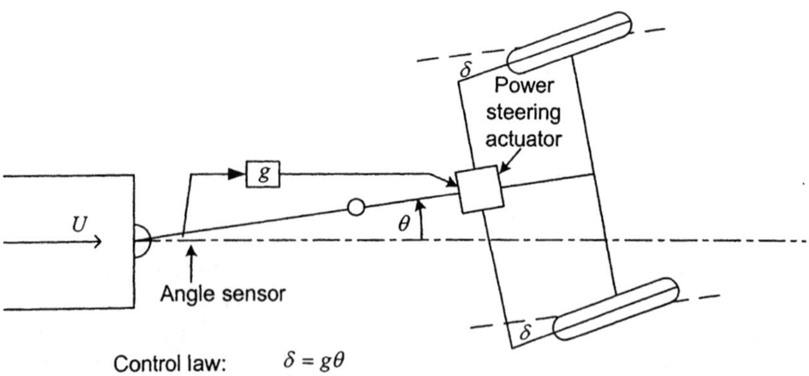

Using the results of the Problem 5.4, modify the expression for the slip angle to include δ and the control law and analyze the system to see if the system would stabilize the trailer when it is backing up.

5.6 The motion of this trailer with a flexible hitch is described with the generalized coordinates x and θ. Show that the tire slip angle α is given by the expression $\theta + (a+b)\dot{\theta}/U + \dot{x}/U$. Assume that $F = C_\alpha \alpha$ and that the flexible hitch has a spring constant k. In a free body diagram, you will have to define a lateral force on the nose of the trailer with a magnitude of kx.

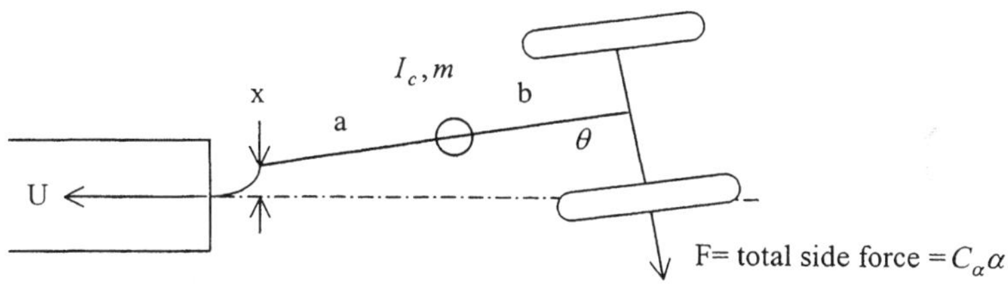

The equations of motion using Lagrange's equations are shown below and your task is to verify these equations by using Newton's laws.

$$\begin{bmatrix} I_c + ma^2 & ma \\ ma & m \end{bmatrix} \begin{bmatrix} \ddot{\theta} \\ \ddot{x} \end{bmatrix} + \begin{bmatrix} C_\alpha(a+b)^2/U & C_\alpha(a+b)/U \\ C_\alpha(a+b)/U & C_\alpha/U \end{bmatrix} \begin{bmatrix} \dot{\theta} \\ \dot{x} \end{bmatrix} + \begin{bmatrix} C_\alpha(a+b) & 0 \\ C_\alpha & k \end{bmatrix} \begin{bmatrix} \theta \\ x \end{bmatrix} = \begin{bmatrix} 0 \\ 0 \end{bmatrix}$$

It is very likely that if you derive two second-order differential equations for x and θ and put them in the matrix form shown above, they will not have the same matrix elements as those resulting from Lagrange's equations.

This can happen even if you have not made a mistake. If you have not made a mistake, then it must be possible, after some manipulation, to achieve the form shown above. Hint: Consider adding one equation to the other to derive a new equation.

5.7 The equations in Problem 5.6 are of the form of Eq. (3.7). Give an expression for the characteristic equation for the system in the form of

Eq. (3.22). Do not attempt to evaluate the determinant unless you have time on your hands.

5.8 Suppose the trailer of Problem 5.6 is assumed to have no side slip at all. Use Newton's laws but with zero lateral velocity as a kinematic constraint. Show that the equations of motion then simplify to

$$\frac{(I_c + mb)^2\,\dddot{\theta}}{b} + mU\ddot{\theta} + \frac{k(a+b)^2\dot{\theta}}{b} + \frac{Uk(a+b)\theta}{b} = 0$$

5.9 Suppose that the trailer of Problem 5.6 were on ice, so that the lateral force on the tires was essentially zero. Modify the equations of motion, Eqs. (5.32), for this situation by assuming $C_\alpha = 0$ and finding the characteristic equation. Check the result with Eq. (5.33). Verify that the trailer could still oscillate under these conditions and give a physical explanation for the mode of motion that this oscillation represents.

5.10 Consider a truck towing tandem trailers at constant speed. The second trailer is hitched directly on the axle of the first. Begin an analysis to the stability of this system using the same types of assumptions used in chapter.

First, sketch the velocity components for the two trailer axles and give expressions for the slip angles.

Using free body diagrams and Newton's laws, write equations of motion for both trailers. This system has only 2 degrees of freedom, but you will have to include forces at both hitch points. These forces can be algebraically eliminated in the final equations of motion.

As an alternative, set up the energy expressions necessary to derive the two equations of motion using Lagrange's equations.

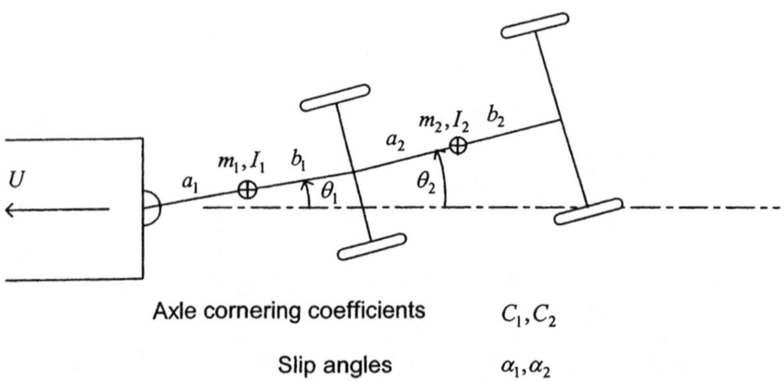

5.11 In Bundorf (1967a), several models of a car and trailer were analyzed for stability. In one of them, the back of the car was idealized as a mass mounted on a single equivalent wheel representing the back axle. The mass, m_c, was assumed to only be able to move sideways with velocity v_c, (with no rotation) under the influence of the trailer. It was assumed that v_c remained small compared to the forward velocity U.

The trailer with mass m_t and centroidal moment of inertia I_t could swing around with the small angle θ. This model is essentially a one and one-half degree-of-freedom model because the position of the car in the y-direction is not important, only the y-velocity, v_c.

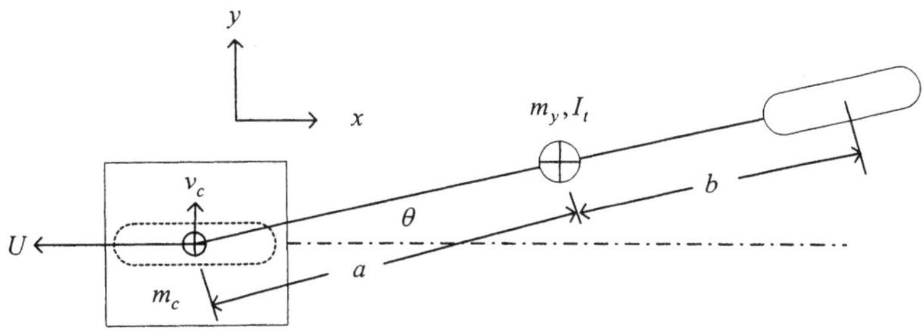

The car axle force is $Y_c = -C_c\alpha_c$, where C_c is a (positive) cornering coefficient and representing the two rear tires of the car and α_c is the slip angle for the car mass. In a similar way, the force on the equivalent trailer wheel is written $Y_t = -C_t\alpha_t$.

Using the sketch of the model shown above, derive the equations of motion as follows:

1. Find expressions for α_c and α_t.
2. Draw free body diagrams for the car mass and the trailer. Include forces on the wheels in the y-direction and the force at the hitch that acts in one direction on the car mass and in the other direction on the trailer.
3. Find the acceleration of the trailer mass center in terms of $\dot{v}_c$ and $\ddot{\theta}$.
4. Write one dynamic equation for the car mass and two for the trailer.
5. See if you have the patience and skill to find a single third-order equation for θ by eliminating the hitch force and $\dot{v}_c$ from the equations.

6. Show how to find the characteristic equation for the system. Assuming you were successful in finding the characteristics equation, how would you determine the condition for stability?

5.12

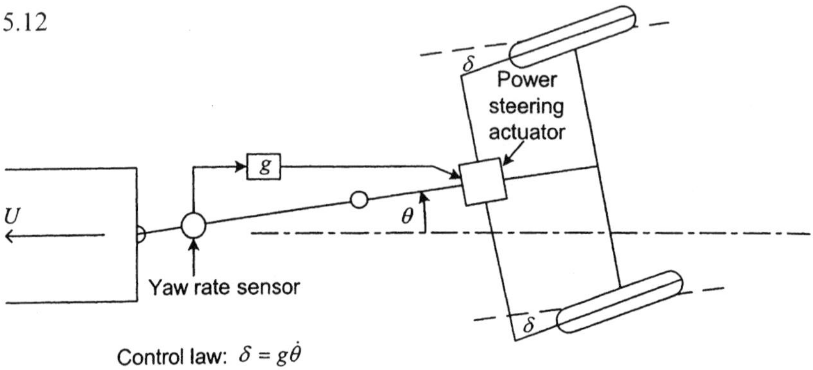

Many trailers exhibit lightly damped oscillations at high speeds, and some actually become unstable for speeds above a critical speed. The active steering system shown is intended to stabilize a trailer by steering the wheels in response to the trailer yaw rate. The steer angle δ is relative to the trailer chassis and is proportional to the measured yaw rate $\dot{\theta}$ with a simple proportional gain g. Modify the analysis presented in the chapter for a single degree-of-freedom trailer to account for the active steering and discuss whether the idea has merit or not. Also, consider whether the control system would stabilize a trailer if it were backing up and what effect the control system would have on a trailer when negotiating a steady turn.

5.13 This is an exercise concerning the concept of static stability discussed in Chapter 3. The idea of static stability involved considering a small perturbation away from the equilibrium position of a system, and then to discover whether there developed forces tending to return the system and subsequently to discover whether there developed forces tending to return the system to its equilibrium position or whether the forces tended to displace the system further away from the equilibrium. The two cases are called statically stable and statically unstable, respectively. As was noted, a statically unstable system usually implies that the system is unstable dynamically, but it is certainly true that statically stable systems are not necessarily stable. A static stability analysis is often much simpler than a dynamic stability analysis and may satisfactorily answer the question of stability. Static stability proves to be useful in aircraft design, as will be seen in Chapter 9.

Consider first the single degree-of-freedom trailer model discussed in the text and shown in Problem 5.2. Noting that $\theta = 0$ is an equilibrium position, show that when θ is given a small value, the trailer is statically stable when moving forward but statically unstable when backing up. (In this case, the static stability analysis agrees completely with the dynamic stability analysis.)

Now, turn your attention to the 2 degree-of-freedom trailer model also discussed in the text and shown in Problem 5.6. For this model, the equilibrium configuration is $\theta = 0$ and $x = 0$. For a static stability analysis, assume that both θ and x have small values but that they are related in such a way that the center of mass remains on the towing vehicle's centerline. In the figure of Problem 5.6, this will mean that if θ is positive, x must be negative.

For a simple static stability analysis, consider only whether the forces on the trailer tend to reduce the angle θ or to increase it. If the forces act to reduce θ, the trailer is statically stable. If the forces tend to increase θ, the trailer is statically unstable.

Once again, consider the forward and reverse motions. By constructing free body diagrams, decide whether the forces are in directions that tend to return the trailer to the equilibrium or to increase the perturbations.

CHAPTER 6

6.1 The figure below is essentially the same as Fig. 6.1.

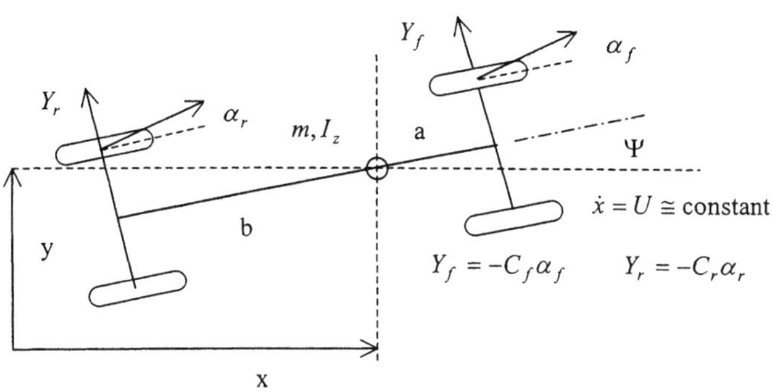

With the definitions shown in the figure, the slip coefficients are positive.

Assume that the slip angles as well as the heading angle Ψ remain small. Also assume that the forward speed U is constant and is essentially the same as the velocity in the x-direction. Derive the following equations of motion for this car model in inertial coordinates y, Ψ, but instead of using Newton's laws as was carried out in the text, try your hand at using Langrange's equations. The kinetic energy is easily expressed and there is no potential energy, thus the only challenge is to find the generalized force for the y-generalized variable and the generalized moment for the Ψ variable.

$$\begin{bmatrix} m & 0 \\ 0 & I_z \end{bmatrix}\begin{bmatrix} \ddot{y} \\ \ddot{\Psi} \end{bmatrix} + \begin{bmatrix} (C_f + C_r)/U & (aC_f - bC_r)/U \\ (aC_f - bC_r)/U & (a^2C_f + b^2C_r)/U \end{bmatrix}\begin{bmatrix} \dot{y} \\ \dot{\Psi} \end{bmatrix}$$
$$+ \begin{bmatrix} 0 & -(C_f + C_r) \\ 0 & -(aC_f - bC_r) \end{bmatrix}\begin{bmatrix} y \\ \Psi \end{bmatrix} = \begin{bmatrix} 0 \\ 0 \end{bmatrix}$$

Derive the characteristic equation using Eq. (3.22). This requires some patience.

Show that this system has two zero-valued eigenvalues, and that the car is always stable if $bC_r > aC_f$. Also show that if the car can be unstable, it will be unstable when $U > U_c$, where the critical speed U_{crit} is determined by the expression Eq. (6.17).

$$U_{crit}^2 = \frac{C_f C_r (a+b)^2}{m(aC_f - bC_r)}$$

6.2 Reconsider the simple car model of Problems 2.2 and 6.1, but use body-fixed coordinates and include a steer angle δ for the front wheels.

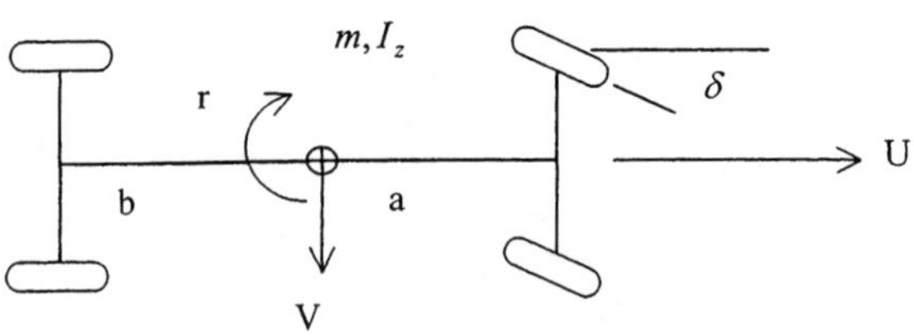

Derive the following equations of motion, Eqs. (6.22) and (6.23):

$$m(\dot{V} + rU) = -(C_f + C_r)V/U - (aC_f - bC_r)r/U + C_f\delta$$
$$I_z\dot{r} = -(aC_f - bC_r)V/U - (a^2C_f + b^2C_r)r/U + aC_f\delta$$

Find the characteristic equation and verify that the same conclusions about stability are found as were reached in Problem 6.1.

6.3 The numerator of the transfer function relating the lateral acceleration of the center of mass of a car to the steering angle at the rear axle is of second order, but with one negative coefficient. It is claimed that this means that transfer function has a right half-plane zero. Using the simplified expression $a_0 s^2 + a_1 s - a_2$ to represent the numerator in Eq. (6.37), show that there will always be a real positive value of s that will make the numerator vanish.

6.4 Using a free body diagram for vertical forces, show that in the absence of aerodynamic forces and longitudinal (x-direction) acceleration, Eqs. (6.58 and 6.59) are correct. How does the situation change if the car has a constant forward acceleration?

6.5 This problem is based on the car model of Problem 6.1. It has been suggested that a steer-by-wire system might be able to stabilize even an oversteering car traveling at a speed greater than the critical speed.

Suppose that the front steer angle δ were to be partly determined by driver input δ_d and partly by a simple control system using sensed yaw rate r as an input. The control law would Be $\delta = \delta_d + Gr$, where G is the gain. Show how the equations in Problem 6.2 would be modified and set up a determinant that would yield the characteristic equation for the steer-by-wire system. Indicate in the determinant where any new terms would arise.

6.6 J. C. Whitehead has examined the steering oscillations that can occur when a car's steering wheel is released in a turn. The two SAE papers "Stabilizing the Steering Weave Mode" (No. 881136) and "Stabilizing Steering Weave with Active Damping" (No. 891981) both use simple vehicle models such as that of Problem 6.2.

His basic vehicle equations in matrix form are given below:

$$\begin{bmatrix} \dot{v} \\ \dot{r} \end{bmatrix} = \begin{bmatrix} -(C_F + C_R)/mu & (C_R l_R - C_F l_F)/(mu) - u \\ (C_R l_R - C_F l_F)/Ju & -(C_R l_R^2 + C_F l_F^2)/Ju \end{bmatrix} \begin{bmatrix} v \\ r \end{bmatrix} + \begin{bmatrix} C_F/m \\ C_F l_F/J \end{bmatrix} \delta_F \tag{1}$$

First, work out the correspondences between Whitehead's notation and that of the model used in Problem 6.2. Then, expand Eq. (1) to Eq. (3) shown below by incorporating the "steering system equation," Eq. (2). Below is a quotation from Whitehead's first paper.

"The fixed control model can be expanded to a torque input model by adding a simple steering system equation

$$b\dot{\delta}_F = \tau - C_F \alpha_F d \tag{2}$$

The viscous rotary damping constant is b, and the two torques on the right-hand side balance the damping torque to determine steering angular rate. The last term is torque due to lateral tire force, where α_F is the front tire slip angle and d is the caster offset distance. Equation (2) applies if steering system damping is sufficiently high relative to rotary inertia that only steering angular velocity needs to be carried as a state variable. Such is the case for a typical passenger car, because damping must be adequate to prevent shimmy, a second-order steering oscillation.

Combining Eqs. (1) and (2) results in

$$\begin{bmatrix} \dot{v} \\ \dot{r} \\ \dot{\delta}_F \end{bmatrix} = \begin{bmatrix} -(C_F + C_R)/mu & ((C_R l_R - C_F l_F)/mu) - u & C_F/m \\ (C_R l_R - C_F l_F)/Ju & -(C_R l_r^2 + C_F l_F^2)/Ju & C_F l_F/J \\ C_F d/bu & C_F L_F d/bu & -C_F d/b \end{bmatrix} \times \begin{bmatrix} v \\ r \\ \delta_F \end{bmatrix} + \begin{bmatrix} 0 \\ 0 \\ \tau/b \end{bmatrix} \tag{3}$$

where α_F has been written in terms of the state variables in the third row of the system matrix. The third-order model is simple enough to examine in detail, and it unequivocally demonstrates that weave is not a steering system oscillation."

In the second paper, Whitehead includes steering system rotary inertia. Incorporate Eq. (8) in the quote given below into the system equations, Eq. (1) above, thus generating a fourth-order matrix equation set analogous to Eq. (3) above. (Of course, this fourth-order system is considerably harder to analyze in general terms than the third-order model.)

"STEERING TORQUE EQUATIONS—In order to model the free control automobile and predict a vehicle's weave mode, the steer angle must be a degree of freedom. Therefore, torque, τ_F, applied through the steering linkage by the driver, becomes the input variable. Other major steering system torques are the product of lateral tire force, f_F, and caster offset, d_F, viscous rotary damping with coefficient b_F, and the weave stabilizer's control torque, τ_A. Taking all quantities to be referenced to the kingpin axis, the torque sum equals the product of steering system rotary inertia and steering angular acceleration:

$$J_F \ddot{\delta}_F = \tau_F - f_F d_F - b_F \dot{\delta}_F + \tau_A \tag{8}$$

If the active torque, τ_A, is zero, Eqs. (1), (2), (3), and (8) form a fourth-order set of differential equations describing the natural response of and automobile to a steering torque input through the steering linkage. If $\tau_F = 0$, the equations represent release of the steering wheel by the driver.

Note that τ_F must be the sum of driver-applied torque and power-assist torque for a vehicle with power steering. Assuming that a power steering system essentially amplifies driver-applied torque, distinguishing between power-assisted steering and nonpower-assisted steering is unnecessary here. In support of this statement, the author has informally observed weave oscillations in numerous automobiles both with and without power-assisted steering."

6.7 Refer to the section on yaw rate and acceleration gains in steady cornering in chapter. Prove that an understeer vehicle has its maximum yaw rate gain at a forward speed equal to its characteristic speed as indicated in Fig. 6.10. Consider Eqs. (6.72), (6.76), and (6.78), and use calculus to derive an expression for the maximum yaw rate again.

6.8 The figure below shows two rear views of a four-wheeled vehicle. In the first instance, the vehicle is running in a straight line with no lateral acceleration. The forces shown as ΣF_{zl} and ΣF_{zr} represent the sum of the front and rear normal forces on the tires at the left and right sides, respectively. The distance between the tires, the track T, is assumed to be the same for the front and rear axles.

In this case, the symmetry of the situation makes it obvious that the total normal forces on the two sides of the vehicle must add up to the weight. Thus $\Sigma F_{zl} = \Sigma F_{zr} = mg/2$.

In the second part of the figure, the vehicle is executing a steady turn of radius R with speed U, so there is a lateral acceleration of $a_y = U^2/R$. There will also be a total lateral force on all the tires of ΣF_y. In this case, the height of the center of mass, h, and the track determine how the normal forces on the outside of the turn increase and the forces on the inside of the turn decrease.

1. Assuming the vehicle is in a steady state with no angular acceleration around the roll axis, show that the normal forces are given by the following expressions:

$$\Sigma F_{zl} = m(g/2 + hU^2/TR), \quad \Sigma F_{zr} = m(g/2 - hU^2/TR)$$

2. Show that the vehicle can achieve a lateral acceleration of 1 "g" only if the center of mass height is equal to or less than half the track.

3. Suppose that the vehicle has suspension springs that allow it to lean a small amount toward the outside of turns. Qualitatively, how would this change the expressions for ΣF_{zl} and ΣF_{zr} in item (1)?
4. In a real vehicle, the suspension roll stiffness at the front and at the rear determines how the total left- and right-side forces given in item (1) are distributed between the front and real axles. For example, if the front roll stiffness is larger than the rear roll stiffness, then the difference between the forces at the front will be bigger than the difference between the forces at the rear although the total left and right forces obey the laws given in item (1). Considering the effect of normal force variation on an axle discussed in Chapter 4, discuss how an increase in front axle roll stiffness would affect the front axle effective cornering coefficient and whether this would tend to promote understeer or oversteer in the vehicle.

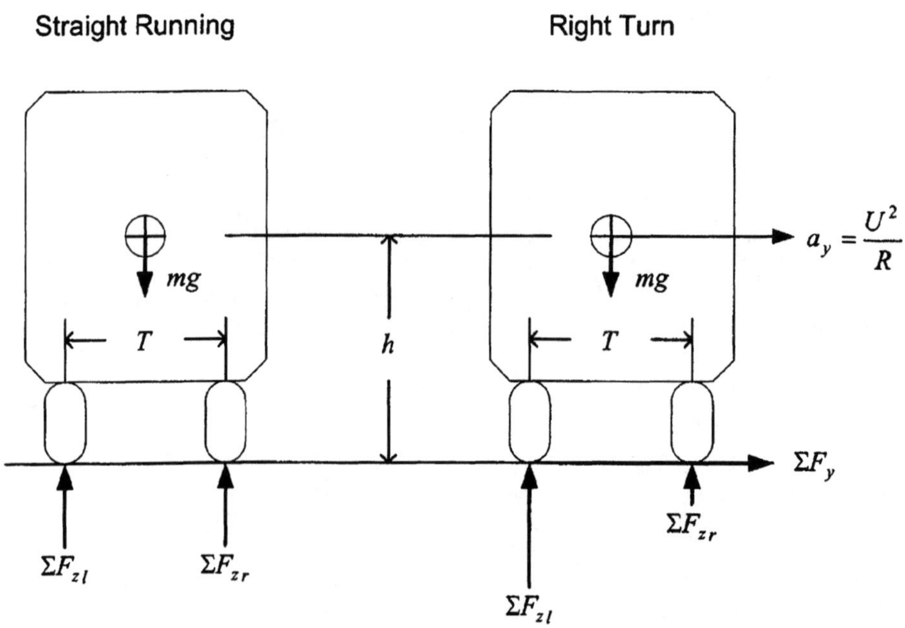

Rear View of Vehicle

6.9 Fig. 6.7 shows a bicycle model of a car with both front-wheel and rear-wheel steering. The positive directions for steer angles, δ_f and δ_r slip angles, α_f and α_r as well as the angle β, which is the angle the velocity vector of the center of mass makes with the center line of the car are shown in the figure.

Fig. 6.8 shows the geometry of a car with only front wheel steering executing a steady turn. In this case, it is seen that the front and rear slip angles, as well as the angle β, are actually negative according to the conventions of Fig. 6.7. This figure is a classic one which appears in Bundorf (1967b), and is used to derive the relation between the steer angle, the slip angles and the vehicle dimensions, Eq. (6.53). The derivation considers the triangles shown with the assumption that the wheelbase is small compared to the turn radius. The text uses a different derivation that avoids the complication of a geometric derivation with negative angle definitions. In this problem, the steady turn results will be generalized to include both front and rear wheel steering.

A careful examination of a figure similar to Fig. 6.8, but including a rear steer angle as shown in Fig. 6.7 yields two expressions:

$$\delta_f = -\alpha_f + \beta + \frac{a}{R} \tag{1}$$

$$\delta_r = -\alpha_r + \beta + \frac{b}{R} \tag{2}$$

The task is to check the correctness of these results by using algebraic manipulations and simpler versions of the turn diagram used to derive them.

1. Show that Eqs. (1) and (2) are consistent with Eq. (6.53) for a front-wheel steering car.
2. Extend Eq. (6.53) to include front and rear steering angles by combining Eqs. (1) and (2).
3. Using simple diagrams, verify Eqs. (1) and (2) for the case that both slip angles are zero and that β is also zero.
4. Make a diagram to verify the results of item (2). for the case of zero slip angles. (This case will appear in Chapter 7.)
5. Derive the following equation for the angle β:

$$\beta = \frac{b}{(a+b)}\delta_f + \frac{a}{(a+b)}\delta_r + \frac{b}{(a+b)}\alpha_f + \frac{a}{a+b}\alpha_r \tag{3}$$

6. How should the rear steer angle be related to the front steer angle if it is desired to keep the angle β to be zero in a steady turn?

Consider the zero slip angle case first and then the case when the slip angles are nonzero.

CHAPTER 7

7.1 Redraw Fig. 7.1 in such a way that the steer angles are small and the turn radius is much larger than the wheelbase. Under these conditions, it should be obvious that the angle between the two lines normal to the wheels shown as dotted lines in Fig. 7.1 is approximately $(a+b)/R$. Then show that this angle is also equal to $\delta_f - \delta_r$ thus proving Eq. (7.2).

7.2 Considering Fig. 7.1, derive an expression for the lateral velocity V in terms of the front and rear steer angles, δ_f and δ_r and the parameters a, b, and U, but without using the small-angle approximations. Check that when the steer angles are assumed to be small, the result becomes Eq. (7.4).

7.3 Derive Eq. (7.12) using Newton's laws by considering the vehicle to be an upside-down pendulum. The mass is m, the moment of inertia about the center of mass is I_l and the center of mass is a distance h above the pivot point. The pivot acceleration to the right is given by $\dot{V}+rU$. (This formula comes from Chapter 2.)

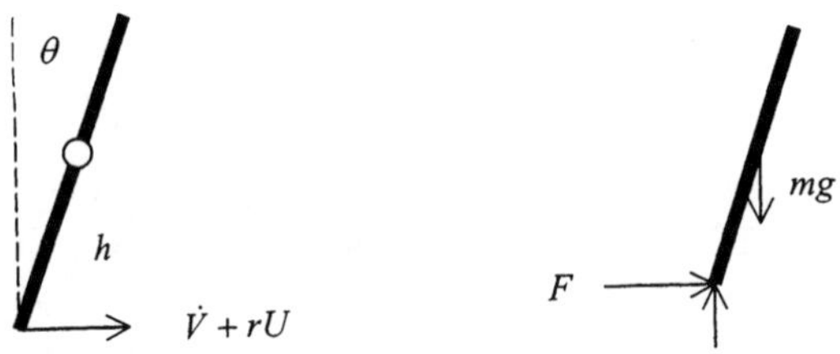

You will need to write two equations involving linear accelerations and one involving angular acceleration. Convince yourself that the vertical force at the pivot is almost equal to mg under the conditions applying to Eq. (7.12). Combine the equations to derive Eq. (7.12).

7.4 The stabilization the of the lean angle of a tilting vehicle using a simple proportional control system acting on either the front wheel steer angle or the rear wheel steer angle is discussed at some length in the chapter. It was shown that front-wheel steering could be made to work quite well except at very low speeds, but that a more complicated control system would have to be devised for rear-wheel steering. The equation for the open loop system in the form of Eq. (7.13) or Eq (7.14) has the influence of both front and rear steer angles.

It has been suggested that there may be advantages to steering both wheels simultaneously. For example, racing motorcycles are conventionally built with short wheelbases so that they respond quickly to steering inputs. However, the short wheelbase combined with a high center of gravity means that the "weight transfer" between the front and the rear limits both acceleration and braking. Excessive acceleration causes "wheelies" and excessive braking causes "stoppies," in which one wheel or the other leaves the ground. A longer wheelbase would ameliorate these tendencies.

Consider a simple scheme in which the rear steer angle is proportional to the front. The rider would steer the front as usual and a servomechanism would steer the rear according to the law

$$\delta_r = \alpha\delta_f$$

Let $-1 < \alpha < +1$, where the proportionality factor α might change with speed or some other variable.

First, find the open loop transfer function θ/δ_f corresponding to Eq. (7.18) by using Eq. (7.14) and the proportional rear steering law above.

Next, using the control scheme of Fig. 7.3, find the closed loop transfer function corresponding to Eq. (7.20) for the proportional rear steering case.

Comment on the possibilities for stabilizing the lean angle for values of the factor α within its range. If the object were to increase the steering response of a long wheelbase motorcycle, should α be positive or negative? (Refer to Fig. 7.1.) Why is there a problem with $\alpha = 1$?

CHAPTER 8

8.1 In Fig.8.2, consider that the pivot housing is fixed in space and that the caster can rotate completely around 360° when it is not touching the ground. Show that the center of the wheel travel is a circle inclined at an angle β to the horizontal. Show the radius of this circle on your sketch. On the basis of this sketch, discuss how far the caster pivot housing would have to move up and down if the wheel remain in contact with the ground. Give an argument justifying the neglect of the up and down movement for small excursions of θ away from the zero position shown in Fig. 8.2.

8.2 Equation (8.41) is an equation with time as the independent variable. But the idea of a relaxation length relates to changes in the force as a function of distance that the tire rolls. Writing the equation in the form $(\sigma/U)\dot{F} + F = C_\alpha\alpha(t)$, change it over to a function of space, x, using the relation $\mathrm{d}/\mathrm{d}t = (\mathrm{d}/\mathrm{d}x)(\mathrm{d}x/\mathrm{d}t) = (\mathrm{d}/\mathrm{d}x)U$. Describe an experiment on a tire rolling at a constant velocity in which the slip angle was changed stepwise at some position and the lateral force was recorded as a function of x. How would one find the relaxation length from this data?

8.3 Sketch a caster that is not moving forward, $U = 0$, but which has a tire with a flexible sidewall with spring constant k. Using a free body diagram, derive Eq. (8.54). Find the natural frequency of the system.

8.4 The text discusses the concept of relaxation length in the context of a caster. In this problem, you are to repeat the analysis for a simpler case that does not involve the caster angle θ. Here, you are to assume that t wheel is simply moving forward with velocity U and with a lateral velocity v_l.

For steady-state conditions, we assume that the lateral tire force, F_{ss}, is related to the slip angle, α, by the approximate relation

$$F_{ss} = C_\alpha \alpha = C_\alpha v_l / U$$

where v_l is the lateral velocity of the wheel center. In bond graph terms, this is a resistance relation with a resistance parameter of C_α/U, as indicated below, which is a simplified version of Fig. 8.6:

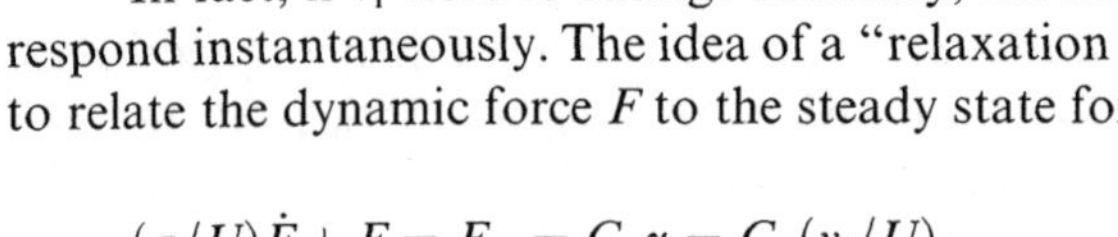

In fact, if v_l were to change suddenly, the actual force F would not respond instantaneously. The idea of a "relaxation length," σ, can be used to relate the dynamic force F to the steady state force F_{ss} as follows:

$$(\sigma/U)\dot{F} + F = F_{ss} = C_\alpha \alpha = C_\alpha (v_l/U)$$

An alternative explanation of this effect is based on the assumption that the tire has a lateral stiffness k and deflection z between the wheel center plane and the contact patch and that the slip coefficient applies to the movement of the contact patch rather than the movement of the wheel itself.

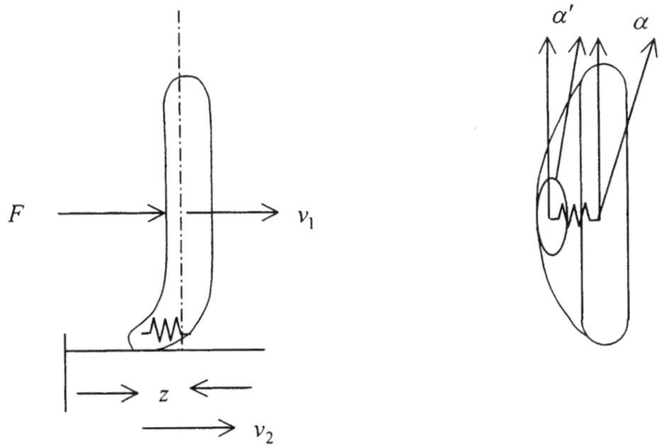

Rear View Top View

Let α' be the contact patch slip angle based on the lateral velocity of the contact patch, v_2. Then $F = C_\alpha \alpha' = C_\alpha v_2/U$. Also, $F = kz$ and $\dot{z} = v_1 - v_2$.

These relations are all expressed by the bond graph shown below which is a simplified version of Fig. 8.7:

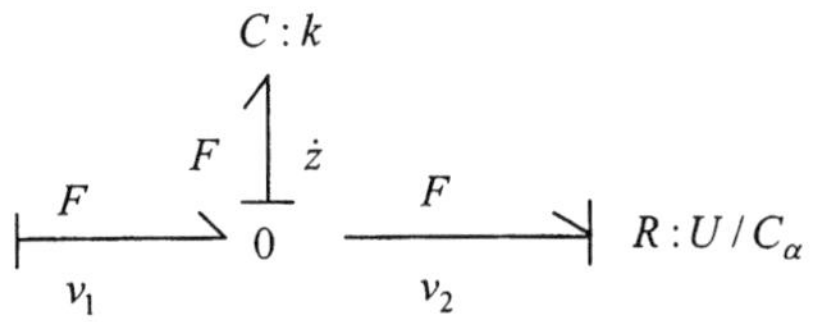

Using these relations, write a dynamic equation for F and show that the relaxation length σ is related to C_α and k by the equation $\sigma = C_\alpha/k$, Eq. (8.51).

8.5 Consider the vertical axis caster shown below. As in Problem 8.4, the tire side force generation is supposed to be dynamic. The mass of the caster system is m and the moment of inertia about the center of mass is $I_c = m\kappa^2$, where κ is the radius of gyration.

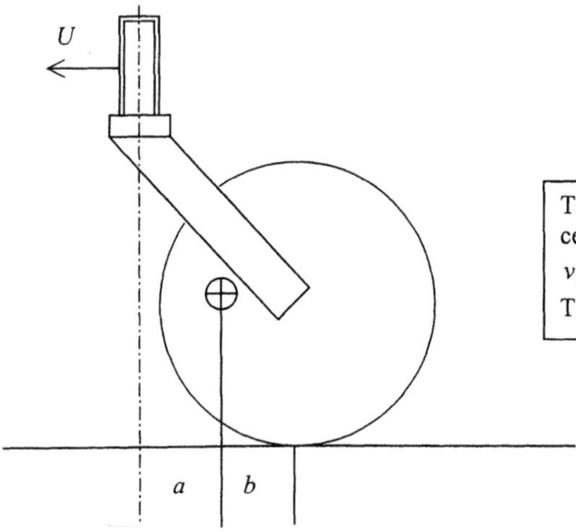

The lateral velocity of the wheel center is given by
$v_1 \cong (a+b)\dot{\theta} + U\theta$
The velocity of the contact patch is v_2

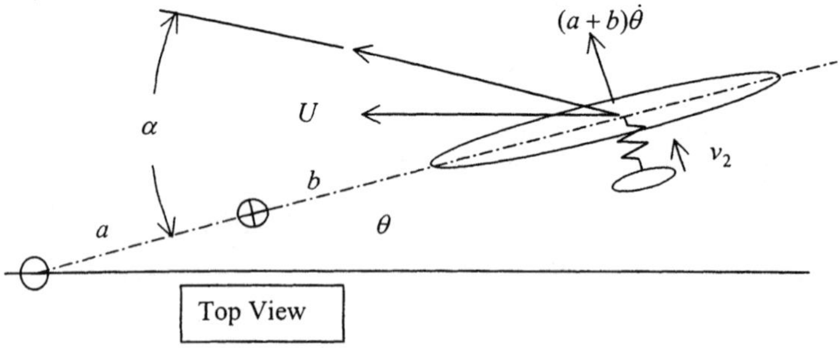

The slip angle α based on v_1 is $\alpha = \theta + (a+b)\dot{\theta}/U$ and the steady state force is $F_{ss} = C_\alpha \alpha$. The equation of motion is $m(\kappa^2 + a^2)\ddot{\theta} = -F(a+b)$, where the dynamic force is given by the equation $(\sigma/U)\dot{F} + F = F_{ss}$, and σ is the relaxation length. From Problem 8.1, $\sigma = C_\alpha/k$. Combine the relations to produce the equation

$$(C_\alpha/Uk)m(\kappa^2 + a^2)\dddot{\theta} + m(\kappa^2 + a^2)\ddot{\theta} + (C_\alpha(a+b)^2/U)\dot{\theta} + C_\alpha(a+b)\theta = 0$$

1. Show that this equation gives correct results for $U \to 0$ and $U \to \infty$.
2. Use Routh's criterion to show that this caster will be stable if $(a+b) > C_\alpha/k$ or $(a+b) > \sigma$.

CHAPTER 9

9.1 The theory of thin airfoils of infinite span yields the force and moment systems shown on the left side of the figure below. The expression $(\rho/2)C_l V_\infty^2 c$ represents the lift force per unit span considered to be applied at the leading edge, and the other expressions represent the (nose up) moment per unit span. The sectional lift coefficient, C_l, varies with angle of attack, α. The constants A_1 and A_2 are dependent upon the camber, ρ is the mass density, V_∞ is the free stream velocity, and c is the chord.

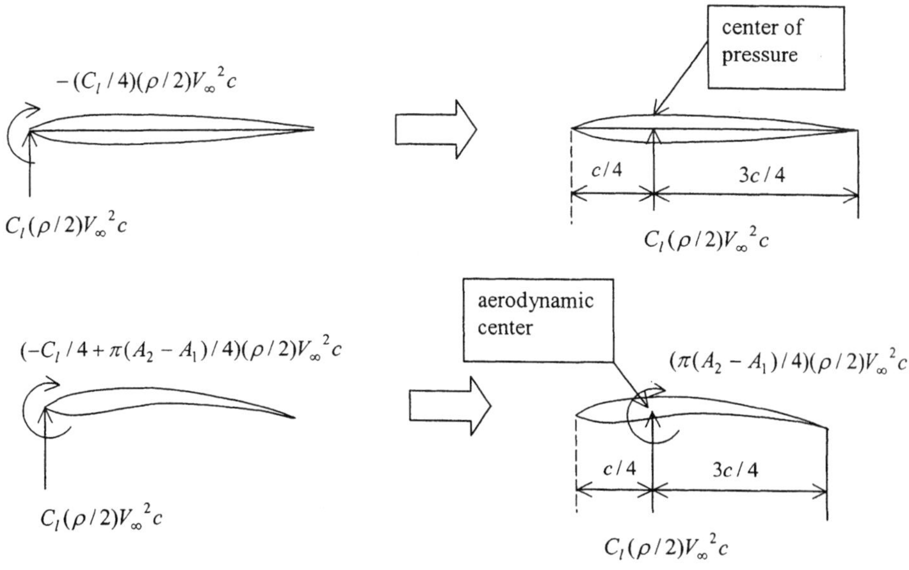

Because C_l varies with α, both the lift and the moment vary for the systems shown on the left.

Show that for the symmetrical wing, if we assume that the lift acts at the *center of pressure* at $c/4$, there is no moment. (That is, show that the force system on the right is statically equivalent to that on the left.)

For a cambered wing, the center of pressure in not convenient because it moves as α varies. However, if we assume that the lift acts at the *aerodynamic center*, again at $c/4$, then there is just a constant moment due to camber. Show that the right-hand force system is also statically equivalent to the left-hand system for the cambered wing.

9.2 Consider a "flying wing" in trimmed flight with $q = W = 0$ and with $\alpha = \alpha_0$.

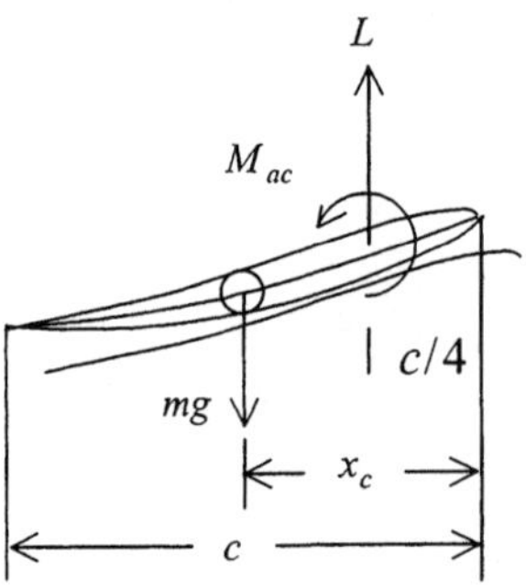

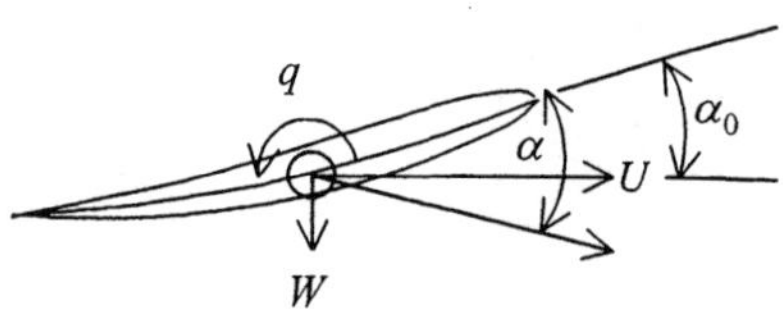

$$\alpha \cong \alpha_0 + W/U$$

$$L \cong \frac{1}{2}\rho U^2 S \frac{\partial C_L}{\partial \alpha}\alpha$$

$$M_{ac} = \frac{1}{2}\rho U^2 S c C_{mac}$$

C_{mac} and $\frac{\partial C_l}{\partial \alpha}$ are assumed constant

Use the equations $m(\dot{W} - Uq) = mg - L$ and $I_{yy}\dot{q} = L(x_c - c/4) + M_{ac}$. Write out the dynamic equations of motion.

1. What relations determine α_0?
2. What are the conditions for dynamic stability?
3. Would a static stability analysis yield the same results as *b*?
4. For stability and trim, where must the center of mass lie and which way should the wing be cambered?

9.3 The following statements are to be explained in words and equations after you figure out what the terms mean:

1. A flying wing can only be stable if the center of mass is ahead of the aerodynamic center.

2. A flying wing can be trimmed if the wing has positive camber, the wing is swept back and it is twisted up so that there is “washout” at the tips. Hint: The tips of the wing can act somewhat as the horizontal tail for a conventional airplane configuration.

9.4 A model glider with a long but very light fuselage is shown. The weight of the glider is mainly made up of the weights of the wing and the horizontal tail.

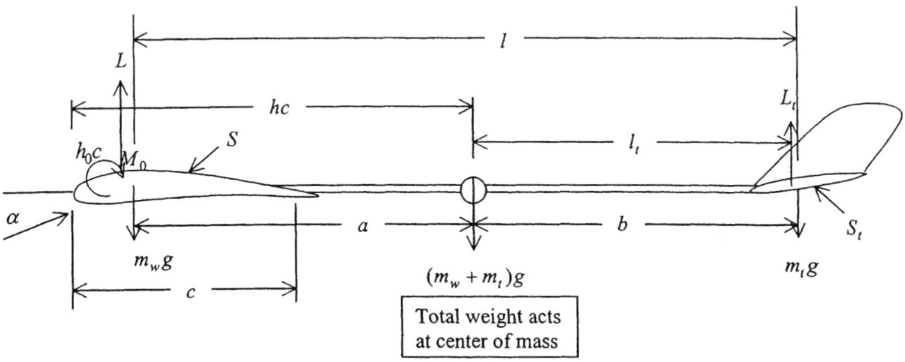

Assume that $(h-h_0)c \cong a$, and that $l_t \cong b$, that is, that the wing weight and the wing lift act nearly at the same point and the tail weight and the tail lift also act at nearly the same point.

1. Let m_w and m_t be the wing and tail masses. Find a and b in terms of m_w, m_t, and l.
2. Using the equation $M_{cg} \cong M_0 + aL - b(L_t + L_{t0})$, where L_{t0} is the tail lift when α is zero, find $dM_{cg}/d\alpha$ in terms of wing and tail areas S, S_t, and lift curve slopes.

3. Assuming equal lift curve slopes for the wing and tail, show that the weight per unit area for the wing must be greater than the weight per unit area for the tail if the glider is to be statically stable.

9.5 Suppose that the wing in Problem 9.4 generates a downwash angle ε at the tail. Discuss how this effect would enter your analysis for stability. Would a given configuration be more or less stable assuming that ε is proportional to α when compared to the case in which downwash is neglected?

9.6 This problem relates to the airplane model of Fig. 9.4 in the chapter.

Some hang gliders are controlled by shifting the center of mass back and forth rather than by the use of elevators on the horizontal tail. (The pilot simply shifts his position back and forth to do this.) Draw a diagram of the moment coefficient vs. the lift coefficient for two locations of the center of mass for such a hang glider. This is the diagram shown in Fig. 9.5 for a single case.

Consider the following questions:

1. Why is the moment coefficient at zero lift independent of the center of mass location?
2. Why can one say that the glider is more stable for the more forward location of the center of mass than for a more rearward location?
3. In which of the two locations does the glider trim at the higher speed?

9.7 Suppose a light airplane is flying at 150 MPH. What is the frequency of the phugoid mode in cycles per second and what is the wavelength of a cycle in meters?

9.8 Suppose the horizontal tail in Problem 9.4 had an adjustable elevator described by the angle η. If the tail lift were given by $L_t = (\rho/2) V^2 S_t(a_1\alpha_t + a_2\eta)$, where α_t is the local angle of attack including downwash and a_1 and a_2 are coefficients, what effect would changes in η have?

CHAPTER 10

10.1 Make a sketch similar to Fig. 10.4, except for the velocity of the contact point of the left-hand wheel in Fig. 10.3. Using the sketch, find the lateral and longitudinal velocities of the contact point and verify that the creepages in Eqs. (10.13) and (10.17) are correct.

10.2 It as been stated that the angular speed of the wheelset shown in Fig. 10.3 around its axle, Ω, is not influenced by the two variables q_1 and q_2. The two variables certainly influence the creepages and through the creep coefficients do influence the forces at the wheel contact points. Show that the influences of q_1 and q_2 on the forces and hence on the moment around the axle are equal and opposite at the two wheels. Thus the angular speed Ω is not influenced by small values of q_1 or q_2.

10.3 The figure below shows an elementary model of a railway truck. The truck frame is considered to be a rigid body on which the axle bearings for two wheelsets are rigidly mounted. There are in principle a number of degrees-of-freedom: the truck frame can move laterally with respect to the rail centerline, it can move longitudinally along the rail, it can yaw with respect to the rails and the two wheelsets can rotate in their bearings. It will be assumed that the longitudinal motion along the track is at a constant velocity V. Furthermore, it can be shown that for a linearized model, the equations of motion for the lateral motion, q_1 and the yaw angle, q_2, actually are uncoupled from the wheelset rotation degrees-of-freedom. This means that the equations of motion for this truck model are very similar to those for the wheelset derived in the text.

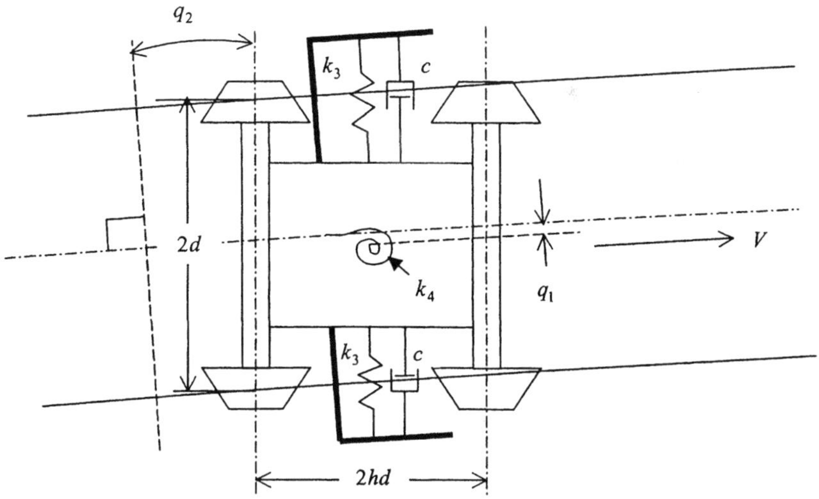

The wheel parameters are the same as those used in the analysis of the simple wheelset in the text. The vertical load from the car body is assumed to be $4N$ and equally distributed among the four wheels. Two springs with constants k_3 and two dampers with damping constants c connect the truck to the car body, which is assumed to move only parallel to the rails. A rotary spring with constant k_4 reacts to the yaw angle of the truck with respect to the car body. The wheels are again a distance $2d$ apart in the lateral direction. The wheelsets are separated a distance $2hd$. The mass of the truck including the wheelsets is m and the centroidal moment of inertia is I.

The derivation for the equations of motion for the simple wheelset is given in some detail in the text. For this model of a two-wheelset truck, the derivation follows a similar pattern. It should be easy to modify the expressions for the lateral and longitudinal creepages at each wheel given for the wheelset wheels to account for the extra distance hd between the mass center and the wheelset axles in the truck model. Similarly, lateral forces due to the gravitational stiffness can be computed for each wheel. The equations of motion by summing forces and moments and equating them to mass times lateral acceleration and moment of inertia times angular acceleration. The result is the following set of equations.

$$m\ddot{q}_1 + \left(\frac{4f_x}{V} + 2c\right)\dot{q}_1 + \left(2k_3 + \frac{4N}{R - R'}\right)q_1 - 4f_x q_2 = 0$$

$$I\ddot{q}_2 + \frac{4f_y d^2}{V}\left(1 + h^2\frac{f_x}{f_y}\right)\dot{q}_2 + \left(k_4 + \frac{4N(hd)^2}{R - R'}\right)q_2 + \frac{4f_y \lambda d}{r_0}q_1 = 0$$

Derive the equations and provide an explanation for each of the terms in the final result.

If you are feeling ambitious, you might attempt to find the characteristic equation and the critical speed. You will have to make the same sort of approximations that were made in the text to make progress. It is also easier if you take out the damping; that is, set $c = 0$. The results, after simplifying the characteristic equation coefficients as was explained in the text are as follows:

$$V_c^2 = \frac{r_0}{\lambda d}\left(1 + h^2\frac{f_x}{f_y}\right)\omega_c^2$$

$$\omega_c^2 = \frac{\dfrac{f_x}{f_y}\left(k_4 + (hd)^2\dfrac{4N}{R - R'}\right) + d^2\left(1 + h^2\dfrac{f_x}{f_y}\right)\left(2k_3 + \dfrac{4N}{R - R'}\right)}{I\dfrac{f_x}{f_y} + md^2\left(1 + h^2\dfrac{f_x}{f_y}\right)}$$

You will notice that the results for the simple truck are quite similar to those for the wheelset. You can also imagine that this model is about as far as one can go in complexity without resorting to purely numerical computations of eigenvalues or, if nonlinear elements appear in the model, to computer simulations of the dynamic equations.

CHAPTER 11

11.1 Consider the state variable feedback steering system shown in Fig. 11.3, but with front wheel steering only. Write the open loop equations in the form

$$\begin{bmatrix} \dot{V} \\ \dot{r} \end{bmatrix} = \begin{bmatrix} A_{11} & A_{12} \\ A_{21} & A_{22} \end{bmatrix} \begin{bmatrix} V \\ r \end{bmatrix} + \begin{bmatrix} B_1 \\ B_2 \end{bmatrix} \delta_{\mathrm{f}}$$

where the A and B coefficients can be seen in Eqs. (11.19) and (11.20).

Write the feedback law in the form

$$\delta_{\mathrm{f}} = \delta_{\mathrm{fref}} - \begin{bmatrix} K_1 & K_2 \end{bmatrix} \begin{bmatrix} V \\ R \end{bmatrix}$$

Combine these relations to find equations for the closed loop system for the response of the lateral velocity and the yaw rate to the reference steer angle supplied by the driver.

Do you think that using the two gains K_1 and K_2 the closed loop system can be given arbitrary eigenvalues (at least in theory)?

11.2 Consider the simple model of the lateral dynamics of an automobile used in Chapter 6 and illustrated in Figs. 6.1 and 6.5. Suppose that a sudden *impulsive force* is applied at the front axle. Show that if $I_z = mab$, the immediate result would be no motion at all at the rear axle, i.e., in this case, the back axle is located at the center of percussion.

Consider the cases $I_z < mab$ and $I_z > mab$. In which case does the back axle have an initial velocity in the same direction as the front axle and in which case does it have an initial velocity in the opposite direction?

Why is it not necessary to consider possible tire forces at the rear axle in these calculations?

11.3 Consider a model reference controller of the general type shown in Fig. 11.1 using only yaw rate error and the relation between steer angle and desired yaw rate given in Eq. (11.35). Construct a block diagram for this controller similar to the diagram in Fig. 11.5 but leaving the controller block unspecified, thus:

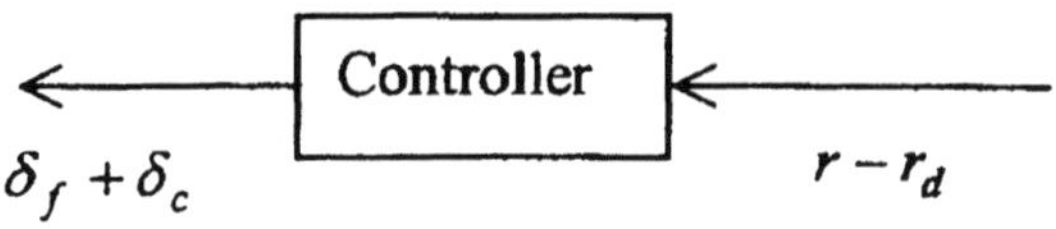

Compare this steering control to the sliding mode control in Fig. 11.5.

11.4 Consider a simple example of a sliding mode controller. Let the state variables be represented by

$$x = \begin{bmatrix} X \\ V \end{bmatrix} \text{ and the state equations be} \begin{bmatrix} \dot{X} \\ \dot{V} \end{bmatrix} = \begin{bmatrix} 0 & 1 \\ 0 & 0 \end{bmatrix} \begin{bmatrix} X \\ V \end{bmatrix} + \begin{bmatrix} 0 \\ 1 \end{bmatrix} U$$

One can think of X as a position, V as a velocity, and U as a control force divided by a mass.

Let the function S, as defined in Eq. (11.5), be

$$S = [c_1 \quad c_2] \begin{bmatrix} X \\ V \end{bmatrix} = c_1 X + c_2 V$$

In this case, $x_d = 0$, so the controller will attempt to force both X and V to be zero.

Show that Eq. (11.8) yields the equation $\dot{S} = c_1\dot{X} + c_2\dot{V} = c_1 V + c_2 U = -\rho S - \alpha \text{ sgn } S$, which gives a formula for the control variable as in Eq. (11.10),

$$U = -\frac{c_1}{c_2}V - \rho\frac{(c_1 X + c_2 V)}{c_2} - \frac{\alpha \text{ sgn}(c_1 X + c_2 V)}{c_2}$$

Plot the line $S = 0$ in the X–V plane, which is the state space for this system, assuming that c_1 and c_2 are positive quantities. Demonstrate that if the system is on the sliding surface, $S = 0$, $\dot{X} = -(c_1/c_2)X$, which represents and exponential motion toward the origin of the state space.

Equations (11.16) and (11.17) have to do with the controlled system with the nonlinear part eliminated by setting $\alpha = 0$. Show that with $\alpha = 0$ the controlled system equations become

$$\begin{bmatrix} \dot{X} \\ \dot{V} \end{bmatrix} = \begin{bmatrix} 0 & 1 \\ -\rho\dfrac{c_1}{c_2} & -\dfrac{c_1}{c_2} - \rho \end{bmatrix} \begin{bmatrix} X \\ V \end{bmatrix}$$

Finally, show that the eigenvalues for the controlled system are $s = -\rho$, as predicted by Eq. (11.17) and $s = -(c_1/c_2)$, which relates to motion in the sliding surface.

11.5 The elementary yaw rate controller shown below could be used to study how to choose a controller transfer function G, such that the actual vehicle yaw rate would closely follow a desired yaw rate.

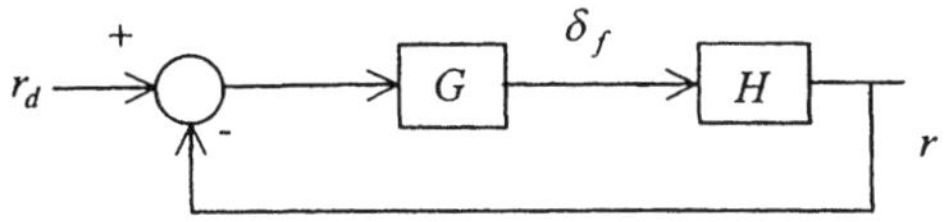

Show that the closed loop transfer function relating the desired yaw rate to the actual yaw rate is given by the expression

$$\frac{r}{r_\mathrm{d}} = \frac{GH}{1 + GH}$$

Assuming that G is a simple gain and that H is given by Eq. (6.33), find the characteristic equation for the closed loop system.

11.6 The block diagram below depicts a yaw rate controller that computes the desired yaw rate r_d from the driver's steer input δ_f according to Eq. (11.35). The controller adds a compensating steer angle δ_c to the driver's steer angle via a differential element in the steering column. As in Problem 11.5, H is a vehicle transfer function such as Eq. (6.33), and G is a controller transfer function.

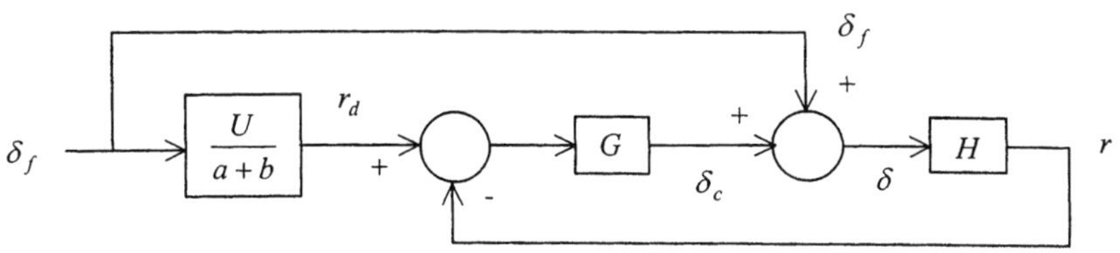

Show that the transfer function relating the yaw rate to the driver's steer input is given by the expression

$$\frac{r}{\delta_\mathrm{f}} = \frac{H(1 + GU/(a+b))}{1 + GH}$$

Would you expect that the characteristic equation for this controller to be different from the one derived in Problem 11.5?

11.7 Fig. 5.5 (in Chapter 5) shows a 2 degree-of-freedom trailer with a flexible connection at the hitch. Derive a static stability criterion for the lateral motion of the trailer analogous to the static longitudinal stability criterion for airplanes.

You will have to consider the type of perturbation away from the basic motion that is appropriate for the static stability analysis and then decide if the trailer will tend to return to the basic motion or not. Discuss why it is possible for the trailer to be statically stable, but dynamically sometimes stable and sometimes unstable.

Bibliography

Anon. Branding Vehicle Dynamics, *Automotive Engineering International*, July 2003; 53–60.

Babister, A.W. *Aircraft Dynamic Stability and Response*; Pergamon Press: Oxford, 1980.

Bakker, E.; Nyborg, L.; Pacejka, H.B. Tyre Modelling for Use in Vehicle Dynamics Studies, SAE Paper 870421, 1987.

Bakker, E.; Pacejka, H.B.; Lidner, L. A New Tire Model with an Application in Vehicle Dynamics Studies, SAE Paper No. 890087, 1989.

Barstow, D. *Car Suspension and Handling*, 2nd Ed.; Pentech Press: London, 1987.

Birch, S. Chassis System Integration, *Automotive Engineering International*, SAE International, June, 2003; 58–62.

Booth, R.A., Jr. A Weave Model for Motorcycle Engineers, SAE Paper 830155, 1983.

Bundorf, R.T. Directional Control Dynamics of Automobile-Travel Trailer Combinations, SAE Paper 670099, 1967a.

Bundorf, R.T. The Influence of Vehicle Design Parameters on Characteristic Speed and Understeer, SAE Paper No. 670078, 1967b.

Cooperrider, N.K. The Lateral Stability of Conventional Railway Passenger Trucks, Proceedings of the First International Conference on Vehicle Dynamics, Swets & Zeitlinger: Amsterdam, 1969; 37–67.

Crandall, S.H.; Karnopp, D.C.; Kurtz, E.F.; Pridmore-Brown, D.C. *Dynamics of Mechanical and Electromechanical Systems*; McGraw-Hill: New York, 1968.

Culick, F.E.C.; Jex, H.R. Aerodynamics, Stability, and Control of the 1903 Wright Flyer, Proceedings of the Wright Flyer: An Engineering Perspective, National Air and Space Museum, Smithsonian Institution, 19–43.

Cumberford, R. Steering: critical path analysis. Automobile Magazine, Nov. 1991; 51 pp.

de Beneto, C.; Borcherts, R.H.; Liang, C.-G.; Walsh, M.R. Dynamically Coupled Steering: A Fail Safe Steer-By-Wire system, 20th International Symposium on Automotive Technology and Automation, (ISATA), Florence: Italy, 1989.

Diboll, W.B.; Hagen, D.H. Lateral stability of road and rail trailers. Transactions of the ASME, Journal of Engineering for Industry, 1969; 1069–1088.

Dole, C.E. *Flight Theory and Aerodynamics*; John Wiley & Sons: New York, 1981.

El-Deen, Y.H.; Seireg, A. Mechatronics for cars: integrating machines and electronics to prevent skidding on icy roads, *Computers in Mechanical Engineering*; January 1987; 10–22.

Ellis, J.R. *Vehicle Dynamics*; Business Books Limited: London, 1969.

Ellis, J.R., *Road Vehicle Dyanmics*; John R. Ellis, Inc.: Akron, OH, 1988.

Etkin, B. *Dynamics of Atmospheric Flight*; John Wiley & Sons: New York, 1972.

Etkin, B.; Reid, L.D. *Dynamics of Atmospheric Flight: Stability and Control*; John Wiley & Sons, 1996.

Fonda, A.G. Steering and Directing Mechanism, United States Patent 3,716,110, 1973.

Ford Motor Company. Steering Gear Actuator, Particularly for Motor Vehicles, British Patent Specification 1 414 206, 1975.

Furukawa, Y.; Yuhara, N.; Sano. S.; Takeda, H.; Mastushita, Y. A Review of Four-Wheel Steering Studies from the Viewpoint of Vehicle Dynamics and Control. *Vehicle System Dynamics*; 1989, *18*, 151–186.

Garg, V.; Dukkipati, R. *Dynamics of Railway Vehicle Systems*; Academic Press: Toronto, 1984.

Garrott, W.R.; Monk, M.W.; Chrstos, J.P. Vehicle Inertial Parameters—Measured Values and Approximations, SAE Paper 881767, 1988.

Gillespie, T.D. *Fundamentals of Vehicle Dynamics*; Society of Automotive Engineers: Warrendale, PA, 1992.

Granda, J.J. *CAMP-G, Computer Aided Modeling Program User's/Reference Manual*; Cadsim Engineering: Davis, CA, 2002.

Grantham, W.J.; Vincent, T.L. *Modern Control Systems Analysis and Design*; John Wiley & Sons, Inc.: N.Y., 1993.

Hoffman, D.D.; Rizzo, M.D. Chevrolet C 5 Corvette Vehicle Dynamic Control System, SAE Paper 980233, printed in Vehicle Dynamics and Simulation, SAE/SP-98/1361, 1998.

Hrovat, D.; Ascari, J. Bond Graph Modeling of Automotive Dynamic Systems, Paper 9438349, International Symposium on Advanced Vehicle Control, (AVEC 94), JSAE, Tsukuba, Japan, 1994.

Hunsaker, J.C.; Rightmire, B.G. *Engineering Applications of Fluid Mechanics*; McGraw-Hill: New York, 1947.

Hurt, H.H. *Aerodynamics for Naval Aviators (NAVWEPS 00-80T-80)*, Office of the Chief of Naval Operations, Air Training Division: Los Angeles, CA, 1960.

Irving, F.G. *An Introduction to the Longitudinal Static Stability of Low-Speed Aircraft*; Pergamon Press: Oxford, 1966.

Jones, D.E.H. The stability of the bicycle. *Physics Today*; April, 1970; 34–40.

Karnopp, D. Bond graphs for vehicle dynamics. *Vehicle System Dynamics*; 1976, *5* (3), 171–184.

Karnopp, D.; Wuh, D. Handling Enhancement of Ground Vehicles using Feedback Steering Control, ASME Monograph DSC 1989, *13*, 99–106.

Karnopp, D.; Wuh, D. *Approximate Inverse Simulation of an Active Steering System Including Steering Wheel Reaction, SAE Paper 912563*; 1991; 11 pp.

Karnopp, D.; Fang, C. A simple model of steering-controlled banking vehicles. ASME, DSC-v.44, Transportation Systems 1992, 15–28.

Kortuem, W.; Lugner, P. *Systemdynamik und Regelung von Fahrzeugen*; Springer-Verlag: Berlin, 1994.

Karnopp, D.C.; Margolis, D.L.; Rosenberg, R.C. *System Dynamics: Modeling and Simulation of Mechatronic Systems*; John Wiley & Sons: New York, 2000.

Lee, A.Y. Vehicle-Stability-Augmentation Systems Design Using Parameter Optimization, ASME Paper 88-WA/DSC-25, 1988.

Li, Y.T.; Meiry, J.L.; Roessler, W.G. An active roll mode suspension system for ground vehicles. Transactions of the ASME, Journal of Basic Engineering, June 1968; 167–174.

Loeb, J.S.; Guenter, D.A.; Chen H.-H.F.; Ellis, J.R. Lateral Stiffness, Cornering Stiffness and Relaxation Length of the Pneumatic Tire, SAE Paper 900129, 1990.

Luenberger, D.G. An introduction to observers. IEEE Transactions on Automatic Control, AC-16, Dec. 1971, 6, 596–602.

Margolis, D.L.; Ascari, J. Sophisticated Yet Insightful Models of Vehicle Dynamics Using Bond Graphs, ASME Symposium on Advanced Automotive Technologies; ASME WAM: San Francisco, 1989.

Margolis, D.; Shim, T. A bond graph model incorporating sensors, actuators and vehicle dynamics for developing controllers for vehicle safety. J. of the Franklin Institute, 2001, *338* (1), 21–34.

Markus, F. Stability-enhancement systems: trick or treat? Car and Driver, June 1996; 109–121.

Martin, G.H. *Kinematics and Dynamics of Machines*; McGraw-Hill: NY, 1982; 58–59.

McCormick, B.W. *Aerodynamics, Aeronautics, and Flight Mechanics*; John Wiley & Sons: NY, 1995.

McRuer, D.; Ashkenas, I.; Graham, D. *Aircraft Dynamics and Automatic Control*; Princeton University Press: Princeton, NJ, 1973.

Milliken, W.F.; Milliken, D.L. *Race Car Vehicle Dynamics*; SAE International: Warrendale, PA, 1995.

Muraca, P.; Perone, G. Applications of variable structure systems to a flexible robot arm. Proc. IMACS 1991, 1415–1419.

Nametz, J.E.; Smith, R.E.; Sigma, D.R. The Design and Testing of a Microprocessor Controlled Four Wheel Steer Concept Car, SAE Paper 885087, 1988.

Nordeen, D.L. Analysis of tire lateral forces and interpretation of experimental tire data. SAE Transactions; 1968, *76*.

Ogata, K. *Modern Control Engineering*; Prentice-Hall, Inc.: Englewood Cliffs, NJ, 1970.

Olley, M. Road manners of the modern car. IAE Proceedings, 1946–47, *41*, 147–182.

Pacejka, H.B. Simplified analysis of steady-state turning behavior of motor vehicles. Part 1. Handling diagrams of simple systems. Vehicle System Dynamics, 1973a, *2*, 161–172.

Pacejka, H.B. Simplified analysis of steady-state turning behavior of motor vehicles. Part 2. Stability of the steady-state turn. Vehicle System Dynamics, 1973a, *2*, 173–183.

Pacejka, H.B. Simplified analysis of steady-state turning behavior of motor vehicles. Part 3. More elaborate systems, Vehicle System Dynamics, 1973b, *2*, 185–204.

Pacejka, H.B. Modelling complex vehicle systems using bond graphs, Journal of the Franklin Institute, 1985, *319* (1/2), 67–81.

Pacejka, H.B. Introduction to the Lateral Dynamics of Road Vehicles, Third Seminar on Advanced Vehicle Dynamics, Amalfi, 1986.

Pacejka, H.B. *Tire Characteristics and Vehicle Dynamics*, course notes, University Consortium for Continuing Education: Washington, DC, Nov 1–3, 1989.

Pacejka, H.B.; Sharp, R.S. Shear force development by pneumatic tyres in steady state conditions: a review of modeling aspects. Vehicle System Dynamics, 1991, *20*, 121–176.

Pacejka, H.B.; Toi, C.G.M. A Bond Graph Computer Model to Simulate the 3-D Dynamic Behavior of a Heavy Truck. Proceedings of the Modeling and Simulation in Engineering Conference, IMACS World Congress on Systems, Simulation and Scientific Computation, Montreal, Canada, 1983.

Puhn, F. *How to Make your Car Handle*; H. P. Books: Los Angeles, CA, 1981.

Sanyal, A.; Karmaker, R. Directional stability of truck-dolly-trailer system. Vehicle System Dynamics, 1995, *24* (8), 617–637.

Schwarz, R. Accident Avoidance Characteristics of Unconventional Motorcycle Configurations, ASME Paper 790258, 1979.

Segel, L. Theoretical Prediction and Experimental Substantiation of the Response of the Automobile to Steering Control. The Institution of Mechanical Engineers: London, 1956; 26–46.

Sharp, R.S. The stability and control of motorcycles. Journal of Mechanical Engineering Science, 1971, *13* (5), 316–329.

Sharp, R.S.; Crolla, D.A. Controlled rear steering for cars—a review. Proceedings of the Institution of Mechanical Engineers, International Conference on Advanced Suspensions, 1988; 149–163.

Slotine, J.-J. Sliding controller design for nonlinear systems. Int. J. of Control 1984, *40* (2), 421–434.

Slotine, J.-J.E.; Sasfry, S.S. Tracking control of nonlinear systems using sliding surfaces with applications to robot manipulators. Int. J. of Control, 1983, *33* (2), 465–492.

Slotine, J.-J.E.; Li, W. *Applied Nonlinear Control*; 1991.

Tanaka,T.; Tani, M.; Yuasa, H.; Mimuro, T.; Yoshida, H. Active Control Technology for Passenger Car, SAE Paper 88105, 1988.

Tran, V.T. Crosswind feedforward control—a measure to improve vehicle crosswind characteristics. Proc. 12th IAVSD Symposium on Dynamics of Vehicles, Lyon, France, 1991; 141–143.

Van Zanten, A.T.; Erhart, R.; Landesfeind, K.; Pfaff, G. VDC Systems Development, SAE Paper 980235, Printed in ABS/Brake/VDC Technology, SAE/SP-98/1339, 1998.

Vincenti, W.G. How did it become 'obvious' that an airplane should be inherently stable? Invention & Technology Spring/Summer, 1988, 51–56.

Whitehead, J.C. Four Wheel Steering: Maneuverability and High Speed Stability, SAE Paper 880642, 1988.

Whitehead, J.C. Stabilizing Steering Weave with Active Torque Versus Semi-Active Damping, SAE Paper 891981, 1989.

Wong, J.C. *Theory of Ground Vehicles*; John Wiley & Sons: New York, 1978.

Zeid, A.A.; Chang, D. A modular computer model for the design of vehicle dynamics control systems. International Journal of Vehicle systems Dynamics, 1989, 201–221.

Index